Learning from the COVID-19 Pandemic

COVID-19 was first identified in Wuhan City in December 2019 and spread throughout Hubei Province and other parts of China. After causing significant morbidity and mortality in China, by February 2020, it had spread to numerous other countries, infecting millions of people and causing a large number of deaths across the world.

The COVID-19 pandemic put a burden on almost all areas of the world including healthcare systems, education, industry, travel, etc. The pandemic revealed the vulnerability of the world's healthcare systems and affected healthcare personnel significantly. The virus is able to attack not only the respiratory tract, but almost all the organs including the brain. Impacts on gut biota have also been noticed. The virus has caused both morbidity and mortality in humans without any geographical, cultural, or religious barriers. The emergence of new variants due to mutations in the virus has aggravated the problem. While the delta variant brought a second wave and killed a large number of people due to various factors such as lowering of saturated oxygen in blood and other physiological emergencies, the omicron variant proved to be less lethal. Though the pandemic has subsided, the emergence of the subvariants BA1 and BA2 and now their hybrids has started to increase the number of cases at exponential levels and has forced new lockdown measures in places such as China. As the conditions laid down to combat the pandemic have been relaxed, the virus may reach other countries and cause additional countries to resort to lockdown again.

COVID-19 became the focus of the scientific community with the aim of developing new drugs, repurposing available drugs to be used against the virus, and developing a series of vaccines in a short time. The mild effect of omicron might have been due to the extensive vaccination programmes carried out in various countries. However, there is genuine fear that newly emerging variants may evade the immune system and cause damage to the body.

This book highlights the impact of COVID-19 on science, industry, and healthcare systems. The chapters included in the volume come from dedicated experts belonging to basic sciences, biotechnology, pharmaceutical sciences, and other fields of sciences. These include discussions on how the virus evolves and attacks various organs in the body. A separate chapter explains the emergence of various strains of virus. The preparedness of hospitals and healthcare workers as well as different agencies such as DRDO to face the challenges posed by virus is also discussed. The way scientists and technologists developed new techniques to detect and control the virus have also been highlighted including a chapter on the development of vaccines to control the pandemic.

This book is a key resource for students, teachers, medical personnel, administrators, and the public as a whole.

Learning from the COVID-19 Pandemic

Implications for Science, Health, and Healthcare

Edited By
R. C. Sobti and Aastha Sobti

CRC Press is an imprint of the
Taylor & Francis Group, an Informa business

Designed cover image: © Shutterstock

First edition published 2023
by CRC Press
6000 Broken Sound Parkway NW, Suite 300, Boca Raton, FL 33487-2742

and by CRC Press
4 Park Square, Milton Park, Abingdon, Oxon, OX14 4RN

CRC Press is an imprint of Taylor & Francis Group, LLC

ISBN: 978-1-032-41603-8 (hbk)
ISBN: 978-1-032-41604-5 (pbk)
ISBN: 978-1-003-35890-9 (ebk)

DOI: 10.1201/9781003358909

Typeset in Times New Roman
by Newgen Publishing UK

Contents

Preface

Coronavirus disease 2019 (COVID-19) first identified in Wuhan City in December 2019 spread throughout Hubei Province and other parts of China. After causing significant morbidity and mortality in China, by February 2020, it had spread to numerous other countries, infecting millions of people and caused a large number of deaths across the world and is, therefore, considered a global pandemic.

The outbreak of COVID-19 pandemic put in a burden on almost all the systems including healthcare, education, life style, aviation throughout the world. This outbreak revealed the world's poor health system and has had a negative influence on healthcare personnel in every way. Because they came in touch with infected people on a regular basis, became more prone to COVID-19 infection than the general population. The virus not only attacks the respiratory tract, but almost all the organs including brain. Impacts on gut biota have also been noticed. The morbidity and mortality by it in humans has been without any geographical, cultural, religious barriers. The emergence of new variants due to mutations in the virus has aggravated the problems. The delta variant that brought in second wave was very disastrous killing a large number of people due to various factors such as lowering of saturated oxygen in blood and other physiological emergencies. The omicron variant has proved to be less lethal. Though the pandemic has subsided, the emergence of sub variants BA1 and BA2 their hybrid and others have created horror feelings in many countries. The increase in the number of cases at exponential level had forced many states in China to repose lockdown. As the conditions laid down to combat the pandemic have been relaxed, and movements from one part of the globe to the other have started, the newer strains are able to spread to other countries and make people to suffer with pandemic.

COVID-19 sparked the scientific community to make all out efforts to delineate the structure and functions and mechanistic approach of virus to cause disease in humans. Besides, attempts to develop new drugs, sincere efforts have been made to repurpose the available drugs to be used against the virus and develop a series of vaccines in a shortest period, not possible in the normal times. The mild effect of omicron and other variants might have been due to the extensive vaccination programmes carried out in various countries. There is, however genuine fear that the new emerging variants may evade the immune system and result in damage to the body system.

The rapidly growing demand on health facilities and health care employees threatened to overburden some health systems and render them unable to function properly. This had a direct impact on health care professionals. All the energies in the nations world over focused on controlling the transmission and curtailing morbidity and mortality due to the pandemic. Despite all the complexities and hardships that health care employees faced, they continued to put in their best to serve the community and society at a large.

The present book highlights the impact of COVID-19 on science, health and health care system. The chapters included in the volume are hard work of dedicated experts in the field. These include the description of the evolution, structure and mode of infection by virus as well as strategies to attack various organs in the body. A separate chapter explains the emergence of various strains of virus. The preparedness of hospitals and health care workers as well as different agencies such as DRDO to face the challenges posed by virus has also been discussed. The way scientists and technologists developed new techniques to detect and control the virus have been highlighted. As the chapter on the development of vaccines to control the pandemic has also found place in this book.

This book will be ready compendium for students, teachers, medical personnel, administrators and public as a whole.

R. C. Sobti
Aastha Sobti

Editor Biographies

Ranbir Chander Sobti is Former Education Consultant Governor of Bihar, Senior Scientist (Indian National Science Academy), Former Vice Chancellor, Babasaheb Bhimrao Ambedkar University, Lucknow (UP) and Panjab University, Chandigarh. Professor R. C. Sobti starting his career as an Animal (including human) Cytogeneticist characterized the karyotypes of animals of groups from protozoa to human. He also looked into the mutagenicity/carcinogenicity of environment pollutants both in *in vivo* and *in vitro* settings. He then moved on to explore the developments in the molecular biology of diseases. He has proactively been involved in cancer biology and focused on finding out novel tumour markers for cancer detection. He has evaluated the role of gene polymorphisms and their expression in the genesis of various cancers and also in COPD, AIDS and metabolic syndrome. He has identified disease susceptible/protective novel genotypes and determined the crucial role of SACS-1 and STAT genes in the genesis of cervical and prostate cancers.

His immense contribution to science includes, inter alia, 350 plus high impact research publications with more than 50 books and 23 Sponsored Research Projects. He has also tried to develop whole organ by tissue/organ culture through custom designed de-cellularization protocol. He has demonstrated that the regulation of stem cell character under *in vitro* conditions is a function of morphological assortment of Solid Lipid Nanoparticles (SLNC) – an observation that has marked significance in the field of stem cell research. At the moment he is involved in meta data analysis for organ dysfunctioning to investigate underlining molecular mechanism: ML approach.

Professor Sobti is a Fellow of the Third World Academy of Sciences (TWAS), National Academy of Sciences, Indian National Science Academy, National Academies of Medical Sciences and Agricultural Sciences and of the Canadian Academy of Cardiovascular Diseases. He is also associated with many other Academic Associations and Institutions in the domain of higher education and research. The litany of honours showered on him includes, among others, the INSA Young Scientist Medal (1978), UGC Career Research Award, Punjab Rattan Award, JC Bose Oration Award and the Life Time Achievement Award of the Punjab Academy of Sciences, Zoological Society of India, and the Environment academy of India, besides many other medals & awards of various reputed National and International Organizations. He was bestowed with Padmashri award by the Government of India in 2009.

He has delivered numerous keynote addresses at reputed National and International forums, based on implications and applications of science and scientific temper for general awareness and improvements of everyday life and lifestyle in a lucid and people-centric manner.

Aastha Sobti is currently carrying out research at the Department of Immunotechnology, Lund University, Lund, Sweden, under Can Faster, Marie SkÅ, odowska-Curie COFUND programme. She has a background in dental sciences and did a specialization in masters in Clinical Dentistry (Oral surgery) from Eastman College, UCL, London. Her core interests are head and neck cancer-based clinical research. In entirety aiding in the field to bring reforms that are required in the present multifarious surgical as well as translational research areas. She has teaching and research experience of about 8 years and published a few papers in international journals of repute. She is a recipient of a large number of prizes and medals for her exceptional work namely the academic certificate of excellence from BAOS, UK and IADR Hatton award (India). Additionally, she has attended and presented papers in international conferences in Hong Kong, Croatia, Brazil, Sweden, UK, Japan, and other countries.

Contributors

Shubham Adhikary
Department of Pharmacology, SPP School of Pharmacy & Technology Management, SVKM's NMIMS, Mumbai, Maharashtra, India

A. S. Ahluwalia
Eternal University, Baru Sahib, District Sirmaur, Himachal Pardesh, India

Sanjay K. Bhadada
Department of Endocrinology, Post Graduate Institute of Medical Education and Research, Chandigarh, India

Gaurav Bhandari
Institute of Renal Science, Sir Gangaram Hospital, New Delhi, India

Shiv Bharadwaj
Department of Biotechnology, Yeungnam University, Gyeongbuk-do, Republic of Korea

Vinant Bhargava
Institute of Renal Science, Sir Gangaram Hospital, New Delhi, India

Brijesh
Punjab and Haryana High Court, Chandigarh, India

Harpal S. Buttar
Department of Pathology & Laboratory Medicine, University of Ottawa, Faculty of Medicine, Ottawa, Ontario, Canada

Neena Capalash
Department of Biotechnology, Panjab University, Chandigarh, India

Munish Garg
Department of Pharmaceutical Sciences, Maharshi Dayanand University, Rohtak, Haryana, India

Rajwant K. Gill
Lumex Instruments Canada, Mission, BC, Canada

Sikander S. Gill
Lumex Instruments Canada, Mission, BC, Canada

Taru Goyal
Department of Biochemistry, All India Institute of Medical Science, Jodhpur, India

Anuragini Gupta
Department of Otorhinolaryngology Head and Neck Surgery, Fortis Hospital, Mohali, India

Ashok Gupta
Department of ENT, Fortis Hospital, Mohali, Punjab, India

Rajesh Harsvardhan
Department of Hospital Administration, SGPGIMS, Lucknow, Uttar Pradesh, India and SSCI&H, Lucknow, Uttar Pradesh, India

Jitendra Kumar Jain
Trishla Foundation, Prayagraj, Uttar Pradesh, India

Varidmala Jain
Trishla Foundation, Prayagraj, Uttar Pradesh, India

Ginpreet Kaur
Department of Pharmacology, SPP School of Pharmacy & Technology Management, SVKM's NMIMS, Mumbai, Maharashtra, India

Guneet Kaur
National Brain Research Centre, Manesar Gurgaon, Haryana, India

Rupinder Kaur
BGJ Institute of Health, Panjab University, Chandigarh, India

Tejinder Kaur
Department of Zoology, DAV University, Jallandhar, Punjab, India

Samander Kaushik
Centre for Biotechnology, Maharshi Dayanand University, Rohtak, Haryana, India

Sulochana Kaushik
Department of Genetics, Maharshi Dayanand University, Rohtak, Haryana, India

Samander Kaushik
Centre for Biotechnology, Maharshi Dayanand University, Rohtak, Haryana, India

Nikhil Kirtipal
Department of Science, MIT, Dhalwala, Rishikesh, Uttarakhand, India

Sukender Kumar
Department of Pharmaceutical Sciences, Maharshi Dayanand University, Rohtak, Haryana, India

Suraj Kumar
Department of Physiotherapy, Faculty of Paramedical sciences, Uttar Pradesh University of Medical Sciences, Saifai, Etawah, UP, India

Vikas Kushwaha
Department of Biotechnology, Panjab University, Chandigarh, India

Sunil K. Lal
School of Science, Monash University, Selangor, Malaysia

Malavika Lingeswaran
Department of Biochemistry, All India Institute of Medical Science, Jodhpur, India

Prakash S. Lohar
KBC North Maharashtra University, Jalgaon, India

Sandeep K. Malhotra
Department of Zoology, University of Allahabad, Prayagraj, India

Shivji Malviya
Department of Zoology, Government H.N.B. Post Graduate College, Naini (Prayagraj), UP, India

Priti Meena
Institute of Renal Science, Sir Gangaram Hospital, New Delhi, India

Geeta Mehra
Department of Food Science, Mehr Chand Mahajan DAV College for Women, Chandigarh, India

Rajneesh Mehra
Amity Global Business School, Mohali, India

Sanjeev Misra
Department of Surgical Oncology, All India Institute of Medical Science, Jodhpur, India

K. P. Mishra
Defence Research and Development Organization (DRDO)-HQ, New Delhi, India

Prasenjit Mitra
Department of Biochemistry, All India Institute of Medical Science, Jodhpur, India

Ahalya Naidu
M. S. Ramaiah Institute of Technology, Bangalore, India

Archna Narula
M. S. Ramaiah Institute of Technology, Bangalore, India

Rimesh Pal
Department of Endocrinology, Post Graduate Institute of Medical Education and Research, Chandigarh, India

Praveen Sharma
Department of Biochemistry, All India Institute of Medical Science, Jodhpur, India

Nikita Pawar
Institute of Renal Science, Sir Gangaram Hospital, New Delhi, India

Gowrishankar Potturi
Department of Physiotherapy, Faculty of Paramedical sciences, Uttar Pradesh University of Medical Sciences, Saifai, Etawah, UP, India

Ranjeev Kumar Sahu
Babasaheb Bhimrao Ambedkar University, Lucknow, Uttar Pradesh, India

Pankaj Seth
National Brain Research Centre, Manesar Gurgaon, Haryana, India

M. C. Sidhu
Department of Botany, Panjab University, Chandigarh, India

A. K. Singh
Defence Research and Development Organization (DRDO)-HQ, New Delhi, India

Gagandeep Singh
Department of Microbiology, All India Institute of Medical Sciences, New Delhi, India

Hardeep Singh Tuli
Department of Biotechnology, Maharishi Markandeshwar (Deemed to be University), Mullana-Ambala, Haryana, India

Sandeep Singh
Department of Biochemistry, Maharshi Dayanand University, Rohtak, Haryana, India

Aastha Sobti
Department of Immunotechnology, Lund University, Sweden

R. C. Sobti
Department of Biotechnology, Panjab University, Chandigarh, India

Immaculata Xess
Department of Microbiology, All India Institute of Medical Sciences, New Delhi, India

Manish Thakur
Department of Microbiology, DAV University, Jalandhar, Punjab, India

Anita Yadav
Department of Zoology, University of Allahabad, Prayagraj, India

Zheng Yao Low
School of Science, Monash University, Selangor, Malaysia

Ashley Jia Wen Yip
School of Science, Monash University, Selangor, Malaysia

1 Impact of Covid-19 on Science, Education, Health Care, Environment, Economy, and Lifestyle

R. C. Sobti[1] *and Aastha Sobti*[2]
[1]Department of Biotechnology, Panjab University. Chandigarh, India
[2]Department of Immunotechnology, Lund University, Sweden

CONTENTS

1.1 INTRODUCTION

The outbreak of Covid-19 (the disease caused by Severe Acute Respiratory Syndrome Coronavirus 2 (SARS-CoV-2)) has been labeled a "Black Swan" event and compared to the economic scenario of World War II. Its effects have been detrimental across a wide range of spheres, including healthcare, education, the environment, lifestyle, economics, and more. Nearly every aspect of human life has been impacted. Covid-19 is the common name for SARS-CoV-2, which has been researched by the Coronavirus Study Group "International Committee on Taxonomy of Viruses" (Baker et al., 2020). The impact of SARS-CoV-2 has been felt around the world for quite some time, especially in the medical sector (Sohrabi et al., 2020).

Given the endless transmission of SARS-CoV-2 and evidence of its transmission to healthcare providers, medical practitioners were the first to be at high risk of infection. Medical examinations, surgical procedures, and even the proximity to infected patients all posed risk of increased pathogen exposure to the healthcare workers. Therefore, patients and medical staff in the outpatient department (OPD) and operating room (OR) were around the clock at risk of infection without proper safety measures being taken. According to the World Health Organization, between 80,000 and 180,000 health and care workers might have died from Covid-19 between January 2020 and May

DOI: 10.1201/9781003358909-1

2021, with a median estimate of 115,500 fatalities (WHO, 2021). Despite adhering to all the Center for Disease Control and Prevention (CDC) and World Health Organisation (WHO) guidelines for management of patients, healthcare workers were impacted not only by pathogen exposure but also by long working hours, psychological distress, fatigue, occupational burnout, stigma, physical and psychological violence, home sickness, and other related issues. Their families were also affected at physical and physiological levels.

The pandemic led governments around the globe to shut down schools, colleges, and universities, separating millions of students and teachers from their institutes. Due to closure of schools in at least 200 nations, more than 91% of the world's students were at home and could not meet their teachers physically during the lockdown imposed because of the Covid-19 pandemic. Because of these unprecedented obstacles, it was imperative that the educational system change and find new ways to keep the students interested in learning. Governments launched online platforms to sustain the educational industry and to help students in their studies. As a result, the educational system witnessed a transition with the distinctive emergence of online teaching and learning processes and a new era of education has begun. This transition has its own challenges. The lockdown presented problems even to wealthier parents in urban spaces: they had to help children set up e-learning stations at home, monitor them around the clock, and deal with additional stress of childcare. People living in rural areas faced unique challenges, such as a lack of information technology infrastructure, poor internet connectivity, and even power outages.

Covid-19 has impacted nearly every aspect of our everyday lives. With more than 600 million infected cases and almost 6.5 million deaths globally, the implications of the Covid-19 pandemic are going to last a long time and affect future generations for centuries to come. Aside from Covid-19 itself and its impact on the healthcare, education, and tourism industries many other sectors were affected leading to an economic recession. Social distancing, quarantine, and travel restrictions forced many employers to lay off their workforce. The demand for consumer goods also reduced significantly. On the contrary, the need and demand for medical supplies and gloves and personal protective equipments (PPEs) has skyrocketed. Many industries are relying more and more on remote workforces requiring newer communicating tools. In response to this unprecedented global pandemic, we attempt to discuss the impact of Covid-19 on the community as a whole.

The year 2020 and 2021 will go down in the history books for many reasons, but most notably because of the Covid-19 pandemic, which has reshaped our lives on a global scale. This Covid-19 pandemic began in Wuhan, China towards the end of 2019 and was named SARS-CoV-2 because of its association with the SARS-CoV, which first appeared in 2002 and re-emerged ten years later under a new name as the Middle East Respiratory Syndrome Coronavirus (MERS-CoV). According to earlier findings on the virus' effects and its origin, it is related to a family of single-stranded RNA viruses known as Coronaviridae (Fehr and Perlman, 2015).

The WHO issued its first alert for the virus on January 16, 2020, declaring it as a Public Health Emergency on January 31, 2020 and as a global pandemic on March 11, 2020. This is not the first time that our civilization has encountered an outbreak like this. During the 6th century we had the notorious plague, the cause of three pandemics in world history. "Black Death," a variation of the plague, took the lives of over 40 million people. Covid-19 is the fifth documented pandemic since the 1918 flue pandemic (Spanish Flu). Fast forward to the 21st century and we are challenged with Covid-19 as never before and it has quickly become one of the deadliest diseases in the world's history having killed a large number of people and infecting many in 188 countries.

Public health experts had warned about the dangers of a pandemic for years so the world should not have been surprised. The pandemic devastated the world economy and international politics and shown the flaws in the global public health infrastructure, educational infrastructure, economy, etc. The fallout was vast and wide ranging. These are unprecedented times and we know nothing of what the future holds and how soon the world will return to some normalcy. The virus is still continuously producing variants due to mutation in its genome. WHO and the related experts continuously monitor these variants which may be characterized as the variants being monitored (VBMs), the variants of Interest (VOIs), the variants of Concern (VOCs) and the variants of high consequence (VOHCs). Some of which such as delta plus have been considered to be of concern. VBMs are the least harmful and pose no risk to public health while the VOHCs are the most dangerous type which significantly impact the disease spread and severity and reduce the effectiveness of diagnosis, therapeutics and vaccination. Alpha, beta, gamma, delta, epsilon, ita, iota, kappa, zeta and mu have been designated as the variants being monitored for the Covid-19 disease. These belong to different lineages and sublineages of the SARS-COV-2 virus and were earlier labeled as the VOIs or the VOCs. The delta variant (B.1.617.2) was earlier classified as the variant of concern on May 11, 2021 because of its higher transmissibility than the alpha variant (B.1.1.7). Omicron (B.1.1.529) is labelled as the variant of concern owing to its increased transmissibility. Omicron is currently the dominant variant circulating globally. Due to the global spread of the Omicron VOC and the associated projected rise in viral diversity, WHO has included "Omicron subvariants under monitoring" to its variant tracking system (WHO, 2023a). BF.7, BQ.1, BA.2.75 and XBB are the omicron variants under monitoring. XBB.1.5, a descendent of the lineage of XBB has been recently (January 25, 2023) upgraded from low to moderate risk assessment (WHO, 2023b). The emerging variants may have the potential to bypass the immunity gained through vaccination. It means there could be fresh waves of infections even in populations that were being considered close to reaching community-level protection, which is what happened in many countries in Europe, particularly in the UK. Even in India, Kerala and Maharashtra saw an unprecedented increase in Covid cases especially during the second wave. Though the first wave waned in early 2020, the second wave wreaked havoc on humanity in countries including India and was much more damaging with a great loss of human lives. After June 2021, there was a reduction in fatalities and new cases, but the third wave hit certain nations including India.

The virus spread at an uncontrollably fast pace but thanks to the scientific and technological advances which made the production of a number of vaccines possible.

Coronaviruses are a family of viruses causing respiratory and gastrointestinal infections. They are enveloped viruses having positive-sense single-stranded RNA and a nucleocapsid with helical symmetry. The genome size of these viruses is 26 to 32 kilobases. They derive their name from an envelope outside them that has protein spikes, giving the appearance of a crown (corona means crown in Latin). It is a zoonotic disease with bats and pangolins as suspected animals for transmission.

The most common symptoms of the disease include fever, cough, difficulty in breathing, loss of appetite, and headache. Some patients also reported sore throat, lesions on skin, sneezing, and runny nose. Many patients also complained of nausea, diarrhea, vomiting, or loss of smell and taste as symptoms of the disease. However, the difficulties that have emerged following the so-called recovery are quite concerning, as are the related symptoms in multiple organs. Covid-19 has a proven human-to-human transmission through large droplets or touching of infected secretions. Researchers have shown that simple human activities like coughing, sneezing, chatting, etc. have a significant influence in the transmission of illness. It also reportedly spreads by touching nose, mouth, or eyes after touching contaminated surfaces and objects, on which the virus has a varying length of viability. It has been reported to survive on plastic for up to 3 days and on cardboard for one day. A person with Covid-19 can spread the virus for 7 to 15 days after infection. The role of asymptomatic carriers is still uncertain.

Certain categories of people such as those with diabetes, cardiovascular disease, chronic respiratory disease, the elderly (especially those above 70 years), hypertensive patients, and those with chronic liver disease are at a high risk of developing the infection.

Since there was no drug and/or vaccine available for about a year, the focus of the governments of various countries was on flattening of the curve, decreasing the infectivity rate, and increasing the efficacy of treatment. The CDC and the WHO advised people to wear masks especially in public places, follow social distancing at crowded places, stay at home unless essential, wash hands frequently, and not to touch body parts with dirty hands so as to reduce the spread of the disease. It was directed by the governments of all nations to wear a good quality mask to avoid the spread of disease by asymptomatic and pre-symptomatic persons. Masks were found to reduce the velocity, volume, and travel distance of droplets from infected patients. In the absence of a mask, use of tissues, use of elbow while sneezing or coughing, and thoroughly washing hands with soap and water are advisable.

The role of social distancing in curbing the disease transmission was also highlighted and people were advised to maintain a distance of atleast 6 feet to reduce the rate of disease spread. For this reason, curfew or lockdown was also imposed by certain countries. There were restrictions in various countries including the closing of malls, schools, gyms, colleges, swimming pools, etc.

For healthcare professionals who were in direct contact with patients, wearing of PPE kits or special safeguard was required. Most importantly, a good personal hygiene with good diet to boost the immunity was also highly recommended. Washing hands with soap and water for at least 20 seconds or use of sanitizer with at least 65% alcohol was recommended by the CDC. A number of frequently touched surfaces also pose a great threat to general public and sanitation of these surfaces frequently was advised as a precautionary measure.

People under home quarantine or institutional quarantine were required to take special precautions and care. They had to self-isolate and avoid sharing utensils, etc., and had to wear masks even at home. Thus, all the precautions mentioned above became a part of our daily lives.

Though development of any vaccine for human use is a long process involving many years of research and development, checking of safety and efficacy, the scientific community has been very innovative to produce about half a dozen vaccines in different countries in less than a year. Hundreds of them are under investigation. Even non-invasive vaccines to be inhaled through the nose are under trial. There are reports that development of certain vaccines in the form of pills are in the pipeline.

1.2 IMPACT OF COVID-19 ON SCIENCE AND TECHNOLOGY

Population increase and industrialization, urbanization as well as anthropogenic activities all over the world have had an impact on the function of the food web, climate cycles, and population dynamics. Because of this ecological imbalance, the majority of diseases identified in one group's organisms begin to spread to the species of other groups. Many diseases such as the plague, rabies, bird flu, swine flu, and others have been discovered by scientists to be slowly spreading from animals to humans, resulting in an increase in death tolls. As mentioned above, towards the end of December 2019, a new zoonotic coronavirus (SARS-CoV-2) was identified in Wuhan, China. In the year 2020, the virus began to spread throughout the world at an amazingly fast pace resulting in 18.6 Cr infected cases and 39.8 L deaths by the end of June 2021. This sparked the greatest scientific revolution all over the world. Thousands of researchers abandoned whatever intellectual difficulties had previously piqued their interest in favour of tackling the pandemic (Ed Yong, 2021). Within a couple of months, most scientific research focused almost completely on Covid-19. Prior to this, there were only a few scientists all over the world who were doing their research on the coronavirus and many did not even know about it. Universities, research institutions, and laboratories offer the perfect environment for the development of new ideas and the extension of future agendas for scientific progress. But the Covid-19 pandemic had a huge negative influence on the scientific community, with universities, research institutions, and laboratories being forced to close. Most scientific activities, such as national and international conferences, symposia, workshops, and training programs, were cancelled, postponed, or shifted to an online format (Drake, 2020). Furthermore, researchers from low- and middle-income countries missed out on numerous opportunities for research awards and travel grants for scientific conferences in 2020. The closure of scientific workplaces or institutions resulted in an extension of research time, and in some cases, even restarting of entire experiments, or the putting of experiments on hold, or the scaling down of experiments to the bare minimum. It is likely that as a result of this, researchers will face an increased economic burden, which will negatively impact their career opportunities, ultimately leading to increased psychological stress, anxiety, tension, or depression, which will ultimately result in decreased scientific output.

As a result of the Covid-19 pandemic a completely new research environment, one that encourages collaboration and communication, was ushered in. Since the release of the first genome sequence of the novel coronavirus in February 2020, scientific progress was exponentially accelerated. Within a year, scientists planned to decode the virus's molecular properties, learn about its mode of propagation and the people most at risk, and create diagnostics, therapies, and vaccinations to battle it. According to the Dimensions Covid-19 data set, researchers published over 38,000 SARS-CoV-2 preprints in the year 2020 (Halford et al., 2021).

In order to develop a safe and effective vaccine for Covid-19, which was a tremendous step forward, scientists and pharmaceutical companies from around the world collaborated and shared their knowledge gained from previous coronavirus research. The development of safe and effective vaccines for any disease generally takes about 8–10 years, but the Covid-19 pandemic accelerated science and scientists to adopt new technologies to discover vaccine candidates in a very short period of time, including Covishield, Covaxin, Sputnik, Pfizer-BioNTech Covid-19 vaccine, moderna mRNA-1273 vaccine, and others. Many of the thousands of clinical trials launched were too small to yield statistically significant results, but the Food and Drug Administration (FDA) granted Emergency Use Authorizations (EUA) for those vaccines (National Center for Immunization and Respiratory Diseases, 2021). Vaccines will not put an end to the pandemic right away. To prevent mortality, millions of doses will need to be manufactured, allocated, and distributed globally. Research is being carried out to develop non-invasive vaccines (e.g., nasal drops and pills). Recently, the Drug Controller General of India (DCGI) gave approval to BBV154 COVID Vaccine, India's first intranasal vaccine for the disease.

As mentioned earlier the Covid-19 virus is transmitted mostly through droplets in the air and social interactions. Wearing a medical mask is one of the preventive methods that can help to minimize the spread of Covid-19 and other respiratory viral illnesses (WHO, 2020). There has been little research into the use of masks made from various materials (e.g., cotton fabric), also known as nonmedical masks, in the research community, but we know with a high level of compliance that mask wearing is effective at reducing the spread of the virus.

The virus that caused the Covid-19 pandemic is still multiplying, mutating, and spreading. Since the outbreak, researchers throughout the world have discovered several coronavirus variants that differ from the "original" strain. In order to identify and learn about the new strains of the coronavirus mutates genome sequencing is used to examine samples from sick people. Sequencing enabled the world to rapidly identify SARS-CoV-2 and develop diagnostic tests and other tools for outbreak management. In the early stages of the pandemic, India began sequencing SARS-CoV-2 genomes from positive samples. In December 2020, the Ministry of Health and Family Welfare (MoHFW) formed the Indian SARS-CoV-2 Genomics Consortium (INSACOG) for centralized genome surveillance in coordination with the Department of Biotechnology (DBT), MoHFW, Indian Council of Medical Research (ICMR), and Council of Scientific and Industrial Research (CSIR), which brought together 28 national laboratories from around the country with a cumulative sequencing capacity of more than 25,000 samples a month (https://dbtindia.gov.in/insacog). However, the situation has changed, and gene sequencing has begun in most of India's medical colleges and government laboratories to investigate virus genome variants.

According to epidemiological statistics recorded from around the world males have a greater morbidity and mortality rate than females. The immunological response of men and women to inflammatory disorders and Covid-19 infection differs significantly. Because of differences in innate immunity, steroid hormones, and characteristics associated with sex chromosomes, women are less susceptible to viral infections than men. Even if one X chromosome is inactive, the presence of two X chromosomes in women boosts the immune system. In females, the immunological regulating genes expressed by the X chromosome result in lower viral load and inflammation than in males (Bwire, 2020; Conti et al., 2020).

The virus has taken a terrible toll on lives, health services, careers, and mental health around the world. Nobody knew how long the coronavirus epidemic would continue, leaving an uncertain future with just one option: adapt and find ways to survive and remain productive. Despite all the above-mentioned negative effects on science, this pandemic has bestowed upon us the gift of a "global perspective," which is a true silver lining. The novel coronavirus is providing us with novel experiences and teaching us how to approach novel obstacles in novel ways.

1.3 IMPACT OF COVID-19 ON HEALTHCARE PROFESSIONALS AND SERVICES

As discussed, Covid-19 hit healthcare hard. Active action plans were required on a priority basis to overcome the sudden challenges. In order to ensure effective health care for society robust, collaborative, and reliable healthcare substitutes were evolved and practiced. Ensuring safety, training available manpower, maintaining social distancing, and procuring and using PPE kits and modified equipment were some of the major challenges in hospitals. Due to the ongoing situation the demand for machine learning and AI-aided electronic device systems was extensively increased. More emphasis was placed on self-care by patients, home-based elderly care, and rehabilitation.

Fear and social distancing decreased the number of patients in almost all hospitals. This resulted in a drastic fall in the income of hospitals. Moreover, the purchase of many needed consumables of the medical sector further added to the running cost of hospitals. This situation resulted in difficulty in importing many consumables and raw materials from other countries when their demand increased because of the steep rise in Covid-19 cases leading exhaustion of resources and infrastructure.

The superfast transmission of the coronavirus disease required contact tracing and this was not possible manually. One of the feasible solutions appeared to be the digitalisation of the data collection systems. The digitization ensured tracking of infected patients or those at high risk in an easy and quicker manner. It provided information about the travellers, both cross- country and within the country therefore helped in timely isolation of the suspected cases. Moreover, the relevant details of the contacts of patients could also be gathered and were useful in contact tracing of a confirmed positive case.

1.3.1 Telemedicine/Telehealth

As Covid-19 spread from nation to nation, health officials urged for the closure of businesses, schools, and cancellation of non-emergency medical visits. Safer at home protocols and social distancing left healthcare providers with a challenge. Covid-19 created an urgent need for telehealth services and what was once a service labeled as low priority soon became a necessity. The healthcare industry had to get creative in ways to help keep their patients healthy, so most facilities moved towards implementing telemedicine. Chatbots or webbots could provide relevant information to people and answer general health questions. Telehealth was particularly useful for home quarantined people who could connect to healthcare workers digitally. The government of India also launched AAROGYA Setu App for this purpose. Digital tools like this further helped healthcare workers in remote areas or villages connect to administrative portals for exchange of information or for data management and resource management solutions.

With an increasing number of Covid-19 cases, the visits of various patients were being discouraged or postponed, if not urgent. To prevent the spread of disease, robots were used for direct examination of patients. To reduce the risk for healthcare workers, carts equipped with cameras, screens, and medical equipment were developed. Various non-invasive monitoring devices were developed that could be connected through Bluetooth, Wi-Fi, etc., to transfer readings to physicians for expert advice.

However, in case of emergency, physical interaction must not be delayed. In this regard, various changes were introduced in the field of diagnosis too, with the aim of providing patient care and protecting healthcare workers. Drive-through stations were promoted where patients could be tested without actually getting out of the vehicle. In addition, walkthrough testing booths were also set up.

Above all, the mode of interaction of physicians amongst themselves as well with the government witnessed a complete change. Virtual meetings and webinars replaced physical meetings and conferences. While this change made it hassle-free to hold webinars and meeting as far as travelling is concerned, the effectiveness of virtual meetings has yet to be determined.

Frontline healthcare workers have been the backbone of the Covid-19 pandemic. When the rest of the world was shutting down, nurses and doctors across the globe were rushing in to do their part and help. If nothing else positive came from the Covid-19 pandemic, the global community has acquired a deeper appreciation for our healthcare workers who have braved long hours with added stress and anxiety during this pandemic.

Healthcare facilities had to implement emergency plans, and some were more prepared than others. When the infection first spread in communities, healthcare facilities started to prepare for when their communities would get hit with an outbreak. Many started to cancel or postpone elective appointments and procedures. The pandemic has rightfully been the focus for much of the medical community as it continues to wreak havoc around the world but it also impacted other areas of health that were postponed or ignored, namely the delay of routine cancer screenings such as mammograms and colonoscopies that require attention.

From the very beginning of this pandemic, healthcare systems all across the globe faced a lack of PPE. Even before the major spikes in various hotspots across the globe, many hospitals struggled to secure enough for their medical staff. The lack of supplies also caused some industries to stop their normal production and switch to manufacturing life-saving equipment in fight against Covid-19. Companies like Ford Motor stopped vehicle production and focused on helping to meet the demand for ventilators (Iyengar et al., 2020). Companies put in lots of time into research and development of 3D printing and how these machines could be utilized to create parts of ventilators and help in manufacturing of some equipment needed to fight against Covid-19.

As the risks of Covid-19 wane, it will be interesting to see in what ways telehealth remains relevant and if it is found to be an efficacious solution. Being able to continue care through virtual visits is likely to revolutionize the healthcare industry. Patients can now have virtual meetings with the providers not only at a lower cost, but also from anywhere with access to the internet (Spinelli and Pellino, 2020).

There are many benefits to using telehealth during emergency situations such as less exposure to illness, saving time, and lower cost since these visits are cheaper. However, there are struggles with using telehealth, mainly related to the digital divide in society. The two big factors that can take away from access to telehealth services are ability to use technology and access to technology (Ramsetty and Adams, 2020).

1.3.2 Psychological Impact on Healthcare Professionals

Healthcare workers dealing with Covid-19 faced increased psychological stress and high rates of psychiatric morbidity similar to the scenario during the SARS and H1N1 outbreaks. Because of the higher risk of viral exposure, frontline physicians, nurses, and healthcare workers were concerned about getting infected and about infecting loved ones and family members, such as elderly parents, newborns, and immuno-compromised relatives. This led to stress and anxiety.

Our healthcare personnel also reported higher stress when dealing with difficult patients who did not follow safety guidelines, as well as feelings of helplessness when dealing with severely sick patients due to the lack of a definite therapy, as well as insufficient intensive care beds and resources (Tsamakis et al., 2020).

A survey of approximately 1,300 healthcare workers treating Covid-19 patients in Chinese hospitals revealed significant rates of stress, distress, anxiety, and insomnia. Guilt, rage, anxiety, fear, embarrassment, and melancholy were all expressed, resulting in resignations and poor work performance. During the Covid-19 pandemic, there were reports of suicides among healthcare personnel in Europe (Lakhani et al., 2020).

Chronic sleep deprivation affected focus, attentiveness, short-term memory, recall capacity, and impaired motor skills and clinical judgment. Chronic stress caused health problems such as backache, exhaustion, headache, irritable bowel syndrome, anxiety, etc. Diabetes, hypertension, and chronic respiratory disorders all increased one's susceptibility to corona-related problems.

To provide healthcare professionals with a sense of security, the Indian government implemented an amendment bill to protect healthcare professionals from the Covid outbreak. This might be seen as a positive impact of Covid-19 on the security of healthcare providers.

1.3.3 Social and Economic Impact on Healthcare Workers

The expected economic impact of Covid-19 on productivity rates has been extensively examined. However, the social and economic impact on healthcare professionals has largely been ignored. That is also an important aspect to consider.

While healthcare workers were busy dealing with Covid-19 patients, hospitals suffered from the decline in other patients. Some primary care practices have reported reductions upto 70%. in the use of other healthcare services The private healthcare sector witnessed an 80% fall in patient visits due to the lockdown and testing volumes and revenue drop of 50–70% during the Covid pandemic (The Hindu Business line, 2020). Many small hospitals and nursing homes, especially in Tier-II and -III cities, were forced to shut down their operations since their cash flows dried up. Salaries of clinical staff were reduced or frozen, and some staff was furloughed. Increase in costs owing to infection control and PPE also needed to be accommodated for.

During the time of this pandemic, hospitals and medical professionals from doctors to nurses to support staff, who are the frontline workers fighting the war against Covid, have faced difficult times. There is an urgent call for action to address the immediate need of the sector and consider the recommendations for financial stimulus for the private healthcare sector too. With an estimated impact of 14,000–24,000 crores in operating losses in the first quarter of the year 2020, the sector required liquidity infusion, indirect and direct tax benefits, and fixed costs.

Covid-19 has had a dramatic influence on the social and emotional wellbeing of people. An increase in post-traumatic stress disorder (PTSD) in the general population along with depression was reported (Torales et al., 2020). Many people felt loneliness due to isolation, anxiety and stress, and other psychological symptoms were also observed in people who were not experiencing mental health issues pre-Covid. Concernswere raised about quarantine being a traumatic event with related sequela. Surveys have shown that 45% of adults in the United States have experienced a decline in their mental health because of the stress and uncertainty in the pandemic. Similar studies have shown that during the state of the outbreak,the sales of alcoholic beverages increased. This is also a clear indication of the rise in stress. It is not only the uncertainty of what Covid-19 brings in terms of health, but also its affect on finances. Many people lost their jobs or had to stay home to take care of children and aging parents. Many families lost their family's earning member too.

A study in China found that females, young adults, migrant workers, the elderly, and those with higher education showed higher rates of reported symptoms due to traumatic stress. There is a concern that since gyms, churches, temples, and structures conducive to coping with stress were shut down, people turned to alcohol and other substances for reprieve (Brown and Schuman, 2020).

1.4 EFFECT OF COVID-19 ON EDUCATION SECTOR

Covid-19 had a significant effect on the education sector due to lockdowns. Many teachers and students had to revert to online learning. When schools did reopen fewer kids returned due to fear of Covid variants and further exposure to the virus. The effects are still ongoing since some families have lost jobs and have had to relocate or have lost family members and thus a loss of income.

Covid also provided several difficulties and possibilities for educational institutions to upgrade their infrastructure (Jena, 2020). Teachers provided assignments to students through the internet and presented lectures through video conferencing utilizing various apps such as Zoom, Google Meet, Facebook, Youtube, and Skype, among others (Jena, 2020). WhatsApp groups were made for guardians, teachers, and students for effective communication. A shift from traditional to digital learning is not easier in developing countries like India, where not all students have access to high-speed internet and digital devices (Jena, 2020). However, many advanced educational institutions in India have digital facilities to deal with the abrupt transition from traditional education to online education. To improve the e-learning experience, educational institutions must follow government standards and suggestions, while also encouraging students to continue learning remotely during the pandemic (Aucejo et al., 2020). Bao (2020) proposed five high-impact criteria for the effective implementation of online education.

While some educational institutions have been able to stay open by changing their standard operating procedures and organizational behaviour to maintain social distancing (e.g., by allowing students to return in shifts to reduce class sizes), it was difficult for teachers and students. When schools were forced to rely on online learning the overall learning process was hampered since students did not have the physical presence of a teacher in the room and might not be able to ask questions or get the help they need.

Equally at loss were the graduate and Ph.D. students, who were unable to complete projects or theses on time due to missing lab work or inability to do research, conduct interviews, etc. As reported by Iivari et al. (2020) school closures forced technological advancements and some schools did not have the necessary equipment and some families also did not have access to computers or the internet. Also, less privileged students, and those in remote areas, lag behind because of poor internet service. However, some countries have found ways to address this issue. For example, in India, Massive Open Online Courses (MOOC) on SWAYAM platforms have been being developed, and chatbots have been used to provide answers to frequently asked questions by students, even during off-hours.

In response to the difficulties the pandemic has created, the government of India, as well as state governments and private sector companies, have adopted appropriate measures. For example, the Ministry of Human Resource Development (MHRD) has created online portals and educational channels via Direct to Home TV and radios. The MHRD's ICT project (eBroucher – https://mhrd.gov.in/ict-initiatives) is a one-of-a-kind portal that incorporates all digital resources for online education.

The implications of closed schools could be severe. Children need interaction for social development. The result has been psychological issues such as frustration, stress, and sadness (Chaturvedi et al., 2021). There is an urgent need to study the effects of the pandemic on student learning (Cook, 2009).

The Covid-19 pandemic has had a greater negative impact on students from lower socioeconomic backgrounds (Aucejo et al., 2020). Reduction in family income, limited access to digital resources, and the high cost of internet connectivity have disrupted the academic life of students. Moreover, changes in daily routine, such as a lack of outdoor activity, abnormal sleeping habits, and social distancing negatively impacted the mental well-being of students. Cao et al. (2020) used a 7-item Generalized Anxiety Disorder Scale (GAD-7) as a diagnostic tool for the assessment of anxiety disorders, panic disorders, and social phobia (Cao et al., 2020). Further, Ye et al. (2020) analyzed mediating roles of resilience, coping, and social support to deal with psychological symptoms (Ye et al., 2020).

The pandemic has also changed the availability of jobs due to the impact on the industrial sector. It is feared that the trend may go on for quite some time. The result is that students are shifting toward vocational rather than traditional courses. More and more students are also staying closer to home instead of attending institutions abroad. Sports activities were hit hard. To avoid undue transmission of the disease, gyms and swimming pools were closed down.

1.4.1 Positive Impact of Covid-19 on Education

Despite the fact that the Covid-19 pandemic has had a detrimental influence on education, educational institutions in India have faced the challenges and are doing their best to give students support services during this difficult time. The Indian education system successfully transitioned from a conventional educational system to a new era. The following factors may be seen as having a beneficial impact (Jena, 2020):

- Easy accessibility from anywhere
- Transition to blended learning
- An increase in the usage of learning management systems (Misra, 2020)
- Increase in the utilization of soft copy learning materials (Jena, 2020)
- Development of collaborative and interdisciplinary work (Misra, 2020)
- An increase in online personal meetings
- Improved digital and information literacy
- Increased use of electronic media for information generation, dissemination, and sharing
- Global exposure
- Improved time management
- Increase in demand for open and distance learning (ODL)

During Covid-19, amidst lockdowns when infrastructures were being expanded to reach more students by using digital tools and content, resources such as the All Indian Institute of Medical Sciences, New Delhi, and PGIMER, Chandigarh, India could be accessed (Ambekar et al., 2019; Bhattacharya et al., 2020). These organizations provided open-access education to the general public as well as healthcare professionals, with a focus on Covid-19-related health concerns. Such efforts showed great potential in resource-constrained situations, but still require additional growth to improve the content in order to fulfill the rising demand for MOOCs.

1.5 THE IMPACT OF COVID-19 ON LIVELIHOODS AND LIFESTYLES

Due to the financial crisis resulting from the pandemic, livelihoods have been lost, emergency savings consumed, and loans borrowed to meet basic requirements leading to excessive financial burdens on earning family members. The only known prophylaxis for the disease is hygiene but developing countries like India have a large sector of population residing in slums with poor waste disposal mechanisms and sanitary conditions. Also, these areas are generally heavily populated. Therefore, growing up in underprivileged or low income areas increases the risk of catching the virus and being a carrier, experiencing underlying health issues, and reduced coverage of vaccination among children. It also affects access to a range of necessities such as good nutrition, quality housing, sanitation issues, space to play or study, and opportunities to engage in online schooling (Golberstein et al., 2019; Nicola et al., 2020; Van Lancker and Parolin, 2020).

A higher level of poverty following any crisis significantly increases the risk of child marriage and teenage pregnancies especially in low-income countries (Marshan et al., 2013; Cas et al., 2014).

There may also be an increase in child trafficking and pornography (Haleemunnissa et al., 2021). The total effect of the Covid-19 pandemic is projected to result in 13 million additional child marriages (Fraser 2020; Giannini and Albrectsen, 2020; Golberstein et al., 2020). Domestic violence services and children's helplines have also reported an increased level of risk for vulnerable children and families (Women's Safety, 2020).

Social media has played a crucial role in spreading awareness and knowledge about public health; however, it has also been misused for encouraging fake news, hatred, and racism (Larson 2018, 2020; Depoux et al., 2020; Kadam and Atre, 2020). Due to existing medico-pleuralism in India, messages containing fake claims about use of herbal and immunity-booster medicines, religious, and spiritual ways for prevention and treatment were widely circulated, which added to the confusion about Covid-19. The confusion was also due to lack of knowledge about non-pharmaceutical interventions like social distancing, quarantine, and isolation.

1.6 IMPACT OF COVID-19 ON ENVIRONMENT AND ECONOMY

1.6.1 Environment

The implications of the Covid-19 pandemic have not all been negative. The global climate saw some improvement throughout the pandemic as a decrease in air pollution and greenhouse gas emissions due to the reduction in transportation and electricity generation was observed (Markard and Rosenbloom, 2020). In addition, with lockdowns leading to business shutdowns, industrial production was also reduced or completely halted.

The Covid-19 pandemic and response measures have significant short- and long-term effects on the macroeconomic activity as well as on the structure of the economy. The structure of the economy plays a key role in how economic effects translate to effects on environmental pressures.

The short-term reductions in environmental pressures are very significant: energy-related emissions declined by 7% and agriculture-related environmental pressures by less (around 2%), and the reduction in the use of non-metallic minerals, including construction materials, reached double digits (Delink et al., 2021).

Long-term changes in environmental pressure depend crucially on their economic drivers and regional impacts. Some sectors (e.g., manufacturing and construction) are more affected than others (e.g., agriculture). Regional differences are also large, with strong long-term effects (e.g., India).

There is a projected long-term – potentially permanent – downward impact on the levels of environmental pressures of 1–3%, depending on the indicator. A slow recovery can double these impacts (Delink et al., 2021).

Due to lockdowns, there was a significant reduction in transportation services, resulting in a decrease in emissions (Venter et al., 2020). For example, the restrictions in Italy were significant enough that the waterways in the city of Venice had clear water for a long time (Braga et al., 2020). In north India there was a significant reduction in carbon monoxide levels because most industries were shut down during this period. China also saw the effects of their shutdown when during the first seven weeks of 2020, they witnessed reduction in their aerosol optical depth (AOD) (Lal et al., 2020). These lockdowns resulted in little or no vehicular movement around the globe, thereby reducing emission rates from vehicular traffic to almost zero. The lockdowns also reduced industrial emissions, which had a significant impact on the global environment and may have allowed time for the ozone layer to revive itself some (Iivari et al., 2020). Due to social distancing, many beaches witnessed fewer crowds and cleaner waters (Zambrano-Monserrate et al., 2020). Some places like Acapulco and Barcelona saw noticeably clearer water along with clean beaches.

While the decrease in pollution has been positive for the environment, the pandemic quarantine also has had a negative impact as more and more people have resorted to online shopping and goods being delivered to their homes. This behaviour has led to an increase in household wastes, particularly packaging and shipping materials (Zambrano-Monserrate et al., 2020).

Another area of increase in waste was from the medical industries as there was an increase in testing and also in hospitalizations due to Covid-19. Hospitals in Wuhan, China saw an increase in hospital waste, averaging about 240 metric tons of waste, almost five times as much as they were producing pre-Covid (Zambrano-Monserrate et al., 2020). With a projected CAGR of 5% from 2020 to 2025, the worldwide waste management market is projected to reach a value of USD 383.83 billion by that year (2021–2026).The amount of biomedical waste being generated from Covid-treating hospitals, quarantine centers, healthcare facilities, and self/home-quarantine has triggered the need for medical waste management (www.researchandmarkets.com).

1.6.2 Economy

Covid-19 has impacted the economy on a global scale due to lockdowns. Lockdowns across the globe interrupted supply chains by altering normal demand and halting supply. The economic impact of the Covid-19 pandemic in India has been largely disruptive. India's growth in the fourth quarter of the fiscal year 2020 went down to 3.1% according to the Ministry of Statistics. Almost all the sectors have been adversely affected including food and agriculture, pharmaceuticals, oil & gas, telecom, aviation, and tourism and other fields as well.

In view of the scale of disruption caused by the pandemic, it is evident that the downturn is fundamentally different from recessions. The sudden shrinkage in demand and increased unemployment is going to alter the business landscape. Social relationships in the context of urban-rural divide, male-female divide, and age-related divide within the family and outside are necessarily being addressed by new perspectives. The social relationship of individuals, families, and institutions now require a culture-specific response, which is a task that respective societies and countries can only handle. In societies, for instance, India, where families have always played the most dominant role for each member's well-being, more considerate ways are expected to be adopted by every member towards each other, especially older men and women and children in a family. The pandemic also unleashed a tsunami of hate and xenophobia, scapegoating, and scare-mongering, which brought changes in international politics that may be irrevocable. The pandemic also amplified a trade war between the United States and the People Republic of China (PRC). Various international organizations, such as SAARC, G20, and EU, attempted to collaborate with the nations to combat the pandemic. The pandemic also caused diplomatic tensions over the distribution of medicines, diagnostic tests, and other hospital equipments.

Covid-19 has impacted the economy on a global scale due to the lockdowns that had many industries, except for essential goods and services, closed for many months. The lockdowns across the globe have interrupted the normal supply chain by altering normal demand and halting supply.

Throughout the globe, fear of an economic recession has caused many to panic. In the United States alone, Bloomberg Economics reported that the full year GDP growth might fall to zero in the worst-case scenario of the Covid-19 pandemic (Orlic et al., 2020). Every industry, from petroleum to finance to hospitality, has been affected by this pandemic.

Due to the epidemic, the economy shrank by 6.6% in fiscal 2021, before showing signs of recovery in fiscal 2022, when it increased by 8.7% (CRISIL, 2022).Unemployment is on the rise, and it was already at a 45-year highest as per NSSO 68th Round. Small businesses have been made unviable. To revive economic activities, the Government of India announced 20 lakh crore relief packages, of which a major part was in the form of liquidity and credit facilities, implemented through different segments of the financial system. Experts advocated measures to support the urban poor, upgrade rural health infrastructure, bolster employment, and protect livelihoods. A research report by the US-based Pew Research Centre in March 2021 revealed that 32 million Indians were moved to the lower income group. Shutdowns and layoffs in factories have pushed employment-providing sectors – construction, retail trade, hospitality, and small-scale manufacturing – to low levels.

1.7 TECHNOLOGICAL IMPACT

The pandemic gave a boost to digital learning, though by necessity. Whether you are working from home, getting an education from home, communicating with friends and family, or just trying to entertain yourself in isolation, technology has had a significant impact on our lives during this pandemic. The use of telecommunications tools like Zoom and Skype has become practically mandatory at this time. One of the major benefits of technology today is that through programs such as Zoom, Microsoft Teams, Google, Skype, and Facebook, we have been able to connect with friends and family and to hold work meetings or online classes.

Most businesses also started using Zoom and other similar services for business meetings to keep their employees safe. Many industries also came to see that even though they were forced into using newer technology it was realized that many things their employees were doing could be done from their homes. Larger businesses extended work from home schedules for their employees.

In brief, Covid-19 has transformed social, political, and economic systems and turned challenges into opportunities in the new normal without discrimination of geographical bounds. In the backdrop of all these facts the present book covers various aspects on the impact of Covid-19 on human civilization as a whole including on lifestyle, family systems, and political and economic systems.

REFERENCES

Alanzi, Turki. "A review of mobile applications available in the app and google play stores used during the COVID-19 outbreak." *Journal of Multidisciplinary Healthcare* 14 (2021): 45. https://doi.org/10.2147/JMDH.S285014

Alsafi, Zaid, Abbas Abdul-Rahman, Hassan Aimen, and Ali Mohamed Adam. "The coronavirus (COVID-19) pandemic: Adaptations in medical education." *International Journal of Surgery* 78 (2020): 64–65. DOI: 10.1016/j.ijsu.2020.03.083

Ambekar, A., A. Agrawal, R. Rao, A. K. Mishra, S. K. Khandelwal, and R. K. Chadda. "On behalf of the group of investigators for the National Survey on Extent and Pattern of Substance Use in India (2019)." *Magnitude of Substance Use in India* (2019).

Aucejo, Esteban M., Jacob French, Maria Paola Ugalde Araya, and Basit Zafar. "The impact of COVID-19 on student experiences and expectations: Evidence from a survey." *Journal of Public Economics* 191 (2020): 104271. https://doi.org/10.1016/j. jpubeco.2020.104271

Baker, Scott R., Nicholas Bloom, Steven J. Davis, and Stephen J. Terry. *Covid-induced economic uncertainty*. No. w26983. National Bureau of Economic Research, 2020.

Bao, Wei. "COVID-19 and online teaching in higher education: A case study of Peking University." *Human Behavior and Emerging Technologies* 2, no. 2 (2020): 113–115. https://doi.org/10.1002/hbe2.191

Bhattacharya, Sudip, Amarjeet Singh, and Md Mahbub Hossain. "Health system strengthening through Massive Open Online Courses (MOOCs) during the COVID-19 pandemic: An analysis from the available evidence." *Journal of Education and Health Promotion* 9 (2020). https://doi.org/10.4103/jehp.jehp_377_20

Braga, Federica, Gian Marco Scarpa, Vittorio Ernesto Brando, Giorgia Manfè, and Luca Zaggia. "COVID-19 lockdown measures reveal human impact on water transparency in the Venice Lagoon." *Science of the Total Environment* 736 (2020): 139612. https://doi.org/10.1016/j.scitotenv.2020.139612

Brewin, Chris R., Nika Fuchkan, Zoe Huntley, M. Robertson, M. Thompson, Peter Scragg, Patricia d'Ardenne, and Anke Ehlers. "Outreach and screening following the 2005 London bombings: usage and outcomes." *Psychological Medicine* 40, no. 12 (2010): 2049–2057.

Brown, Sabrina, and Donna L. Schuman. "Suicide in the time of COVID-19: A perfect storm." *The Journal of Rural Health* (2020). Accessed January 6, 202. www.npr.org/2021/01/06/953254623/massive-1-year-rise-in-homicide-rates-collided-with-the-pandemic-in-2020

Bwire, George M. "Coronavirus: Why men are more vulnerable to covid-19 than women?" *SN Comprehensive Clinical Medicine* 2, no. 7 (2020): 874–876. https://doi.org/10.1007/s42399-020-00341-w

Cao, Wenjun, Ziwei Fang, Guoqiang Hou, Mei Han, Xinrong Xu, Jiaxin Dong, and Jianzhong Zheng. "The psychological impact of the COVID-19 epidemic on college students in China." *Psychiatry Research* 287 (2020): 112934. https://doi.org/10. 1016/j.psychres.2020.112934

Cas, Ava Gail, Elizabeth Frankenberg, Wayan Suriastini, and Duncan Thomas. "The impact of parental death on child well-being: Evidence from the Indian Ocean tsunami." *Demography* 51, no. 2 (2014): 437–457.

Chaturvedi, Kunal, Dinesh Kumar Vishwakarma, and Nidhi Singh. "COVID-19 and its impact on education, social life and mental health of students: A survey." *Children and Youth Services Review* 121 (2021): 105866. https://doi.org/10.1016/j.childyouth.2020.105866

Cohen, Alison K., Lindsay T. Hoyt, and Brandon Dull. "A descriptive study of COVID-19–related experiences and perspectives of a national sample of college students in spring 2020." *Journal of Adolescent Health* 67, no. 3 (2020): 369–375. https://doi.org/ 10.1016/j.jadohealth.2020.06.009

Conti, Pio, and A. Younes. "Coronavirus COV-19/SARS-CoV-2 Affects Women Less than Men: Clinical Response to Viral Infection." *Journal of Biological Regulators and Homeostatic Agents* 34, no. 2 (2020): 339–343. https://doi.org/10.23812/Editorial-Conti-3

Cook, David A. "The failure of e-learning research to inform educational practice, and what we can do about it." *Medical Teacher* 31, no. 2 (2009): 158–162. https://doi.org/10.1080/ 01421590802691393

CRISIL. Indian Economy: States of Aftermath. July 29, 2022. www.crisil.com/en/home/our-analysis/reports/2022/07/crisil-insights-indian-economy-states-of-aftermath.html#:~:text=The%20economy%20contracted%206.6%25%20in,%2C%20when%20it%20grew%208.7%25

Dellink, R., et al. "The long-term implications of the COVID-19 pandemic and recovery measures on environmental pressures: A quantitative exploration," OECD Environment Working Papers, No. 176, OECD Publishing, Paris (2021). https://doi.org/10.1787/123dfd4f-en

Depoux, Anneliese, Sam Martin, Emilie Karafillakis, Raman Preet, Annelies Wilder-Smith, and Heidi Larson. "The pandemic of social media panic travels faster than the COVID-19 outbreak." (2020): taaa031. https://doi.org/10.1093/jtm/ taaa031

Di Pietro, Giorgio, Federico Biagi, Patricia Costa, Zbigniew Karpiński, and Jacopo Mazza. *The likely impact of COVID-19 on education: Reflections based on the existing literature and recent international datasets*. Vol. 30275. Publications Office of the European Union, 2020.

Drake, Nadia. "How the Coronavirus is Hampering Science." *Scientific American*. March 10, 2020. www.scientificamerican.com/article/how-the-coronavirus-is-hampering-science/

Ed Yong. "How science beat the virus," *The Atlantic*, February 2021. www.theatlantic.com/magazine/archive/2021/01/science-covid-19-manhattan-project/617262/

Fehr, Anthony R., and Stanley Perlman. "Coronaviruses: An overview of their replication and pathogenesis." *Coronaviruses* (2015): 1–23.

Fraser, Erika. "Impact of COVID-19 pandemic on violence against women and girls." *UKAid VAWG Helpdesk Research Report* 284 (2020). www.sddirect.org.uk/media/1881/vawg-helpdesk-284-covid-19-and-vawg. pdf

Giannini, Stefania, and Anne-Birgitte Albrectsen. "Covid-19 school closures around the world will hit girls hardest." *ErişimTarihi* 3 (2020): 2020.

Golberstein, Ezra, Gilbert Gonzales, and Ellen Meara. "How do economic downturns affect the mental health of children? Evidence from the National Health Interview Survey." *Health Economics* 28, no. 8 (2019): 955–970. https://doi.org/10.1002/ hec.3885

Golberstein, Ezra, Hefei Wen, and Benjamin F. Miller. "Coronavirus disease 2019 (COVID-19) and mental health for children and adolescents." *JAMA Pediatrics* 174, no. 9 (2020): 819–820.

Gupta, Bhavna, Sukhminder Jit Singh Bajwa, Naveen Malhotra, Lalit Mehdiratta, and Kamna Kakkar. "Tough times and Miles to go before we sleep-Corona warriors." *Indian Journal of Anaesthesia* 64, no. Suppl 2 (2020): S120.

Haleemunnissa, S., Siyaram Didel, Mukesh Kumar Swami, Kuldeep Singh, and Varuna Vyas. "Children and COVID19: Understanding impact on the growth trajectory of an evolving generation." *Children and Youth Services Review* 120 (2021): 105754.

Halford, Bethany, Laura Howes, and Andrea Widener. "How Covid-19 Has Chanrged the Culture of Science." *C&EN*, January 25, 2021. https://cen.acs.org/biological-chemistry/infectious-disease/How-COVID-19-has-changed-the-culture-of-science/99/i3.

Hasan, Najmul, and Yukun Bao. "Impact of 'e-Learning crack-up' perception on psychological distress among college students during COVID-19 pandemic: A mediating role of 'fear of academic year loss'." *Children and Youth Services Review* 118 (2020): 105355.

Howard, Jeremy, Austin Huang, Zhiyuan Li, Zeynep Tufekci, et al. "An evidence review of face masks against COVID-19." *Proceedings of the National Academy of Sciences* 118, no. 4 (2021). https://doi.org/10.1073/pnas.2014564118; www.researchandmarkets.com/reports/5125736/medical-waste-management-global-market- report? The Business Research Company.

Iivari, Netta, Sumita Sharma, and Leena Ventä-Olkkonen. "Digital transformation of everyday life–How COVID-19 pandemic transformed the basic education of the young generation and why information management research should care?." *International Journal of Information Management* 55 (2020): 102183. Accessed December 2020. https://doi.org/10. 1016/j.ijinfomgt.2020.102183

International Council of Nurses. "High proportion of healthcare workers with COVID-19 in Italy is a stark warning to the world: Protecting nurses and their colleagues must be the number one priority." (2020).

Iyengar, Karthikeyan, Shashi Bahl, Raju Vaishya, and Abhishek Vaish. "Challenges and solutions in meeting up the urgent requirement of ventilators for COVID-19 patients." *Diabetes & Metabolic Syndrome: Clinical Research & Reviews* 14, no. 4 (2020): 499–501. https://doi.org/10.1016/j.dsx.2020.04. 048.

Jena, Pravat Kumar. "Online learning during lockdown period for covid-19 in India." *International Journal of Multidisciplinary Educational Research (IJMER)* 9 (2020): 82–92.

Johnson, Nicole, George Veletsianos, and Jeff Seaman. "US faculty and administrators' experiences and approaches in the early weeks of the COVID-19 pandemic." *Online Learning* 24, no. 2 (2020): 6–21.

Kadam, Abhay B., and Sachin R. Atre. "Negative impact of social media panic during the COVID-19 outbreak in India." *Journal of Travel Medicine* 27, no. 3 (2020): taaa057.

Khan, M., S. F. Adil, H. Z. Alkhathlan, M. N. Tahir, S. Saif, M. Khan, and S. T. Khan. "COVID-19: A global challenge with old history, epidemiology and progress so far. *Molecules* 23;26(1) 2020: 39. DOI: 10.3390/molecules26010039.

Lakhani A, E. Sharma E, K. Gupta, S. Kapila, S. Gupta. "Corona virus (COVID-19) and its impact on health care workers." *Journal of the Association of Physicians of India* 68, no. 9 (2020): 66–69. PMID: 32798348.

Lal, Preet, Kumar Amit, Kumar Shubham et al. "The dark cloud with a silver lining: Assessing the impact of the sars covid-19 pandemic on the global environment." *Science of the Total Environment* 732 (2020): 139297. https://doi.org/10.1016/j.scitotenv.2020.139297

Larson, Heidi J. "Blocking information on COVID-19 can fuel the spread of misinformation." *Nature* 580, no. 7803 (2020): 306–307.

Larson, Heidi J. "The biggest pandemic risk? Viral misinformation." *Nature* 562, no. 7726 (2018): 309–310.

Lee, Joyce. "Mental health effects of school closures during COVID-19." *The Lancet Child & Adolescent Health* 4, no. 6 (2020): 421.

Maldives, U. N. I. C. E. F. "Indoor play ideas to stimulate young children at home." (2020).

Markard, Jochen, and Daniel Rosenbloom. "A tale of two crises: COVID-19 and Climate." *Sustainability: Science, Practice and Policy* 16, no. 1 (2020): 53–60. https://doi.org/10.1080/ 15487733.2020.1765679.

Marshan, Joseph Natanael, Mohammed Fajar Rakhmadi, and Mayang Rizky. "Prevalence of child marriage and its determinants among young women in Indonesia." In *Child Poverty and Social Protection Conference*. SMERU Research Institute, 2013.

Misra, Kamlesh. "Covid-19: 4 negative impacts and 4 opportunities created for education." *Retrieved on May* 25 (2020): 2020.

"National Center for Immunization and Respiratory Diseases (NCIRD), Division of Viral Diseases." March 25, 2021. www.cdc.gov/coronavirus/2019-ncov/vaccines/distributing/steps-ensure-safety.html).

Nicola, Maria, Zaid Alsafi, Catrin Sohrabi, Ahmed Kerwan, Ahmed Al-Jabir, Christos Iosifidis, Maliha Agha, and Riaz Agha. "The socio-economic implications of the coronavirus pandemic (COVID-19): A review." *International Journal of Surgery* 78 (2020): 185–193.

Orlic, T., J. Rush, M. Cousin, and J. Hong. Coronavirus could cause the global economy $2.7 Trillion. *Bloomberg* (2020). www.bloomberg.com/graphics/2020-coronavirus-pandemic-global-economic-risk/#xj4y7vzkg

Ramsetty, Anita, and Cristin Adams. "Impact of the digital divide in the age of COVID-19." *Journal of the American Medical Informatics Association* 27, no. 7 (2020): 1147–1148. https://doi.org/10.1093/jamia/ocaa078

Sohrabi, Catrin, Zaid Alsafi, Niamh O'neill, Mehdi Khan, Ahmed Kerwan, Ahmed Al-Jabir, Christos Iosifidis, and Riaz Agha. "World Health Organization declares global emergency: A review of the 2019 novel coronavirus (COVID-19)." *International Journal of Surgery* 76 (2020): 71–76.

Spinelli, A., and G. Pellino. "COVID-19 pandemic: Perspectives on an unfolding crisis." *Journal of British Surgery* 107, no. 7 (2020): 785–787. https://doi.org/10.1002/bjs.11627.

Torales, J., M. O'Higgins, and J. Mauricio. "Castaldelli-Maia, & Ventriglio, A. (2020)." *The Outbreak of COVID-19 Coronavirus and Its Impact on Global Mental Health. Pubmed* 66, no. 4: 317–320. https://doi.org/10.1177/0020764 020915212

Tsamakis, Konstantinos, Emmanouil Rizos, Athanasios J. Manolis, Sofia Chaidou, Stylianos Kympouropoulos, Eleftherios Spartalis, Demetrios A. Spandidos, Dimitrios Tsiptsios, and Andreas S. Triantafyllis. "[Comment] COVID-19 pandemic and its impact on mental health of healthcare professionals." *Experimental and Therapeutic Medicine* 19, no. 6 (2020): 3451–3453.

Van Lancker, Wim, and Zachary Parolin. "COVID-19, school closures, and child poverty: A social crisis in the making." *The Lancet Public Health* 5, no. 5 (2020): e243–e244.

Venter, Zander S., Kristin Aunan, Sourangsu Chowdhury, and Jos Lelieveld. "COVID-19 lockdowns cause global air pollution declines." *Proceedings of the National Academy of Sciences* 117, no. 32 (2020): 18984–90. https://doi.org/10.1073/pnas.2006853117.

Women's Safety, N. S. W. "New domestic violence survey in NSW shows impact of COVID-19 on the rise." *Women's Safety NSW*, April 3 (2020).

World Health Organization. "Naming SARS-CoV-2 variants." (2023a). Tracking SARS-CoV-2 variants (who.int).

World Health Organization. "Situation Report-17 SITUATION IN NUMBERS total and new cases in last 24 hours." (2020).

World Health Organization. "The impact of COVID-19 on health and care workers: A closer look at deaths." World Health Organization (2021). https://apps.who.int/iris/handle/10665/345300

World Health Organization. "XBB.1.5 Updated Rapid Risk Assessment, 25 January 2023." (2023b). 25012023XBB.1 (who.int).

Xie, Xinyan, Qi Xue, Yu Zhou, Kaiheng Zhu, Qi Liu, Jiajia Zhang, and Ranran Song. "Mental health status among children in home confinement during the coronavirus disease 2019 outbreak in Hubei Province, China." *JAMA Pediatrics* 174, no. 9 (2020): 898–900.

Ye, Zhi, Xueying Yang, Chengbo Zeng, Yuyan Wang, Zijiao Shen, Xiaoming Li, and Danhua Lin. "Resilience, social support, and coping as mediators between COVID-19-related stressful experiences and acute stress disorder among college students in China." *Applied Psychology: Health Well Being*. 12 (4): 1074–1094 (2020). doi: 10.1111/aphw.12211.

Zambrano-Monserrate, Manuel A., María Alejandra Ruano, and Luis Sanchez-Alcalde. "Indirect effects of COVID-19 on the environment." *Science of the Total Environment* 728 (2020): 138813. https://doi.org/10.1016/j.scitotenv.2020.138813

2 Lessons Learned from Covid-19

Health, Education, and Environment

Tejinder Kaur[1], *Manish Thakur*[2], *and R. C. Sobti*[3]
[1]Department of Zoology, DAV University, Jalandhar, Punjab, India
[2]Department of Microbiology, DAV University, Jalandhar, Punjab, India
[3]Department of Biotechnology, Panjab University, Sector 14, Chandigarh, Punjab, India
tkkalra@gmail.com

CONTENTS

2.1 INTRODUCTION

"Coronavirus" is the most popular word used since December 2019 because this seemingly miniature organism has turned the whole world upside down. As of May 2, 2021, the disease had infected 152,788,755 people all over the world and claimed 3,205,786 lives (Worldometer, 2021). The United States is the worst hit country with 99,974,387 infected cases and 1,100,631 deaths. India is second in the list with 44,667,744 patients infected with the disease and 530,535 deaths (Worldometer, 2021). The situation in India got even worse and a huge number of infections and deaths were reported during the second wave of the pandemic which was far more serious as compared to the first wave because of the breathlessness experienced by greater number of patients and the number of ventilators and oxygen cylinders fell short. The fatalities have grown to the extent that the space in cremation grounds is falling short and people have to wait in long queues to cremate the dead bodies of their loved ones. But how did this deadly virus enter our otherwise normal lives and what lessons have we learned from the pandemic? These are a few questions to consider. Let us first get ourselves acquainted with the virus.

DOI: 10.1201/9781003358909-2

2.2 CORONAVIRUSES

The word "coronavirus" is derived from the Latin word "corona," meaning "crown" or "wreath" (Webster, 2020). It refers to a group of associated RNA viruses that cause diseases in mammals and birds. Coronaviruses belong to the subfamily Orthocoronavirinae (Fan et al., 2019), one of two subfamilies in the family of Coronaviridae (de Groot et al., 2011). These are single-stranded positive-sense RNA viruses (Cherry et al., 2017). They have distinctive club-shaped spikes that project from their surface, which produce an image reminiscent of the solar corona, from which their name derives (Almeida et al., 1968). They are grouped into four genera: alphacoronavirus, betacoronavirus, gamma-coronavirus, and delta-coronavirus (Pal et al., 2020). Mammals are infected by alphacoronaviruses and betacoronaviruses, while gammacoronaviruses and deltacoronaviruses mostly infect birds (Wertheim et al., 2013). They cause respiratory tract infections in humans and the pathogenicity varies from mild to lethal. Following are the details of recent infections caused by these viruses.

2.2.1 Severe Acute Respiratory Syndrome (SARS)

In February 2003, the World Health Organization (WHO) reported a new and previously unknown illness, severe acute respiratory syndrome (SARS). It is caused by a coronavirus named SARS-CoV or SARS-CoV-1. SARS originated in November 2002 across Guangdong Province in southern China, where it was initially thought to cause atypical pneumonia. Within a short span of time, it spread to many countries like the United States, Canada, Singapore, Hong Kong, and European countries. More than 8,000 reported infections were caused by the global outbreak, which eventually resulted in the deaths of around 800 people (Reed, 2018). Bats of the genus *Rhinolophus* were found to be the reservoirs of SARS and SARS-like coronaviruses (Wang and Crameri, 2014). Similar to the common cold and influenza, the disease spreads by tiny droplets of saliva. SARS incubation time is usually 2–7 days, but can be as long as 10 days. Fever (>38°C), which is sometimes severe and often associated with chills and rigors, is normally the first symptom of the disease. Other symptoms can also follow it, including headache, malaise, and muscle pain. There is no SARS cure or vaccine and care should be compassionate and focused on the symptoms of the patient (WHO, 2020).

2.2.2 Middle East Respiratory Syndrome (MERS)

Another coronavirus named MERS-CoV is responsible for an acute respiratory disease called Middle East Respiratory Syndrome (MERS) (Zaki et al., 2012). It is a viral respiratory disease identified in Saudi Arabia for the first time in 2012. Symptoms of MERS include fever, cough, shortness of breath, and pneumonia rarely. There have also been cases of gastrointestinal problems, including diarrhea. Some patients were also reported to be asymptomatic. The fatality rate was between 40% and 50%, and about 35% of patients with confirmed MERS-CoV infection died (Wang and Crameri, 2014). The origin of the virus is not well known, but it is thought that it may have originated in bats and was then transmitted to camels, which became a source of infection to human beings.

2.2.3 Coronavirus Disease (COVID-19)

A recently identified coronavirus is responsible for Coronavirus disease (Covid-19). It is caused by a coronavirus, named SARS-Cov-2. It is named so because of its genetic similarity to SARS-Cov-1 (Chen et al., 2020). The first reports of the infection came from Wuhan, China in December, 2019 (Lu et al., 2020; Xu et al., 2020). The virus has high potential to mutate, which makes it adaptable to a large number of hosts like bats, chickens, civets, dogs, rodents, etc. SARS-CoV-2 is a more virulent variant of SARS-CoV-1 and MERS (Guarner, 2020). Because of its high virulence and mutation rate, the virus infected people all over the world and the Covid-19 outbreak was declared a global pandemic by the WHO on March 11, 2020 (Cucinotta and Vanelli, 2020). The virus has been creating havoc all over the world since December 2019 when its first case was reported. Recent statistics of infection and mortalities reported were mentioned in the introduction. The virus is transmitted by direct or indirect contact of mucous membranes with infectious respiratory droplets. The outbreak of SARS-CoV-2 has badly affected the world not only health wise but economically also.

2.3 LESSONS LEARNED

All episodes in life whether good or bad teach us something. Covid has taught us many things. Some of the lessons learned are discussed here.

2.3.1 Animals

For ages, animals have served as trusted companions to human beings. But human-animal interactions are beyond companionship only. Be it the daily dependence for milk on cows or buffaloes, for eggs on poultry birds, for wool on sheep, for leather on skin of different animals, or many animals for meat; animals give a lot to human beings, but this interaction sometimes leads to serious consequences in the form of zoonotic diseases. These are the infections transferred from animals to humans and sometimes the repercussions are serious. In fact, most of the pandemics in the past were caused due to inter-species transmission of pathogens from animals to humans (Jones et al., 2008; Wolfe et al., 2007). The current Covid-19 pandemic also seems no exception to this. The novel coronavirus, SARS-Cov-2, is also a zoonotic infection that has been linked to bats and pangolins (Xiao et al., 2020; Flores-Analis et al., 2020). Sometimes human beings also become a source of infection to animals. An example of this is the isolation of SARS-Cov-1 of human origin from a pig during the SARS epidemic (Chen et al., 2005). Although animals are a source of many infectious diseases, they are not to be blamed. Each and every animal species is important for maintaining the ecological balance and the proper functioning of food chains and food webs. But the species called "Homo sapiens" definitely needs to change its attitude toward animals. The health of animals should be the top priority of breeders whether they are bred in laboratories for experimentation purposes or reared for the purpose of meat or for the sake of companionship as pets. Regular heath checkups by trained veterinarians should be carried out for early detection of diseases so that disease transmission is controlled at the source station only. Only those animals should be used for eating purposes, which can be bred in controlled and hygienic conditions. Hunting wild animals for the purpose of meat or any other use should be strictly prohibited. For ages, animals have sacrificed themselves for the human race. Now is the time to return their debt. Nurturing animals with love and care will not only benefit them but will also profit the human race because animals are unconditional givers.

2.3.2 Environment

The health of human beings and that of the environment are in direct relation with each other. Continuous deforestation and habitat destruction make way for zoonotic diseases. One such example is the origin of Nipah virus epidemic in 1998, which spread from bats to pigs and then to humans (Chua et al., 2002). Therefore, in order to prevent wildlife-to-human transmission of disease, we need to stop all the activities that disturb the ecological niches of animals. The Covid-19 pandemic has also made us understand the impact of urbanization and industrialization on pollution. Never were the urban skies as clear and air as fresh as in the days when lockdown was imposed in many countries to break the human-human transmission chain. Sadly, we could not enjoy the freshness of pollution free air because fear of the disease locked us inside our homes and when outside we had to wear face masks. Last but not least Covid-19 taught us the importance of oxygen generators—the trees. The second wave of coronavirus disease was so severe, especially in India, that many patients required ventilators and oxygen cylinders to breathe. These cylinders come with a price tag and sometimes poor people can not afford it. But trees provide us with oxygen throughout their life without asking for anything in return. The only way we can repay the dept of these unconditional givers of oxygen is to plant more and more trees. This will also profit the human race more than any other living organism.

2.3.3 Health

"Health is wealth" is an old saying but many of us are so busy with our daily chores that we forget to focus on our health. The results are very damaging, which was proven by Covid. The same organism behaved differently in different people. Some were asymptomatic, some had mild symptoms, some were hospitalized due to severe infections, and some lost their battle with the disease. Undoubtedly the virus has immense potential to mutate and hundreds of variants have already been reported from all over the world. Many of these variants are far more infectious than others. But the different physiological responses to the disease is not only due to variants but also due to the different health status of individuals. Patients with co-morbidities like hypertension, diabetes, AIDS, etc., are more prone to complications than healthy individuals. Many of the co-morbidities like obesity, hypertension, and diabetes are the result of unhealthy lifestyles. Sedentary habits, unbalanced diet, intake of junk food, and lack of exercise are some of the reasons for these diseases. Therefore, a daily routine that includes a balanced diet and at least some exercise is very important. Mental health is equally as important as physical health. Many people suffered from depression and anxiety because of the pandemic. A few minutes of daily meditation and breathing exercises may help in such cases.

2.3.4 Science

The ongoing Covid-19 pandemic has once again highlighted the importance of science to mankind. Never in the history of medical science have vaccines been produced and rolled out to the general public as fast as in this case. In just a span of one year, more than 300 vaccines were in trials and 11 vaccines have been granted emergency use listing (EUL) by WHO. These include the inactivated first-generation vaccines (Covaxin, CoronaVac, etc.); second-generation vaccines based on protein subunits (RBD-Dimer and EpiVacCorona); third-generation RNA-based vaccines (Pfizer–BioNTech and Moderna); and viral vector-based vaccines (Sputnik V, Oxford–Astra Zeneca/Covishield, Johnson & Johnson, etc.). Vaccination drive is at its peak in most countries. The WHO is working in cooperation with scientists, businesses, and global health organizations through the ACT Accelerator to accelerate the response of the pandemic and for the distribution of these vaccines to protect people in all countries (WHO, 2022).

2.3.5 Education

Education is an ever-growing sector that needs to change with the changing times. The novel corona virus outbreak has also created new requirements for education in schools, colleges, and universities. First of all, there is a need to educate students about their physical and mental health. Secondly, the lockdown and online classes have put forward many challenges for educators. The students sitting at their homes are distracted by many things that they find more interesting than regular studies. This disconnects them from the teacher, which impacts their academic performance. The teachers need to brush up on their creative skills to increase the interest of students. They need to device new methods of pedagogy. The current situation requires a revolution in the education sector and needs a complete shift from traditional methods of teaching to the latest ones. Lastly, both teachers and students need to be aware of and trained on information and communications technology (ICT) tools available for teaching and creating lessons.

2.3.6 International Collaborations

One very important lesson taught by Covid-19 is the need for international collaborations for the production and distribution of medical equipment, vaccines, diagnostic kits, drugs, etc. No country has the potential to deal with this pandemic alone. The WHO has played a commendable role in bringing countries together to achieve the higher objective of making the world corona free. The exchange of the genomic sequence of the virus among all the countries as soon as it was decoded was the first step in this direction. After this the cross-country collaborations for providing help in the form of PPE kits, diagnostic kits, drugs for treatment, and now the vaccines wherever required proves that all countries around the globe are united in the fight against corona. If these inter-country collaborations continue, the world will get be rid of this deadly virus soon.

2.4 CONCLUSIONS

The world is going through one of the most difficult times in history. But this is not the first time that it has experienced a pandemic. Many viral epidemics and pandemic outbreaks have been reported in the past. The need today is to learn from the past experiences and prepare for the future. If we do not learn now, we will land in the same situation time and again.

REFERENCES

Almeida, J.D., D.M. Berry, C.H., Cunningham et al. 1968. Virology: Coronaviruses. *Nature*. 220:650.

Chen, W., M. Yan, L. Yang et al. 2005. SARS-associated coronavirus transmitted from human to pig. *Emerg Infect Dis.* 11:446–448.

Chen, Y., Q. Liu and D. Guo. 2020. Emerging coronaviruses: Genome structure, replication, and pathogenesis. *J Med Virol.* 92:418–423.

Cherry, J., G.J. Demmler-Harrison, S.L. Kaplan et al. 2017. Feigin and Cherry's textbook of pediatric infectious diseases. *Elsevier Health Sciences*. PT6615.

Chua, K.B., B.H. Chua, and C.W. Wang. 2002. Anthropogenic deforestation, El Nino and the emergence of Nipah virus in Malaysia. *Malays J Pathol.* 24:15–21.

Cucinotta, D., and M. Vanelli. 2020. WHO declares COVID-19 a pandemic. *Acta Biomed.* 91:157–160.

de Groot, R.J., S.C. Baker, R. Baric et al. 2011. "Family Coronaviridae." In *International Committee on Taxonomy of Viruses, International Union of Microbiological Societies*. Virology Division (Eds.). King, A.M., Lefkowitz, E., Adams, M.J., and Carstens, E.B. Ninth Report of the International Committee on Taxonomy of Viruses. Oxford: Elsevier. pp. 806–828.

Fan, Y., K. Zhao, Z.L. Shi et al. 2019. Bat Coronaviruses in China. *Viruses.* 11:210.

Flores-Alanis, A., L. Sandner-Miranda, G. Delgado et al. 2020. The receptor binding domain of SARS-CoV-2 spike protein is the result of an ancestral recombination between the bat-CoV RaTG13 and the pangolin-CoV MP789. *BMC Res Notes.* 13:398.

Guarner, J. 2020. Three emerging coronaviruses in two decades. *Am J Clin Pathol.* 153:420–421.

Jones, K.E., N.G. Patel, M.A. Levy, et al. 2008. Global trends in emerging infectious diseases. *Nature* 451:990–993.

Lu, R., X. Zhao, J. Li, et al. 2020. Genomic characterization and epidemiology of 2019 novel coronavirus: Implications for virus origins and receptor binding. *Lancet.* 395:565–574.

Pal, M., G. Berhanu, C. Desalegn et al. 2020. Severe acute respiratory syndrome coronavirus-2 (SARS-CoV-2). *An Update Cureus* 12:7423.

Reed, K.D. 2018. Viral zoonoses. *Reference Module in Biomedical Sciences.* B978-0-12-801238-3.95729-5.

Wang, L.F., and G. Crameri. 2014. Emerging zoonotic viral diseases. *Rev Sci Tech.* 33:569–581.

Webster, M. 2020. Coronavirus. In *Merriam-Webster.com Dictionary*. www.merriam-webster.com/dictionary/coronavirus (accessed May 2, 2021).

Wertheim, J.O., D.K. Chu, J.S. Peiris, et al. 2013. A case for the ancient origin of coronaviruses. *J Virol.* 87:7039–7045.

WHO. 2022. https://covid19.trackvaccines.org/agency/who/ (accessed Nov 16, 2022).

WHO. 2020. www.who.int/health-topics/severe-acute-respiratory- syndrome#tab=tab_1 (accessed May 2, 2021).

Wolfe, N.D., C.P. Dunavan, and J. Diamond. 2007. Origins of major human infectious diseases. *Nature.* 447:279–283.

Worldometer. 2021. www.worldometers.info/ coronavirus/ (accessed May 2, 2021).

Xiao, K., J. Zhai, and Y. Feng et al. 2020. Isolation of SARS-CoV-2-related coronavirus from Malayan pangolins. *Nature.* 583: 286–289.

Xu, X.W., X.X. Wu, X.G. Jiang et al. 2020. Clinical findings in a group of patients infected with the 2019 novel coronavirus (SARS-Cov-2) outside of Wuhan, China: Retrospective case series. *BMJ.* 19:368: m606.

Zaki, A.M., S. van, Boheemen, T.M. Bestebroer et al. 2012. Isolation of a novel coronavirus from a man with pneumonia in Saudi Arabia. *N Engl J Med.* 367:1814–1820.

3 Science and Research during Pandemic

Covid-19

R. C. Sobti[1], *Archna Narula*[2], *and Ahalya Naidu*[2]
[1]Department of Biotechnology, Panjab University, Chandigarh, India
[2]M S Ramaiah Institute of Technology, Bangalore, India
rcsobti@pu.ac.in

CONTENTS

3.1 INTRODUCTION

The causative agent of Covid-19, SARS-CoV-2, was first discovered in Wuhan, China, in early December 2019. It promptly spread throughout the world from China with efficacious human-to-human transmission and has now circumnavigated the globe, becoming a worldwide pandemic (WHO, 2020). The World Health Organization (WHO) first declared it a public health emergency and subsequently a pandemic. The virus affected the normal life of many people around the world and had negative effects on all aspects of human life—effects that were unprecedented for most people.

Since the outbreak of the pandemic, many countries have been promoting the research and development (R&D) of medicines, vaccines, and testing reagents for the novel coronavirus. The Covid-19 pandemic has underscored the pressing need for countries to focus more on elevating science, technology, and innovation (STI) in both policy and practical terms.

Scientific and technological innovation have played a crucial role in fighting this "war without smoke," and scientific culture has also demonstrated its power to inspire and unite people to overcome difficulties together. Through concrete action, scientific professionals around the world have demonstrated a spirit of patriotism, innovation, truth-seeking, dedication, and teamwork and shown great compassion and shared humanity. In that sense, the Covid-19 pandemic shows the importance of developing scientific culture (Han, 2020).

The present work reviews the efforts of the scientific community in discovering the different variants of the virus, diagnostic tools for its detection, preventive and curative measures. The work also presents other challenges faced by the research community during the pandemic.

DOI: 10.1201/9781003358909-3

3.2 UNDERSTANDING THE VIRUS AND ITS VARIANTS

Researchers have collaborated in an undeclared joint effort to understand and target various aspects of the virus as a means to limit the infection, stop its spread, and eventually devise a potential cure for Covid-19. The aspects looked into include the virus's structure, its mechanism of infection, its molecular biology, the origins of its nucleic acid sequence, and the epidemiology of disease spread. Like other coronaviruses, SARS-CoV-2 particles are spherical and have proteins called spikes protruding from their surface. These spikes latch onto human cells, then undergo a structural change that allows the viral membrane to fuse with the cell membrane. The viral genes can then enter the host cell to be copied, producing more viruses. Recent work has shown that SARS-CoV-2 spikes bind to receptors on the human cell surface called angiotensin-converting enzyme 2 (ACE2) (Wrapp et al., 2020).

All viruses, including SARS-CoV-2, the virus that causes Covid-19, change over time. Most changes have little to no impact on the virus' properties. However, some changes may affect the virus's properties, such as how easily it spreads, the associated disease severity, or the performance of vaccines, therapeutic medicines, diagnostic tools, or other public health and social measures (WHO, 2020).

Viruses mutate all the time and most changes are inconsequential. But others can make the disease more infectious or threatening—and these mutations tend to dominate.

Viruses with the most potentially concerning changes are called "variants of concern" and are identified as follows:

1. **Delta** (B.1.617.2), which currently accounts for 99% of new Covid-19 cases in the UK.
2. **Alpha** (B.1.1.7), first identified in the UK but which has spread to more than 50 countries.
3. **Beta** (B.1.351), first identified in South Africa but which has been detected in at least 20 other countries, including the UK.
4. **Gamma** (P.1), first identified in Brazil but which has spread to more than 10 other countries, including the UK.

The variants of concern have all undergone changes to their spike protein—the part of the virus that attaches to human cells. Delta has some potentially important ones (such as L452R) that might make it spread more easily. As of March 2021, there were several new variants in the picture with more transmissibility, disease severity, immune escape, diagnostic or therapeutic escape, and were identified to cause significant community transmission resulting in risk to global public health. These variants of interests (VOI) are identified as:

1. Eta (B.1.525): First identified on 17 March 2021
2. Iota (B.1.526): First identified on 24 March 2021
3. Kappa (B.1.617.1): First identified on 4 April 2021
4. Lambda (C.37): First identified on 14 June 2021
5. Omicron (B.1.1.529): First identified on 24 November 2021

Since the outbreak of Covid-19, scientists in many countries have largely collaborated under the principle of "open science"—where knowledge, methods, data, and evidence are made freely available and accessible to everyone. Collaborative arrangements of open science, especially the mapping of the virus's genome, helped in the development of the Covid-19 vaccines being administered in various countries.

3.3 DIAGNOSTIC MEASURES

Science and technology have played important roles in Covid-19 prevention and control. Computed tomography (CT) scanning—which used to be a high-end technique—became a mandatory examination for every patient and even suspected patients. Imaging technologies like CT scans/chest X-rays were used to rule out false negatives in symptomatic patients.

Tremendous efforts have been employed to develop highly accurate diagnostic testing for Covid-19 since January 2020 (Sheridan, 2020). Together, the WHO and Foundation for Innovative New Diagnostics (FIND) collaborated to independently validate tests from different manufacturers across the globe (Sheridan, 2020; Cheng et al., 2020). The U.S. Food and Drug Administration (FDA) initiated fast-tracked FDA review and approval of Covid-19 tests through emergency use authorization (EUA) in mid-March (Cheng et al., 2020). According to the data from FIND COVID-19 resource center (www.finddx.org/covid-19/) on 9 June 2020, over 400 molecular and serological antibody tests had been approved by different countries' and regions' agencies of certification.

After the discovery of patients with severe pneumonia of unknown origin in China, many scientists took a week to isolate the causative virus and accurately sequence its genome, exhibiting a level and efficiency that was far higher than during the 2003 severe acute respiratory syndrome (SARS) outbreak. Moreover, scientific researchers rapidly produced original, significant research results (Han, 2020).

In this regard, several platforms for highly specific as well as sensitive, rapid diagnostic tests have been reported globally. A polymerase chain reaction (PCR)-based technique has been the primary platform for detection of Covid-19, globally as well as in India. RT-PCR is fast, specific, sensitive as well as cost-effective in diagnosis, where time is of the essence. Next-generation sequencing can also play a very significant role in high-throughput and highly specific detection of SARS-CoV-2, although it involves higher costs. Very recently, a targeted gene-editing technique-based detection of the virus has also been reported, which is highly sensitive and faster than RT-PCR.

In India, Covid KAVACH ELISA was developed by the National Institute of Virology, Pune, India, by isolating virus from tested positive patients. The kit is manufactured by Zydus-Cadila, which is the leading global pharmaceutical company in India (www.indianexpress.com/article/india/icmr-clears-1st-batch-of-key-elisa-antibody-testing-kits-made-in-india-6410431). The Indian Council of Medical Research (ICMR) kit is reported to have 98.7% sensitivity and 100% specificity. GCC Biotech India Pvt. Ltd. has developed an indigenous low-cost Covid-19 test kit after rigorous clinical trials. It has high accuracy and can deliver results in < 90 min. The firm has received the approval of ICMR (www.newindianexpress.com/videos/good-news/2020/may/15/only-rs-500-per-test-kolkata-firm-develops-low-cost-indigenous-covid-19-testing-kit-108432.html). Mylab Discovery Solutions in June 2021 has rolled out low cost self-test kits for COVID-19 and it can be ordered online (www.thehindu.com/news/national/indigenous-self-test-kits-for-covid-19/article34722048.ece).

Overall, in a short period of time, the scientific community developed several methods useful for correctly diagnosing a suspected case of Covid-19 infection.

3.4 THERAPEUTICS USED FOR COVID-19

In the race to find solutions to combat the SARS-CoV-2 virus, several possible therapeutics have been suggested and approved for the market. The broad groups of therapeutic agents include those that: (i) block viral entry to host cells; (ii) block viral replication and survival in host cells; and (iii) downplay the host immune response (Shetty et al., 2020).

The scientific community engaged with the WHO to handle Covid-19 has prepared online recommendations for the use of therapeutics in the treatment of Covid-19. The five versions of the WHO guideline now contain seven recommendations, including a new recommendation regarding interleukin-6 (IL-6) receptor blockers, including both tocilizumab and sarilumab. This latest update of WHO *Therapeutics and Covid-19* includes:

1. To use IL-6 receptor blockers (tocilizumab or sarilumab) in patients with severe or critical Covid-19 (published 6 July 2021);

2. Not to use ivermectin in patients with Covid-19 except in the context of a clinical trial (published 31 March 2021);
3. Strong recommendation against hydroxychloroquine in patients with Covid-19 of any severity (published 17 December 2020);
4. Strong recommendation against lopinavir/ritonavir in patients with Covid-19 of any severity (published 17 December 2020);
5. Conditional recommendation against remdesivir in hospitalized patients with Covid-19 (published 20 November 2020);
6. Strong recommendation for systemic corticosteroids in patients with severe and critical Covid-19 (published 2 September 2020);
7. Conditional recommendation against systemic corticosteroids in patients with non-severe Covid-19 (published 2 September 2020).

Other Covid-19 therapeutics that are currently under consideration by the WHO include colchicine, monoclonal antibodies, and anticoagulants (www.who.int/publications/i/item/WHO-2019-nCoV-therapeutics-2021.2).

Due to the severity of the disease, alternative medicines have also come to the forefront. Ayurvedic formulations and other Indian alternative medicines have also been studied and used for their potential utility in various kinds of viral infections (Dhama et al., 2018). However, none of such natural products have actually been tested to treat Covid-19 (Shetty et al., 2020). The presence of a variety of phytochemicals such as flavonoids, tannins, triterpenes, phenolic acids, alkaloids, saponins, lignins, proteins, and peptides provide a plethora of functions to such natural products and extracts that have been demonstrated to modulate various aspects of viral infection including virus entry, viral gene expression, and replication (Ganjhu et al., 2015). Combinations of natural products like curcumin (Avasarala, 2013), *Azadirachta indica* (neem plant) (Pooladanda, 2019), *Withania somnifera* (Ashwagandha) (Ganguly et al., 2018), etc., may have the potential to be used as adjunct therapy to treat infected individuals.

3.5 VACCINE DEVELOPMENT

In an endeavour to seek a cure for the SARS-CoV-2 virus, the scientific research community started work in early 2020. Success was not promised as the challenges were immense. The fastest any vaccine had previously been developed, from viral sampling to approval, was four years, for the mumps in the 1960s. To hope for a curative/preventive vaccine even by the summer of 2021 seemed highly optimistic.

But by the start of December 2019 the developers of several vaccines had announced excellent results in large trials, with more showing promise. And on December 2, 2019, a vaccine made by drug giant Pfizer with German biotech firm BioNTech became the first fully tested immunization to be approved for emergency use. Today there are several vaccines available tha are WHO approved and have undergone Phase 3 trials. They are as follows:

1. **Pfizer:** BNT162b2/COMIRNATY Tozinameran (INN)
2. **Moderna:** mRNA-1273
3. **Janssen (Johnson & Johnson):** Ad26.COV2. S
4. **Two versions of AstraZeneca:** AZD1222
5. **Sinophar:** SARS-CoV-2 Vaccine (Vero Cell), Inactivated (lnCoV)
6. **The Serum Institute of India:** Covishield (ChAdOx1_nCoV19)

Covaxin, also known as BBV152, was authorized for emergency use by India's Central Drugs and Standards Committee (CDSCO) on January 3, 2021, even though Phase 2 clinical trials were unpublished, and larger Phase 3 trials were still ongoing and awaiting WHO approval. Outside of

India, Covaxin has also been approved for emergency use in 15 countries, including Iran, Zimbabwe, Mexico, the Philippines, Guatemala, and Botswana.

The world was able to develop Covid-19 vaccines so quickly because of years of previous research on related viruses and faster ways to manufacture vaccines, enormous funding that allowed firms to run multiple trials in parallel, and regulators moving more quickly than normal.

3.6 ROLE OF RESEARCH, ML, AND AI IN COMBATING THE PANDEMIC

The need to promote scientific culture has been highlighted by the Covid-19 pandemic. It has provided a focal point for popular science and healthcare education. With the pandemic not showing signs of abating, people are more concerned about public health than ever before. With knowledge of how SARS-CoV-2 is transmitted, the general public has consciously worn masks, practiced social distancing, and strictly adhered to home quarantine guidelines. People have gained a deeper appreciation for establishing good living habits and maintaining harmony between humans and nature. Public recognition of vaccines has increased rapidly, and many people have joined voluntary vaccine trials (Han, 2020).

Machine learning (ML)-based technologies are playing a substantial role in the response to the Covid-19 pandemic. Experts are using ML to study the virus, test potential treatments, diagnose individuals, analyze public health impacts, and more (Vaishya et al., 2020). The application of advanced information technology made epidemiological investigation much more efficient by rapidly identifying new and suspected cases and close contacts, ensuring the safety and mobility of healthy people via digital tracking, and increasing the reliability of epidemic trend prediction through big data.

Even before the world was aware of the threat posed by the coronavirus (Covid-19), artificial intelligence (AI) systems had detected the outbreak of an unknown type of pneumonia in China. As the outbreak became a global pandemic, AI tools and technologies have been employed to support efforts of policymakers, the medical community, and society at large to manage every stage of the crisis and its after effects (Figure 3.1): detection, prevention, response and recovery (www.oecd.org/coronavirus/policy-responses/).

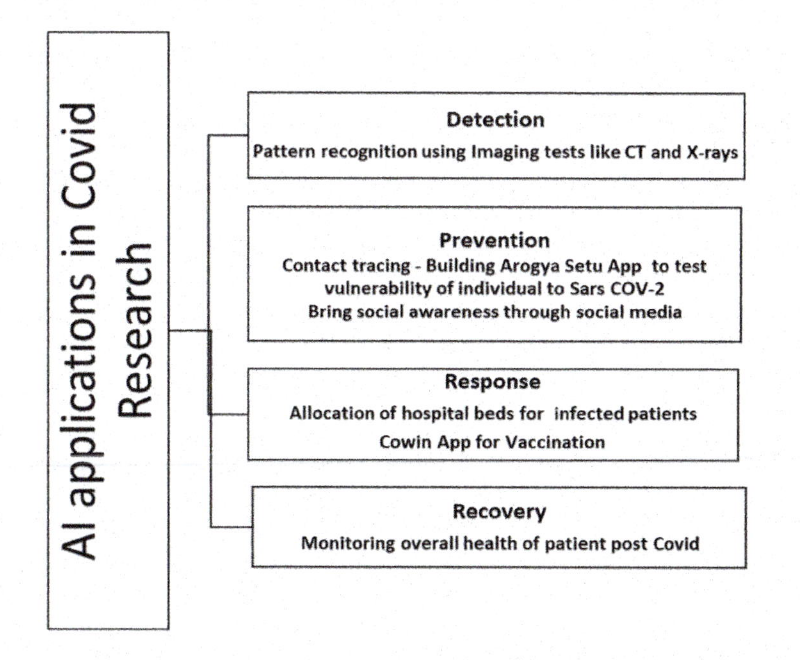

FIGURE 3.1 AI applications at different stages of the COVID-19 crisis.

3.7 OTHER CHALLENGES IN RESEARCH DURING COVID-19

One of the areas most affected by the pandemic is research, which has slowed drastically due to lockdowns and social distancing. Research institutions have been closed across countries, and conventional research has been disrupted (UNESCO, 2020). Covid-19 has yielded unprecedented challenges to experimental research with laboratories being shut and fieldwork largely suspended. As we navigate through and beyond this pandemic, which will have a long-lasting impact on our world, it is important to recognize the challenges and limitations faced by researchers.

SARS-CoV-2 virion, a few nanometers in diameter, disrupted the conduct of research and forced scientists to adapt their work to continue its progress. The genetic structure of the virus (SARS-CoV-2) was sequenced within weeks of its discovery due to the integration of scientists and researchers working in the field of genetics, electrical, mechanical, computer, and chemical engineering. The research community united to solve the challenges in the areas of development of applications for tracking and prevention of this disease.

Funding Priorities: The first concern is about research funding priorities being largely focused on COVID-19 solutions and less emphasis on non-COVID related research. Researchers across the world are working around the clock to find a way to neutralize COVID-19 and reduce its virulent impact on humanity. The pandemic has shifted research endeavors and focus away from other equally important issues. The confidence of researchers in applying for grants that are not focused on COVID-19 has become undermined and researchers are concerned that the extreme focus on the pandemic means less funding and attention for other priority areas.

Time Constraints: Frequent lockdowns and time lag in implementing rigorous methodology-based research is causing serious crisis. Validation of research data and solutions with limited resources pose challenges for the research community. Prof. Dipshikha Chakravortty at the Department of Microbiology and Cell Biology, Indian Institute of Science, Bangalore said that disruptive effects of SARS-CoV-2 were felt on every sector, but especially in experimental research. Research requiring experimental setups including animals, cell lines, microbes, etc., were halted due to the frequent lockdowns. These perturbations proved to be detrimental for the research community. Research students in the middle of important findings became discouraged and an enormous amount of time was lost. There is no way in which the scientific community can emerge unscathed from this pandemic.

Remote Research and Funding: The third challenge is about the ability to conduct research remotely. Research involving established databases is easier than empirical evidence collected via interviews, which may require a change in research technologies; archival research is clearly more problematic. The shift to online interactions and mathematical modelling or usage of application software created a new dimension in research. The commitment to providing everything through computational software helped researchers to plan work with new tools and with different time constraints.

Prof. Jyant Kumar at the Department of Civil Engineering, Indian Institute of Science, Bengaluru, India reported that engineering researchers engaged in laboratory and field work had to halt their work during Covid-19. The researchers engaged in computational research were able to progress steadily. In a very short time, lab based discoveries and research had to be scaled up for the mass production of vaccines and other related equipment's to address the challenges posed by the pandemic. Engineers required laboratories to pilot the scientific findings with regards to therapeutics and vaccines and lockdowns posed a challenge. The pandemic interrupted projects and curtailed travel, which intensified the challenges to the engineering research community.

Human Resources: Covid-19 has had an array of impacts on researchers, including faculty members, technicians, postdocs, and graduate students. It has impacted their progress, career development, and work-life balance. Doctoral students are facing delays due to disruptions in laboratory and field research. Some researchers may no longer be on track to complete work within the stipulated time frame that external funding typically covers. While Covid-19 may not have caused disparities within the research enterprise, it has exacerbated pre-existing inequalities.

Publications: The Covid-19 crisis has activated a massive influx of publications. More than 20,000 papers have been published since December 2019, many in prestigious journals (Harper et al., 2020). There is also an increasing number of studies being uploaded to preprint servers, such as BioRxiv, for rapid dissemination. There is a need for increased caution in the wake of this massive influx of submissions, especially since we are increasingly seeing these results being picked up by the media and diffused to an inharmonious audience. On June 15, 2020, *The New York Times* highlighted potential lapses in the peer review process affecting major scientific journals (New York Times, 2020).

Gender Inequality in Research: Another significant challenge during the pandemic is gender inequality within the field of research (Dorn et al., 2020). There has been a disproportionate adverse effect on female contributions, which have declined in comparison to their male peers. This might be due to the fact that more women than men are overworking to take care of families and children (Lammare et al., 2020; The Guardian News, 2020).

3.8 CONCLUSION

Science and research during the pandemic have significantly contributed to our understandinig of the virus and its variants and to the development of diagnostic tools, therapeutics, and vaccines. In addition, AI tools have played an important role in detection, prediction, response, and recovery. While the pandemic has wreaked havoc with research timelines and raised questions about possible funding sources, there have been some positive outcomes as well. One of the biggest opportunities is that the pandemic has made researchers rethink traditional ways of doing research and consider new ways that are beneficial for everyone involved. The pandemic has made the research community work collaboratively to develop effective approaches to tackling the virus. This pandemic has bought cohesion between the scientific community and has promoted collaborative research across institutions.

REFERENCES

Avasarala S, Zhang F, Liu G, Wang R, London SD, London L. Curcumin modulates the inflammatory response and inhibits subsequent fibrosis in a mouse model of viral-induced acute respiratory distress syndrome. *PLoS One*. 2013; 8:e57285.

Cheng MP, Papenburg J, Desjardins M, et al. Diagnostic testing for severe acute respiratory syndrome-related coronavirus 2: A narrative review. *Ann Intern Med*. 2020; 172:726–734.

Covid-19: ICMR Clears First Batch of Key ELISA Antibody Testing Kits Made in India. www.indianexpress.com/article/india/icmr-clears-1st-batch-of-key-elisa-antibody-testing-kits-made-in-india-6410431

Dhama K, Karthik K, Khandia R, Munjal A, Tiwari R, Rana R, et al. Medicinal and therapeutic potential of herbs and plant metabolites/extracts countering viral pathogens-current knowledge and future prospects. *Cure Drug Metab*. 2018; 19:236–263.

Dorn AV, Cooney RE, Sabin ML. COVID-19 exacerbating inequalities in the US. *Lancet*. 2020; 395(10232):1243–1244.

Ganguly B, Umapathi V, Rastogi SK. Nitric oxide induced by Indian ginseng root extract inhibits infectious bursal disease virus in chicken embryo fibroblasts *in vitro*. *J Anim Sci Technol*. 2018; 60:2.

Ganjhu RK, Mudgal PP, Maity H, Dowarha D, Devadiga S, Nag S, et al. Herbal plants and plant preparations as remedial approach for viral diseases. *Virusdisease*. 2015; 26:225–236.

Han Q. Introduction: The COVID-19 pandemic calls for the strengthening of scientific culture. *Cultures of Science*, 2020;3(4): 223–226.

Harper L, Kalfa N, Beckers GMA, et al. The impact of COVID-19 on research. *J Pediatr Urol*. 2020; 16(5):715–716. doi: 10.1016/j.jpurol.2020.07.002

Lamarre Vincent P, Sugimoto CR, Larivière V. The decline of women's research production during the coronavirus pandemic. *Nature Index*. 2020, 19. www.theguardian.com/education/2020/may/12/womens-research-plummets-during-lockdown-but-articles-from-men-increase.

Only Rs 500 Per Test: Kolkata Firm Develops Low-cost Indigenous COVID 19 Testing Kit. www.newindianexpress.com/videos/good-news/2020/may/15/only-rs-500-per-test-kolkata-firm-develops-low-cost-indigenous-covid-19-testing-kit-108432.html

Pooladanda V, Thatikonda S, Bale S, Pattnaik B, Sigalapalli DK, Bathini NB, et al. Nimbolide protects against endotoxin-induced acute respiratory distress syndrome by inhibiting TNF-alpha mediated NF-kappaB and HDAC-3 nuclear translocation. *Cell Death Dis.* 2019; 10:81.

Raju V, Mohd Javaid, Khan IH, Haleem A. Artificial Intelligence (AI) applications for COVID-19 pandemic., *Diabetes Metab. Syndr.: Clin. Res. Rev.*, 2020; 14(4):337–339, ISSN 1871-4021, https://doi.org/10.1016/j.dsx.2020.04.012; www.oecd.org/coronavirus/policy-responses/using-artificial-intelligence-to-help-combat-covid-19

Sheridan C. Fast, Portable tests come online to curb coronavirus pandemic. *Nat Biotechnol.* 2020; 38:515–518.

Shetty R; Ghosh A; Honavar SG; Khamar P; Sethu S, Therapeutic opportunities to manage COVID-19/SARS-CoV-2 infection. *Indian J. Ophthalmol.*: May 2020; 68(5):693–702 doi: 10.4103/ijo.IJO_639_20

The pandemic claims new victims: Prestigious medical journals. *New York Times*, June 2020.

UNESCO. COVID-19 educational disruption and response Paris. France: UNESCO; 2020. https://en.unesco.org/covid19/ education response

WHO, 2020, World Health Organization. Coronavirus disease (COVID-2019) situation reports, Geneva. 2020. www.who.int/emergencies/ diseases/novel-coronavirus-2019/situation-reports/

Wrapp D, Wang N, Corbett KS, Goldsmith JA, Hsieh CL, Abiona O, Graham BS, McLellan JS. Cryo-EM structure of the 2019-nCoV spike in the prefusion conformation. *Science*. 19 Feb 2020. pii: eabb2507. doi: 10.1126/science.abb2507. [Epub ahead of print]. PMID:32075877.

4 Molecular Biology Lessons Learned from the Spiky Intruder SARS-CoV-2-19

Sikander S. Gill[1], *Rajwant K. Gill*[1], *and R. C. Sobti*[2]
[1] Freelance
[2] Department of Biotechnolog, Panjab University, Chandigarh, India

CONTENTS

DOI: 10.1201/9781003358909-4

4.1 INTRODUCTION

On December 31, 2020 the national authorities of China informed the local office of the World Health Organization (WHO) about an outbreak of acute respiratory pneumonia of unknown etiology in Wuhan City, Hubei Province of China (WHO, 2019). On January 3, 2020 at 1:30 pm, Dr. Zhang Yongzhen, a professor and virus sequencing expert at Shanghai Public Health Clinical Center & School of Public Health, Fudan University of China, received a metal box enclosing a test tube packed in dry ice that contained swabs from a patient in Wuhan (Campbell, 2020). Little did he know this box would eventually turn out to be "Pandora's" box. He and his team started sequencing the genome of the pathogen in these swabs. By 2:00 am on January 5, 2020, the virus genome was sequenced and eventually uploaded to the National Center for Biotechnology Information (NCBI). On January 7, 2020, the causative agent of the outbreak was identified as a new type of spiky coat virus (coronavirus) and was named 2019-novel coronavirus (2019-nCoV), also called human coronavirus 2019 (HCoV-19 or hCoV-19) (Wong et al., 2020; Andersen et al., 2020). On January 13, 15, and 20, 2020, the first entries of 2019-nCoV were reported in Thailand, Japan, and Republic of Korea, respectively. Realizing its potential for global spread, the WHO declared the outbreak a Public Health Emergency of International Concern on January 30, 2020 (WHO, 2020a).

After debating the nomenclature of this virus, on March 2, 2020, the Coronaviridae Study Group (CSG) of the International Committee on Taxonomy of Viruses renamed it. The new name, Severe Acute Respiratory Syndrome Coronavirus 2 (SARS-CoV-2), became a global scare because of its intrusion capabilities (Coronaviridae, 2020; Zhu et al., 2019). Genetic analysis of this virus showed it to be most closely related to the group of spiky viruses, SARS-CoV and SARS-related bat and civet coronaviruses within the family Betacoronavirus. Eventually, the outbreak caused by this spiky intruder destabilized the whole world and transformed the entire globe into a massive scale of infection forcing the WHO to declare it a pandemic on March 11, 2020 (WHO, 2019).

4.2 THE PANDEMIC AND MOLECULAR BIOLOGY

Professor Yongzhen's team's success in discovering and publishing the virus genome (updated version deposited on GenBank on March 18, 2020, accession # MN908947) allowed scientists to instantly design COVID-19 tests to fight the pandemic (Robin, 2020; Zhang, 2020). In fact, for identifying this virus in patient samples, molecular biology played a great role globally since it has been revolutionizing the diagnostic pathology for the detection and characterisation of pathogens. In recent years, it has led the way into this new era by allowing rapid detection of micropathogens, and development of vaccines. It provided a testing solution for the specimens that were previously known as difficult or impossible to detect by conventional microbiological methods. In addition to the detection of fastidious microorganisms, more rapid detection by molecular methods has now become possible for pathogens of public health importance. The molecular methods have now progressed beyond identification to offer strain characterisation by genotyping. Treatment of certain pathogens has also been improved by viral load testing for the monitoring of responses to antiviral therapies.

Molecular biological methods, especially polymerase chain reaction (PCR) techniques, turned out to be a widely used and accepted as a reliable tool for detecting this virus in clinical laboratories with excellent performance (Lamoril et al., 2007; Matic et al., 2021). With the advent of multiplex real-time reverse transcription polymerase chain reaction (RT-PCR) and improvement in efficiency through automation, the costs of molecular methods have been decreasing, further increasing their popularity. Thus, the clinical utility of molecular methods performed in the clinical laboratory have changed laboratory diagnosis and the management of infectious diseases (Speers, 2006). In knowing and managing this pandemic as well, molecular biology lessons learned have played a great role in understanding the genome of SARS-CoV-2 in clinical samples and to know whether the genome of this pathogen is evolving in the form of variants from the alpha to very unusual ones such as PANGO lineage B.1.1.529, the omicron variant named after the 15th letter of the Greek alphabet on November 26, 2021 following its discovery on November 24, 2021 (www.who.int/news/item/28-11-2021-update-on-omicron). Among the variants of SARS-CoV-2, the delta variant was the most dominant one, followed by eight including epsilon, zeta, eta, theta, iota, kappa, and lambda that mostly fizzled out. However, the WHO while naming omicron by the next Greek letter on the list, Nu and then Xi were skipped to avoid the Nu easily being confused with "new" and Xi being a common surname in certain communities (Ingrid, 2021; Patel 2021; Robitzski, 2021).

The emergence of the novel pathogens was in fact reflected during the analysis of scientific data and possibility of mutations in the spike region of progenitor viral genome published 17 and 13 years ago, and during the first semester of 2020 (Cheng et al., 2007; NCBI, 2021). Therefore, the novel pathology of this virus was all set to throw new challenges and teach new lessons in all fields including science, health, sociology, economics, political, and logistics. Even though after more than a year since its emergence, many questions are answered but new questions are still emerging, and many questions still have been hovering unanswered. Poor understanding and lack of knowledge about this novel pathogen hampered the development of reliable, specific, sensitive, and rapid detection tools.

The lessons being learned on immediate deficiencies in terms of biological materials, harmonized methodologies, sharing practical experiences of encountering this pathogen will help in tackling many new viruses enlisted by the WHO that have the potential of causing pandemics in the future (WHO, 2020c; Taubenberger et al., 2010). The lessons ranging from management of resources to sorting of supply bottlenecks would also help in preparing and managing the crises of the future.

4.3 SARS-COV-2 DETECTION BY THE RT-PCR METHOD

A week after the publication of the genome sequence of the SARS-CoV-2, on January 17, 2020, a group of German scientists developed the PCR-based first diagnostic method for COVID-19 in nasopharyngeal swab samples that was later selected by the WHO (Corman et al., 2020a). The RT-PCR is a powerful and gold standard tool that amplifies the unique sequences of the target genome for the detection by employing primers and probe sequences specific for the target to be detected. Therefore, to support the potential public health emergency response to COVID-19, the US Center for Disease Control and Prevention (US-CDC) worldwide developed and validated a RT-PCR panel of targets based on this SARS-CoV-2 genome sequence (US-CDC, 2020b). The panel targeted the nucleocapsid protein (N) gene of SARS-CoV-2. This RT-PCR panel was validated under the Clinical Laboratory Improvement Amendments for the US-CDC use for diagnosis of SARS-CoV-2 from respiratory clinical specimens (CMS, 2021).

On January 20, 2020, the US-CDC RT-PCR panel confirmed an early case of COVID-19 in the United States (Holshue et al., 2020). To enable emergency use of the US-CDC's RT-PCR panel as an *in vitro* diagnostic test for SARS-CoV-2, on February 4, 2020, the US Secretary of Health and Human Services initiated the US Food and Drug Administration (US-FDA) to issue an Emergency Use Authorization (EUA) of the test. The EUA authorizes a device if it is reasonably believed that the device is safe and effective for detection of the virus, rather than waiting to grant full approval. This authorization played a significant role in rolling out the testing spree for SARS-CoV-2 (Lu et al., 2020).

In addition to contributing to the availability of an increasing number of samples and understanding of molecular biology of the virus and the molecular tests, the EUA also evolved in terms of refining the terms of authorization. Eventually, it defined the use of samples of known SARS-CoV-2 status ranging from low to high viral load, recommended test controls, required post-authorization validation, and determination of relative limits of detection (LODs) leading to establishing true and accurate performance. In view of the burdensome nature of EUA validation, in the week of February 24, 2020, various groups of laboratories proposed to use laboratory-developed tests (LDTs) without obtaining an EUA.

Thus, access to SARS-CoV-2 genome isolated from COVID-19 patients and eventually the availability of primers and probes of various test kits on the website of the WHO revolutionized the rapid development of the RT-PCR tests (WHO, 2020d; Razvan et al., 2021). The RT-PCR method used to detect SARS-CoV-2 is very often multiplexed for co-amplification of several targets. In the RT-PCR methods, the Ct values of the RT-PCR are often taken at their face value to determine the viral load except at low viral loads (Cheng et al., 2020). However, the models of viral load evolved over time for both symptomatic and asymptomatic people became more reliable than Ct values (Lippi et al., 2020; Kucirka et al., 2020).

4.4 THE TEST METHOD PERFORMANCE CHARACTERIZATION

It is a well-known fact that the RT-PCR tests are prevalent and preferred over other methods for the detection of SARS-CoV-2. However, as with any diagnostic method, the PCR meets certain criteria of the analytical and diagnostic sensitivity like LOD, analytical and diagnostic specificity, amplification efficiency, repeatability, reproducibility, accuracy of results, robustness, cost, and other aspects (Belouafa et al., 2017). For an overall effective diagnosis test, specificity and sensitivity are two fundamental requirements in addition to the user friendliness of the testing method.

It is also important to determine the difference between the accuracy of the method validated in specific conditions as outlined by the kit developer and the compliance of the kit with standards itself. For example, the quality of the reagents and other materials depends upon various factors including storage conditions and performance within the life cycle of the test kit.

4.5 VALIDATION OF RT-PCR METHODS FOR THE DETECTION OF SARS-COV-2

The RT-PCR methods for the detection of SARS-CoV-2, although quite popular and straightforward, have been the subject of several controversies (Rahman et al., 2020; Stanley, 2020). A concern is also anticipated by the variation found in specificity through bioinformatics analysis (*in silico*) through the primer specificity evaluation procedures and experimental approach (Yip et al., 2020). Through this *in silico* procedure, the absence of cross-reactions with other related coronavirus strains, human coronavirus, or other respiratory virus strains is ensured. Therefore, another lesson on the validation of the test kits or test methods is the lack of reference test sample materials that suffers due to the heterogeneity in the samples and sample types used for method validation. It is especially important to ensure that the quality and diversity of the reference samples for method validation are representative of the samples routinely tested in terms of same sample type, and similar composition for example samples from positive patients be made available (ANSES, 2015).

4.6 PERFORMANCE OF THE TEST IN CURRENT PANDEMIC

In the ongoing pandemic, the real-time RT-PCR methods have emerged as the standard method for the detection of SARS-CoV-2 (Kubina and Dziedzic, 2020; Razvan et al., 2021). With the availability of primer and probes from the WHO or the CDC, many companies and laboratories have developed RT-PCR kits for the detection of this novel virus. The differences that exist among various RT-PCR methods are because of different molecular target genes, variable analysis durations, differences in reagents used, and differences in performance in terms of sensitivity indicated by LOD values (Table 4.1).

A comparative study of such tests indicates significant differences in sensitivity, displaying LOD of 200 to 600 copies/ml (Yip et al., 2020; Lai et al., 2020; Iglói et al., 2020). This difference can be critical and the difference in performance between various methods is dependent on several of the following factors.

TABLE 4.1
Differences among Various RT-PCR Test Kits

Factor	Problems	Solution	Reference
Reporter dyes (fluorochromes)	Signal-to-background noise ratio can be variable.	1. Background noise to be low. 2. Background noise should not mask fluorescence. 3. Thresholds set to detect low but significant Ct values.	Ishige et al., 2020
Quality of the reagents	1. Quality of the kit reagents seriously impacts on the reliability of the test results. 2. Composition of the kits by the suppliers is kept confidential. 3. Evaluation of the quality of reagents over time or batch to another difficult.	4. The amplification conditions play a significant role. 5. Certain additives can improve detection tests. 6. Bovine serum albumin improves detection of E and RdRP genes. 7. Evaluation of additives on the analytical sensitivity is needed.	Chen et al., 2020
Reaction volume	1. The reaction volume on the performance of SARS-CoV-2 test. 2. A lower reaction volume may interfere with the sensitivity of the test.	A test demanding less reagent offers cost effective (Figure 4.3).	Nelson et al., 2020; 25, Gill et al., 2021
Enzyme pair in RT-PCR	1. Efficiency of Reverse transcriptase and *Taq* Polymerase varies pair to pair.	The enzyme pair should be selected with due care.	Alcoba-Florez et al., 2020
RNA Extraction method	2. The extraction stage clears PCR inhibitors. 3. The extraction reagents can be in short supply.	Alternative extraction methods: 1. Direct heating without additives. 2. Saliva Direct using proteinase K and heating to 95°C. 3. Processed saliva sample. 4. No extraction methods.	Alcoba-Florez et al., 2020 Vogels et al., 2021 Gill et al., 2021 Fukumoto et al., 2020

4.6.1 Molecular Targets Selected

The first RT-PCR based detection test for SARS-CoV-2 detected the presence of envelop (E) gene of the virus, which codes for the envelope of the viral coat and the second gene for the enzyme RNA-dependent RNA polymerase (RdRp). This test was initially developed from genetic similarities between SARS-CoV-2 and its close relative SARS, and later refined using the SARS-CoV-2 genome data to target viral genes unique to the newly discovered virus. However, different test kits have their own molecular targets that the test kit primers and probe detect in the genome of the virus present in the test sample (Bezier et al., 2020). The most frequently tested gene targets for the SARS-CoV-2 detection include the following targets while some gene targets cross-react with SARS-CoV-1 (even though it is no longer in the human population) and endemic coronaviruses associated with the common cold.
Structural genes:

1. Envelope protein: S and E (E and RdRp)
2. Transmembrane: M
3. Helicase: H
4. Nucleocapsid: N (N1, N2, and N3)

Accessory genes:

1. Enzymatic machinery: RdRp (E and RdRp)
2. Haemagglutinin esterase: HE
3. Open reading frame: ORF (1a/1b)

If a test kit used more than one target, the target sequences may be present in two different genes or inter-genes or other genomic regions or within the same gene. The concomitant detection of more than one target gene is often preferred to optimize the detection of SARS-CoV-2, enhancing the sensitivity of the test. For example, the US-CDC recommended EUA test kit initially having two targets N1 and N2 present within the nucleocapsid (N) gene, the main target gene that was positive in the repositive cohort samples of the Wuhan patient tested retrospectively (Zhang et al., 2020a). However, the EUA test kit recommended by the US-CDC initially optimized with N1 and N2 targets eventually dropped the N2. Similarly, detection of the E gene by itself is not recommended by the EU-CDC but with the RdRP gene (Colton et al., 2020; Corman et al., 2020b). It also appears that the genes that are efficiently transcribed in host cells (e.g., M gene) also prove to be good targets for the test kit development (Toptan et al., 2020). Different copy numbers of the molecular targets in the genome also impact the efficiency of diagnostic test.

4.6.2 Limitations of SARS-CoV-2 Diagnostic Tests

The COVID-19 disease caused by this virus has become the deadliest pandemic worldwide. Part of the attribution to this impact of COVID-19 is the poor diagnosis, absence of timely diagnosis, absence of timely treatment regime, and asymptomatic transmission of its causative agent SARS-CoV-2. The available rapid tests for antigen and antibody are error prone. The RT-PCR tests, although the gold standard at this point of time, suffer from the following limitations.

4.6.3 Sampling Limitations

The reliability of the diagnostic process depends upon the sensitivity of the tests to detect SARS-CoV-2. In return, the sensitivity of the test also depends upon the following factors.

4.6.4 COVID-19 Viral Load in Sample Collection

The sample taken with higher viral load offers higher probability of isolating viral particles in the sample leading to higher diagnostic sensitivity. The studies performed on different samples—saliva, nasal swabs from both nostrils, sputum, or phlegm, and bronchoalveolar lavage (BAL) indicate that the viral load is greater in the lower respiratory tract with significant difference in LOD (Bezier et al., 2020; Vogels et al., 2021). Taking samples from the upper and lower respiratory tract is recommended by the WHO especially if a sample from the upper respiratory tract is negative in light of strong clinical observations of infection (WHO, 2020b). This recommendation is meant to improve the reliability of the diagnosis. Unfortunately, it involves logistical concerns since BAL is a complex medical procedure and poses biosafety hazard for health professionals taking the samples in addition to the time and cost it takes. Therefore, simpler procedures are more applicable for mass screening (Wang et al., 2020). Nasopharyngeal swab sample collection is the most common sample type for mass testing, despite this region not having the most significant viral load (Tsang et al., 2021).

4.6.5 Diagnostic Timing and Risk of False Negatives

The SARS-CoV-2 load in samples taken during different stages of infection and on viral load kinetics from patient to patient affect the performance of molecular tests (Loeffelholz and Tang, 2020). The RT-PCR like all virological tests has imperfect sensitivity over the viral load during course of the disease (Poggiali et al., 2020). Evolution of viral load in terms of time is one of the important parameters that influence the reliability of SARS-CoV-2 mass testing of the population. Ever since the application of the diagnostic method for the detection of SARS-CoV-2, the test kits have been lacking sensitivity and thus false negatives result from samples with low viral loads. Thus, some patients receive a positive result then a negative result and vice versa in just a short duration of time. Moreover, a significant number (~25%) of asymptomatic subjects test positive after about 2 weeks, indicating the persistence of SARS-CoV-2 in the nasal passages (Zhang et al., 2021). Such cases can result in transmission during the latent period and provide evidence of the insufficiency of a single negative result (Poggiali et al., 2020; Jarvis and Kelley, 2021).

This is a key lesson for public health agencies that RT-PCR might be highly accurate in detecting the virus, but only at certain times the course of infection. However, serological tests by measuring circulating antibodies against the virus offer an alternative approach (Zhang et al., 2021; Saurabh et al., 2020). In contrast to virological tests, serological tests can detect infected people months after viral clearance. Thus, in view of public health safety, a combination of RT-PCR and serological tests is the most accurate determination of public health.

In symptomatic patients, the COVID-19 infection viral load has been modeled. On average, the incubation period appears to be 5 to 6 days followed by symptoms lasting for 12 days (Wang et al., 2020). The viral load evolves in parallel with symptoms over 5 to 10 days in patients reaching its maximum the 10th day after infection. The graph presents two sectors on the course of infection where the low viral load in a patient becomes a false-negative result. The false-negative sectors are supported by the statistical data on the variation of viral load over time in symptomatic individuals (Lippi et al., 2020; Kucirka et al., 2020). During this course, the model can also help in predicting the rate of false negatives each day from the day of contacting the virus to the 21st day. The results of this study on the appearance rate of false negatives with RT-PCR versus number of days following exposure are shown in Figure 4.1. This figure offers only a general picture of the evolution of false

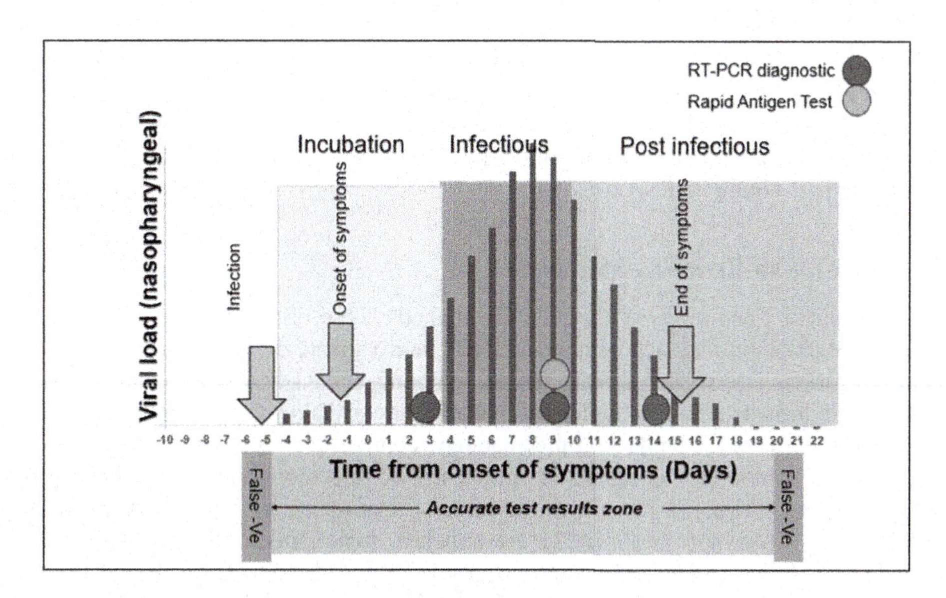

FIGURE 4.1 Course of infection and false negative zones of the RT-PCR tests.

negatives. In the first 3 days after exposure, the probability of a false negative is ~100% (asymptomatic period, where the viral load is low due to early period of infection) and the rate falls to the minimum at ~7 days (symptomatic period, where the viral load is at its peak). This model also shows that the appropriate time to take a test sample from nasopharyngeal, oropharyngeal, or saliva is from 7 to 10 days from the day of exposure (Loeffelholz and Tang, 2020). Thus, appropriate timing of the testing is an important limiting factor for molecular testing.

4.6.6 Quality of Samples for Accurate Testing

Appropriate sample collection, transportation, and storage is essential for the reliability of test results. The important factors that contribute to negative impact on test results include (US-CDC, 2021c):

1. Quickness in sample collection.
2. Lack of sensitivity for the collected sample.
3. Poor transporting of samples (compromising cold chain and transit duration).
   a) To be tested within 72 h: Rapid storage between 2°C and 8°C
   b) To be tested after long period: Storage at −18°C or, ideally at −70°C
4. Presence of substances in samples that inhibit the RT-PCR.
5. Contamination of samples by the RNA or the virus.
6. Inactivation of sample with means other than viral transport medium or by lysis buffer containing guanidinium to better conserve RNA in the sample. For example, heat inactivation of the samples generates higher Ct in the RT-PCR indicating degradation of the sample (Pan et al., 2020).

4.7 TECHNICAL LIMITATIONS OF THE DIAGNOSTIC TESTS

No diagnostic test is perfect since many factors play a role in the outcome of the test. Diagnostic tests suffer from the following limitations.

4.7.1 Methodological Limitations

The current RT-PCR-based tests suffer from inadequately harmonized test methodologies, limited and inadequate biological reference materials, and limited information on diagnostic test systems. To dissipate this limitation to some extent, following the Minimum Information for Publication of Quantitative RT-PCR Experiments (MIQE) recommendations, the publication of research articles on the RT-PCR method validation should include complete information on evaluation parameters (Bustin, 2017). Accordingly, the end user of clinical laboratories can intelligently choose the diagnostic tools for performing SARS-CoV-2 tests.

4.7.2 Limitation of Reference Materials

Despite the popularity and utility of the RT-PCR methods, these tests have certain limitations like false-positive/negative results, lack of reproducibility, and variable detection limits of the test kits depending on various aspects such as assay design, sampling, and RT-PCR reagent activity (Benzigar et al., 2021). While some reference biological materials are available, they are not yet standardized and lack quality in terms of production methodology, composition, and storage conditions. Thus, the lack of access to adequate quantities of sufficiently diverse and reference biological materials has been hampering testing. To improve the outcome of measurements, it is important to use well-referenced materials (Akyurek et al., 2021). Nevertheless, some guidelines on standardization and validation of RT-PCR for SARS-CoV-2 detection have been uploaded on relevant websites. The JIMB (Joint Initiative for Metrology in Biology) and European Virus Archive-GLOBAL have started sharing reference materials for the validation of methods for SARS-Cov-2 detection by RT-PCR that include harmonized practices to increase confidence levels in RT-PCR tests and thus eased out the situation (Akyurek et al., 2021; EVA, 2021; US-CDC, 2021d; JIMB, 2021). In addition, molecular controls in terms of SARS-CoV-2 RNA to be employed as positive control in the test kits have become available (Gill et al., 2022).

4.7.3 Inter-Laboratory Comparisons

Comparisons of inter-laboratory tests, methodologies, and results on the detection of SARS-CoV-2 have been done despite the limitations on sample homogeneity, storage, and transportation of samples (Chik et al., 2021). Such inter-laboratory comparisons result in concerns about disparity of practices, accuracy, and confidence levels of detection tests in practical terms (Charki and Pavese, 2019). It would be an important lesson learned to develop inter-laboratory comparisons that can move us forward towards consistency of reliable test results across regions. For example, lack of access to detailed information on exact composition of the commercial extraction kit and test kit reagents and sequences of primers and probes limits the appreciation and comparison of the performance of the RT-PCR test kits. Unfortunately, in the absence of this information, the end-user laboratory must spend time and resources in validating the test kit before it can be adopted for routine test service. This wastage of resources and time is unacceptable in the wake of pandemic scenario.

4.8 CONCEPTUAL LIMITATIONS OF SARS-COV-2 DIAGNOSTIC TESTS

Besides the technical limitations, insufficient knowledge of the molecular biology of the virus affects the confidence levels in the detection tests. This is a significant limitation in view of the emergence of new variants. Therefore, to run a successful testing strategy, knowledge of the following has added importance in the accuracy of the detection test.

4.8.1 Lack of Knowledge about the Virus Genome

The isolation of SARS-CoV-2 is not being performed in accordance with standard practice as per current guidelines (US-FDA, 2021). A correlation between the diversity of symptoms among patients and specific symptoms with the presence of SARS-CoV-2 RNA is difficult to make. Thus, an element of uncertainty hovers over the conserved and variable regions of the viral genome, molecular targets, the primers, and probes of RT-PCR kits. However, more and more sequencing and its biology information has poured in from different regions of the world since a large number of viral genomes are being sequenced (Umair et al., 2021). This data is being utilized to refine the conserved region for improving molecular tests for more reliable diagnosis. But in light of the fact that new variants are constantly emerging the PCR primers being used in clinical laboratories need to be reviewed constantly to ascertain the validity. Since this virus evolves fast, every new variant must be sequenced for its mutations to be fully characterized to develop strategies in diagnostics and management. For example, the omicron variant has 60 mutations (50 nonsynonymous mutations, 8 synonymous mutations, and 2 non-coding mutations) in comparison to the Wuhan strain taken as reference (Haseltine, 2021a; Ingrid, 2021). The S-protein alone contains 32 mutations (Haseltine, 2021b). The total set of mutations also includes a snippet of genetic material from a common cold causing HCoV-229E virus (Lapid, 2021; Venkata et al., 2021). Therefore, SARS-CoV-2 keeps perpetuating in the presence of other human viruses, the emergence of more variants is to be expected. Moreover, to predict future mutations in the genome of this virus, bioinformatics tools may be used to improve specificity of primers in the conserved regions of the viral genome (Yip et al., 2020).

4.8.2 Genetic Diversity of SARS-CoV-2

For designing a specific and sensitive test kit, primers and probes must be constructed to detect and amplify all strains of SARS-CoV-2 but excluding those of other viruses. However, the known genetic diversity of SARS-CoV-2 makes this problematic. It is a well-known fact that viruses mutate, and SARS-CoV-2 is no exception. There have been a number of circulating viral strains over the course of the COVID-19 pandemic. SARS-CoV-2 has been mutating at a rate of about 1–2 mutations per month (Callaway, 2021c). These mutations occur as a by-product of viral replication, but the RNA viruses make relatively more mutations than DNA viruses. On the other hand, the coronaviruses have an enzyme that corrects some of the errors of replication and have lower mutation rate than most RNA viruses (Lauring et al., 2021). But co-infections of SARS-CoV-2 with other human viruses can generate variants like omicron with genetics insertions from the co-infecting viruses (Lapid, 2021; Venkata et al., 2021).

Having too many mutations in its genome is not always beneficial for a virus since mutations may counter or complement one another. The fate of a new mutation is determined by natural selection, which favors the mutation that confers a competitive advantage in viral replication or in transmission or evasion from immunity. But the mutation that reduces viral fitness naturally tends to get eliminated (Speers, 2006; Lauring et al., 2021). Based on the data collected between December 24, 2019 and May 29, 2020, the mutation rate in this virus was evaluated at 0.80–2.38 × 10–3 nucleotide substitutions per site per year reflecting the genetic diversity of SARS-CoV-2 around the world (Álvarez-Díaz et al., 2020). However, this mutation rate is half that of influenza and one fourth that of HIV (Hagan, 2021). Thus, a significant contribution to this genetic diversity is from both the intrinsic polymorphism and the selection pressure exerted by the human immune system (Shen et al., 2020).

Although the effect of this genetic diversity would be premature on the viral phenotype but can allow distinction only between conserved and variable regions but also between more conserved and more variable regions in the viral genome. Accordingly, meta-transcriptome sequences of SARS-CoV-2 have shown that many variants can be found in the same patient at the same time (Shen et al., 2020). Unfortunately, this data also warns about the possible impact of this genetic diversity on the validity of molecular, antigen, and serology tests due to the inherent design differences of each test. To determine the validity of RT-PCR detection tests it is imperative to recognize the existing variants. But this also implies that it is not the quantum of mutations that matters but their position in the genome whether in the conserved regions or in the variable regions of the genome. Since test kits are normally designed on the conserved regions of the genome, a high mutation rate in a variable region is not a point of concern in view of the validity of the test kits. On the other hand, a low number of mutations in a conserved region can have serious consequences leading to a serious problem with sensitivity and false negatives in the population.

In a study of 31,000 SARS-CoV-2 genome sequences taken from nasopharyngeal samples from 30 patients, 99% were identical in the target regions of the primers included in detection tests (Álvarez-Díaz et al., 2020). The 1% of non-homogenous sequences posed a mismatch between the genome of SARS-CoV-2 and the commercial primer for the genes of the detection tests recommended by the WHO. Rather two sites of genetic variability were presented in the US-CDC recommended sequence of primers (US-CDC, 2021b). This variability had a critical influence on the validity of tests involving RdRP gene. Depending upon the position of the discrepancies, primer binding to the target sequences may not be affected, whereas others are critical and may lead to false negatives from patient samples. If the mutations occur in the 3' region that critically decides the primer mismatch, an absence of amplification resulting fromfalse-negative test results will occur (Álvarez-Díaz et al., 2020). This implies that the sensitivity and reliability of detection tests depends upon the understanding and exploitation of the genetic diversity of SARS-CoV-2.

4.9 VARIANTS OF SARS-COV-2

As expected, multiple variants of SARS-CoV-2 have been documented in the United States and globally throughout this pandemic (Lauring et al., 2021). The US FDA first alerted clinical laboratory staff and healthcare providers on January 8, 2021 of the following impacts:

1. False-negative results from tests involving only one genetic target of this virus.
2. Tests involving multiple genetic targets of this virus partly affected.
3. Do not ignore negative results and review with clinical observations.
4. Test with kits involving different genetic targets if still suspected after reviewing negative test result.
5. In laboratory studies, specific monoclonal antibody may be less effective due to variants with the L452R or E484K substitution in the receptor binding domain (RBD) of the spike protein that binds to human lung receptor sangiotensin-converting enzyme-2 (ACE-2). This event also accompanies priming and activating transmembrane protease serine 2 (TMPRSS2), also known as serine protease leading to the eventual internalization of the virus.

A variant of SARS-CoV-2 may have one or more mutations that make it different from other variants. However, understanding the variants requires in-depth knowledge of SARS-CoV-2 molecular biology, its evolution, and the genomic epidemiology. Accordingly, a classification system of SARS-CoV-2 variants was developed by a SARS-CoV-2 Interagency Group (SIG) of the US Department of Health and Human Services (HHS) and accordingly the SIG and the US-CDC have grouped the variants in three categories. The WHO also classifies variant viruses as Variants of Concern with the omicron as the latest variant as of November 26, 2021, and Variants of Interest. The WHO classification may slightly differ from that of the US classifications since the importance of variants may differ by location. The variant status in one category may escalate or deescalate based on scientific data (WHO, 2021a). Therefore, the list of variants gets updated as needed to show the variants that belong to each class (Table 4.2).

4.10 DOMINANT DELTA VARIANT VERSUS UPCOMING OMICRON VARIANT

Adequate information on the molecular biology of the dominant delta variant indicates that the worldwide dominant delta variant is between 40% and 60% more transmissible than the alpha variant. Its two-fold higher transmissibility than the Wuhan's SARS-CoV-2 virus is thought to contribute to its dominance over the other strains. Additionally, it can replicate its genome much faster, producing about 1,000 virus particle than its predecessors which is a key factor in viral fitness. Another factor is the contribution of mutation, especially L452R, in the spike protein leading to more efficient binding of the spike protein to the ACE-2 receptor in patients (Ingrid, 2021; Robitzski, 2021).

The immune evasion and the COVID-19 vaccine evasion capability conferred upon by the delta variant mutations including K417N proved another factor in its dominance. The severity of disease-causing hospitalization of more people and higher death rates were the result of the P681R mutation on its spike protein improving its ability to enter cells.

In view of this, it is too early to determine if the omicron variant would out compete the delta variant, although both the variants share some mutations. But the omicron variant also possesses quite different mutations in addition to the 10 mutations in the receptor-binding domain of the spike protein that interacts with the ACE-2 receptor compared with just two mutations in the delta variant (Callaway, 2021c).

Omicron may spread globally if it becomes more transmissible with immune evasion, which seems likely based on its spread to over 38 countries in a few days of its being declared as a Variant

TABLE 4.2
Categories of SARS-CoV-2 Variant

Category	PANGO Line	Region	Dated	Performance Markers
Variant of Interest (VOI)	B.1.526 B.1.526.1 B.1.525 and P.2	NY (USA) NY (USA) UK/Nigeria Brazil	Nov 2020 Oct 2020 Dec 2020 Apr 2020	1. Changes in receptor binding. 2. Reduced interaction with antibodies raised against previous infection or vaccination. 3. Reduction in efficacy of treatments. 4. Potentially evading of diagnostic kit. 5. Predicted severity in transmissibility or disease.
Variant of Concern (VOC)	B.1.1.7 B.1.351 P.1 B.1.427, and B.1.429 B.1.1.529 variant	UK South Africa Japan/Brazil CA (USA) CA (USA) South Africa	Nov 2020 Nov 2020 Jan 2021 Jan 2020 Jan 2020 Nov 2021	1. Increase in transmissibility. 2. Increase severity in disease. 3. Decreased neutralization by antibodies raised during prior infection or vaccination. 4. Reduced efficacy of treatments or vaccines. 5. Failure of diagnostic tests.
Variant of High Consequence (VOIC)	No variants identified			1. Failures of prevention measures. 2. Failures of clinical countermeasures. 3. Failure of diagnostic tests. 4. Failures of vaccine effectiveness. 5. Failures of vaccine-induced protection. 6. Failures of multiple EUA or approved therapeutics. 7. Severe clinical disease.

of Concern by the WHO (www.aljazeera.com/news/2021/12/3/new-york-becomes-fourth-us-state-to-confirm-omicron-live). More variants may emerge that upon natural selection and adaptation may result in more variants with higher severity. It is apparent that molecular biology has learned new lessons while dealing with the fast-pace evolution of variants.

4.11 MOLECULAR TESTS IMPACTED BY SARS-COV-2

The mutations emerging in the genome of the SARS-CoV-2 variants were expected to affect the specificity and efficiency of the test kits if sequences of their primers or probe corresponded with these mutated sequences. Normally, the test performance for most molecular tests for SARS-CoV-2 is not expected to be impacted by a single-point mutation; however, the Cepheid tests were affected by a single-point mutation in the target sequence of the test as analyzed by the FDA (Hasan et al., 2021; US-CDC, 2021a). Mutations in the E gene (C26340T) and N gene (C29200T) were reported to impact the detection of target genes of two commercial assays that were common single nucleotide. Some reports indicated that two independent single-point mutations reduced the test sensitivity of the N2 target. The E target of the test kit (the Xpert Xpress SARS-CoV-2 and Xpert Xpress SARS-CoV-2 DoD) still detected a positive signal in cases where enough virus was present in the sample. Detection of the E target without detecting the N2 target will be reported as "positive" in the Xpert Omni SARS-CoV-2 (Hasan et al., 2021). These mutations happen to be in those regions of the virus including the N1, N2, and E genes that do not seem to offer much of a selection advantage to the virus.

Despite the fact that mutations lead to generation of variants, the diversity of the virus and the number of tests currently available for testing patient samples can still be reliably utilized to detect the variants. There are more than 246 molecular diagnostic tests on the list of the FDA's EUA, and the majority have targets other than the spike gene (West et al., 2021; US-CDC, 2020b). Therefore, these tests should still be valid for these variants and are not expected to generate false negatives

TABLE 4.3
Test Kits Possibly Affected by the Variants

Test Kit	Outcome of Test	Genetic Variant at Position	Potential Impact	Analysis – *in silico*
Accula SARS-CoV-2 Test (Mesa Biotech Inc.)	1 of the 2 targets of the test reduced sensitivity.	28881 (GGG to AAC) is tested.	1. Demands attention. 2. Overall test sensitivity not impacted.	1. Primer binding may be affected. 2. Must be sequenced to confirm the mutant(s).
Linea COVID-19 Assay Kit (Applied DNA Sciences, Inc.)	1 of the 2 targets of the test reduced sensitivity.	B.1.1.7 variant (UK VOC-202012/01).	1. Detects multiple genetic test targets. 2. Overall test sensitivity not impacted. 3. Identifies new variants of the virus.	4. 1 positive target and 1 negative target display the S-gene dropped the test kit sensitivity. 5. Must be sequences to confirm the mutant.
TaqPath COVID-19 Combo Kit (Thermo Fisher Scientific, Inc.)	1 of the 3 targets of the test reduced sensitivity.	B.1.1.7 variant (UK VOC-202012/01).	6. Detects multiple genetic test targets. 7. Overall test sensitivity not impacted. 8. Identifies new variants of the virus.	1. 2 positive targets and 1 negative target display the S-gene dropped the test kit sensitivity. 2. Must be sequences to confirm the mutant.
1. Xpert Xpress SARS-CoV-2 2. Xpert Xpress SARS-CoV-2 DoD 3. Xpert Omni SARS-CoV-2 (Cepheid)	1 of the 2 targets of the test reduced sensitivity.	B.1.1.7variant (UK VOC-202012/01).	1. Detects multiple genetic test targets. 2. The mutation does not lead to false negative. 3. Mutation leads to "presumptive positive" outcome.	4. Test instructions are important. 5. Mutation leading to "Presumptive Positive" results. 6. Must be sequenced to confirm the mutant.

from the sample with a variant having mutations in the spike gene. For example, the test kits having multiple targets within the genome of SARS-CoV-2 like ORF1ab and N genes in addition to the spike gene. Therefore, these kits can still give accurate results. Of the four test kits identified by the US-CDC on its website that involve target S gene, the sensitivity of the test would be affected and thus may or may not clearly identify a patient infected with a spike gene variant (Table 4.3).

From this perspective, S-gene drop-out like other variants is also observed for the omicron variant, and thus these tests cannot be used for diagnostic or as an identifier of the omicron variant. Obviously, such S-gene drop-outs samples are the candidates for sequencing. Since the delta variant is the primary variant circulating in major regions of the globe, the kits using S-gene target may indicate the S-gene drop-out and potentially be omicron variants and should be prioritized for confirmation by sequencing.

4.12 WHAT SHOULD TEST KIT MANUFACTURERS AND TESTING LABS DO?

To ensure the accuracy of the tests in view of the variants of SARS-CoV-2, the manufacturers of test kits or testing laboratories must confirm the following:

1. The sequences of the primers and probes of the test must be determined to confirm analytical sensitivity of the test at least fortnightly.

2. In the event of such an outcome of results, the primer sequences of the virus present in the test sample should be compared to the variant sequences as suggested by the CDC in the guidelines.
3. It would be advisable and vital for the accuracy of tests that the testing labs and the test kit manufacturers watch for notices from the local CDC or the regulatory agency of their region regarding the validity of the EUA diagnostic tests (US-CDC, 2020a).
4. Follow up on failed tests to investigate potential variants of SARS-CoV-2. Thus, the test labs must remain informed of any potential impacts on test efficacy (West et al., 2021).
5. Validate alternative diagnostic tests to proactively keep alternative tests kit ready for deployment.
6. Develop tests for variant detection if possible using primers and probes for the qualitative detection of emerged variants.
7. The confirmation of the variant must be performed with Whole Genome sequencing.

4.13 RAPID SCREENING OF VARIANTS WITH TEST KITS

It is amply clear by now that the mutations happening in the test kit target sequences may or may not affect the overall sensitivity and specificity of the test depending upon the number of targets of the test kit (US-CDC, 2020b). However, indirectly, such mutations happening in the sequences area of the test kit are indicative of the possible variant of the virus. For example, a commercial RT-PCR test kit of SARS-CoV-2 experienced a hit from the UK variant, B.1.1.7, having a deletion (69/70) in the spike gene of the SARS-CoV-2 present in the patient sample. Similarly, some other kits were also hit by the occurrence of variants in their target sequences. But the mutations within the spike receptor binding domain have displayed a potential for enhanced transmission, thus being classified as VOCs (Nelson et al., 2020; Peterson et al., 2021). Moreover, the VOC also demonstrated in vitro tests a significant reduction in antibody neutralization by the K417N/T, E484K, and N501Y substitutions in the spike protein (Liu et al., 2021; Xie et al., 2021; Feldman et al., 2021; Matic et al., 2021). These findings suggest that early identification of VOCs is incredibly important for containment of the virus.

Interestingly, the use of test kits in the detection of possible variants of the virus emerged as a tool for surveillance for the detection of this variant in the population (EU-CDC, 2020). The surveillance data collected through this tool from the United Kingdom indicated a rapid increase in SARS-CoV-2 cases in September 2020 concerning the B.1.1.7 variant (Public Health England, 2021; Matic et al., 2021). This variant eventually predominated in most of of Europe (Matic et al., 2021, Galloway et al., 2021). Another VOC, the B.1.1.28/P.1 variant in Japan and Brazil, later emerged as the predominant variant in parts of Brazil and North America (Matic et al., 2021). Strangely, B.1.1.28/P.1 was found to have mutations other than 69/70 deletion that were found in the B.1.1.7 variant necessitating the demand for broad VOC surveillance to detect multiple mutations.

Detecting variants using test kits is not the gold standard, of course, but comprehensive variant detection in a genome including single nucleotide variants through whole-genome sequencing (WGS) or whole exome sequencing or sequencing of random subsets of SARS-CoV-2 samples (Koboldt, 2020) is laborious, costly, time consuming, and complex.

On March 16, 2021 Roche launched the cobas® SARS-CoV-2 Variant Set 1 Test to detect and differentiate variants that originated in the UK (B.1.1.7), South Africa (B.1.351), and Brazil (P.1). The test involves respective primers and probes in the ready-made 384-test cassette for screening these variants. Moreover, multiple RT-PCR test kits and the RT-PCR-based algorithm-based strategies have been designed to detect variant mutations in specimens testing positive for SARS-CoV-2 in clinical samples that can play a great role in rapid VOC surveillance (Matic et al., 2021; WHO, 2021b). For example, a research team based in British Columbia, Canada identified a cluster of B.1.1.28/P.1 cases, a variant not previously known to circulate in British Columbia, Canada (Matic et al., 2021).

4.14 SURVEILLANCE OF VARIANTS

While certain mutations in the SARS-CoV-2 genome are acceptable, it is essential to monitor key mutations in important regions of the virus genome. For example, the mutations that alter the genes involved in infection and evade the immune system of the host.

4.14.1 Genomic Surveillance of Variants

Regulatory agencies like the WHO and the US CDC regularly survey for variants by sequencing and characterizing the virus since mutations occur frequently. For example, a variant of SARS-CoV-2 generated a D614G substitution in the spike gene between late January and early February 2020. A few months later this mutation replaced the wild type of Wuhan city and became the dominant global strain.

The CDC's National SARS-CoV-2 Strain Surveillance (NS3) characterizes about 750 samples weekly through sequencing and evaluation (www.cdc.gov/coronavirus/2019-ncov/cases-updates/variant-surveillance.html). In addition, the US CDC also expands its surveillance program by:

1. Collaborating with commercial diagnostic laboratories.
2. Collaborating with seven universities across the country.
3. Supporting regional and tribal health departments.

4.14.2 Wastewaters Screening of Variants

In recent years, wastewater-based epidemiology has been used to monitor the presence of drugs or pathogens like poliovirus in many countries (Ivanova et al., 2015). Traces of SARS-CoV-2 in community wastewaters have been detected in diverse countries using RT-PCR (Bosch, 1998; Wu et al., 2020; Jorgensen et al., 2020). Thus, wastewater monitoring can serve as a potential point of monitoring public health concerning the SARS-CoV-2 virus, since it can indicate potential outbreaks and offer a cost-effective way of surveillance for determining burden of disease in a community. However, it is still unclear if sequencing of the RT-PCR tested samples can be used to detect the viral variants in community sewage. The COVID-19 Wastewater Coalition is investigating the usefulness of this strategy (Chik et al., 2021; Canadian Water Network, 2021).

4.15 IMPACT OF GLOBALIZATION ON VARIANTS

The scenario of the last pandemic of the Spanish flu of 1918 cannot be compared with the current pandemic of COVID-19. The impact of the Spanish flu was no doubt massive, and has been shadowed by the consequences of the current pandemic (Jarvis and Kelley, 2021; Pandey et al., 2020). At the time of the Spanish flu, the globe was not a village. However, globalization has played a great role in the current pandemic. Connectivity through globalization offers a positive selection pressure for the mutants in getting selected and thus plays a role in spreading the mutants of the potential pathogens. This also implies that the next epidemic could be just a flight away since infectious diseases are everywhere (Dyer, 2021). There are 1.67 million unknown viruses on this planet out of which between 631,000 and 827,000 viruses have the ability to infect people. However, in the scientific literature only about 263 viruses are known that can infect people. This implies that little is known about the potential pandemic threats (Kessler, 2021). A list of viruses is presented by the WHO whose molecular biology is unknown. Thsu, there may not be effective vaccines or medicines to counter potential pandemics caused by such viruses (Pennisi, 2020).

Many factors contributed to the current pandemic. The frequent mixing of different animal species has been thought to facilitate the creation of some coronaviruses (Lauring et al., 2021; Jasper et al., 2015). In addition, other factors like air travel, trade of goods, urbanization, and climate change may have played a significant role in generation and spread of mutated viral genomes and emergence of pathogenic viruses (Davies et al., 2021).

The single-stranded RNA (ssRNA) nature of viral genetic material has an influence on the propensity for emergence (Davies et al., 2021). The high mutation and genome recombination rates in such viruses may allow them to adapt to new hosts across species barriers and global regions (Lauring et al., 2021). Due to these genomic traits, the coronaviruses are known to be highly rapidly evolving viruses (Kumar et al., 2021). As an example, the SARS viruses have the capability of direct transmission from animals to humans (Davies et al., 2021; Kumar et al., 2021; Hedman et al., 2021; Made et al., 2021).

Despite the progress in development of diagnostics, vaccines, and treatments, the emergence of new infectious diseases is a threat to human health and global stability. Many diseases that are identified by the WHO normally exist in animals, but due to globalization, the probability of outbreaks is increased since animals and humans are in frequent contact (WHO, 2021b). This scenario is aggravated when snippets of genetic materials from co-infected human virus are picked up as additional molecular weapons as likely happened with the omicron variant (Ingrid, 2021).

4.16 THROUGHPUT OF DETECTION TESTS

The rapid spread of COVID-19 resulted in the development of molecular methods to detect the virus and manage it, but also strained health care systems worldwide (Kreier, 2021). Unable to operate effectively in wake of the unprecedented inflow of infected cases, even excellent healthcare systems in developed countries struggled to keep up. High-throughput RT-PCR has helped meet the needs of both patients and healthcare facilities.

The use of RT-PCR tests in 96-well formats in centralized laboratories is the standard. However, saving costs on testing, and increasing the throughput, 384-well formats using lower volume has also been introduced for SARS-CoV-2 testing LOD down to five copies/microliter (Nelson et al., 2020; Giri et al., 2021). The granting of EUA for diagnostic test systems on February 4, 2020 by the FDA enabled molecular diagnostic tests to be developed, validated, and deployed at war scale within a short time (Jeffery et al., 2018).

4.17 IMPACT OF PANDEMIC BOTTLENECKS ON TESTING STRATEGIES

The pandemic has caused a number of bottlenecks including a shortage of supplies such as reagents, lab plasti-wares, extraction kits, and RT-PCR kits. Thus multiple options for test kits, methods, testing platforms, and other accessories are needed. The current RT-PCR test kits are costly, and hence some countries with limited resources cannot afford enough test kits to adequately screen their people. This lack of resources leaves important gaps in screening of asymptomatic people during the incubation phase of the disease. This shortage is being experienced globally (Giri et al., 2021).

COVID-19 testing platforms are operating at full capacity putting stress on key reagent supply chains and thus creating bottlenecks. These supply bottlenecks have resulted in the testing of only symptomatic individuals even though there is an urgent need to test asymptomatic and pre-symptomatic populations as well (Shental et al., 2021). In parallel, however, pandemic shortages have encouraged the development of less resource demanding, rapid, reliable, and low-cost diagnostic methods like microchip-based RT-PCR test systems (Figure 4.2). The microchip test system allows up to 4-plex testing in a miniaturized reaction volume of only 1.2 microliters offering PCR runtime of 32 minutes for 45 samples with LOD of < 5 copies/microliter. The reagents of the test kit are lyophilized in the microwells of the microchips and therefore the end-user needs to add sample RNA following the layout of the test kit making it very user-friendly (Razvan et al., 2021; Gill et al.,

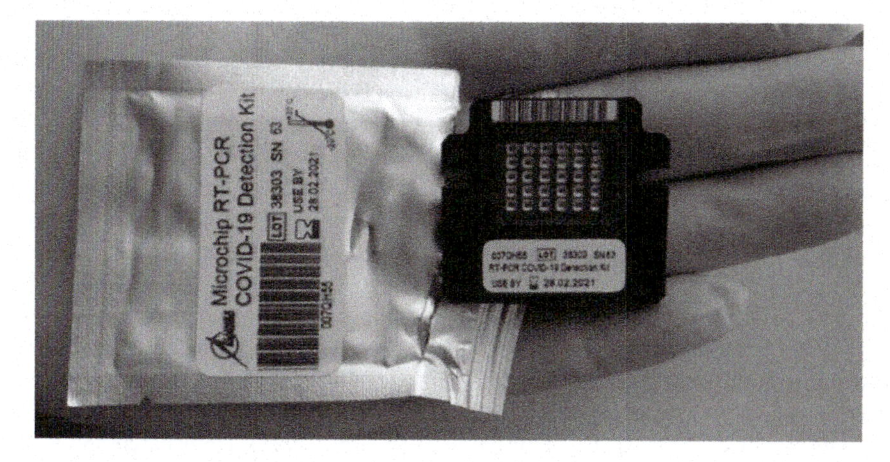

FIGURE 4.2 Microchip RT-PCR Test System (30 or 48 microreactors).

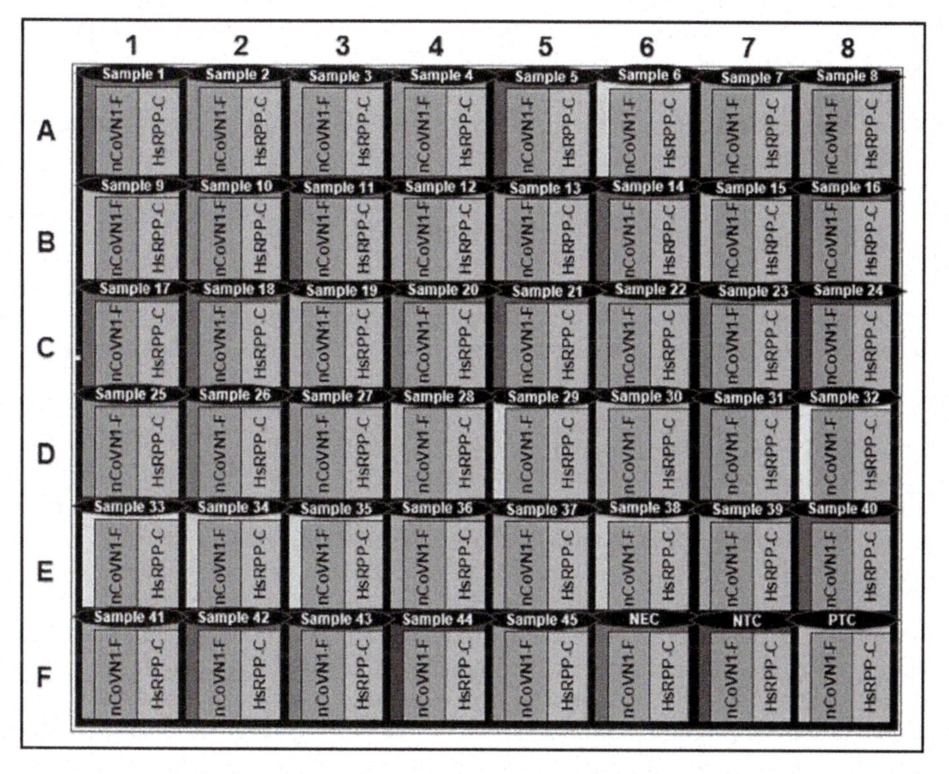

FIGURE 4.3 Microchip 48 microreactor layout.

2021). The technology works with RNA extracted from nasopharyngeal swab samples and with extraction-free saliva samples for SARS-CoV-2 detection as research use only (RUO) or laboratory developed test (LDT). The lyophilized microchips are good for at least six months in storage offering a longer-lasting solution while maintaining the RT-PCR based diagnosis of SARS-CoV-2 as the benchmark among testing methods for diagnosis of the disease (Gill et al., 2021).

To meet the rising demand for additional testing molecular diagnostic tools are constantly emerging. Moreover, the limitations of RT-PCR can also be potentially overcome by whole-genome sequencing of SARS-CoV-2 by detecting fragments of the virus even if the complete genome is not there in the sample. The SARS-CoV-2 genome being free of repeats is conducive for sequencing since short construction of RNA library and short sequence reads can make the testing faster and cost effective. The genome sequencing may also overcome the false-negative results due to sidewise confirmation of the sequences. However, cost, technical complexities, data analysis, and 24–48 h turnaround time limit its use in clinical testing (Chandler-Brown et al., 2020). Another emerging method with potential is clustered regularly interspaced short palindromic repeat (CRISPR), a new gene editing tool-based technology that performs simultaneous RT-PCR and loop-mediated isothermal amplification (RT-LAMP). It is a single-tube technique for the amplification of DNA and a low-cost alternative to detect certain diseases with LOD of 10 copies/microliter with duration of < 40 min (Nguyen et al., 2021).

An innovative pilot protocol that has benefits of massive scale-up potential due to the use of DNA molecular barcodes (Udugama et al., 2020; Pardo-Seco et al., 2021), DNA barcoding is a method that can be used to uniquely identify a pathogen just like a scanner recognizing the black stripes of the UPC barcode. This approach involves screening the swabs of SARS-CoV-2 to determine whether the gene matching the reference DNA barcode is present in the swabs or not. The approach can screen thousands of samples per day speeding up the testing process and reducing cost after its limitations are addressed (Pardo-Seco et al., 2021). Furthermore, a handheld, portable, low cost, and rapid device built by Conservation X Labs that can detect the presence or absence of any species with an available genetic sequence has potential for SARS-CoV-2 testing (https://conservationxlabs.com/dna-barcode-scanner).

4.18 RNA-BASED VACCINE DEVELOPMENTS

Currently available antiviral vaccines can be put into two categories: Protein-based vaccines like the polio vaccine that delivers stimulating antigen to the immune system (live or attenuated virus or subunit antigen) and tene-based vaccines that carry genetic instructions of the antigen (e.g., spike protein of the virus) to instruct the host cell to make the antigen.

While it is alarming to witness the generation of variants with increasing potency of transmission, an effective vaccine seems the only solution to contain the spread. The published genome sequencing by Professor Yongzhen's team allowed scientists to quickly begin developing COVID-19 vaccines (Robin, 2020; Zhan, 2020). How effective current vaccines will be in the face of new mutations and variants remains to be seen (Callaway et al., 2021b).

To develop the mRNA vaccine, a new approach on vaccine development was taken, even though the idea of applying RNA in vaccine development has been around since the first report of *in vitro* transcribed (IVT) mRNA into mice models in 1990 (Wolff et al., 1990). Key molecular biology lessons were learned in 2002 from the SARS-coronavirus and then from the MERS-coronavirus in 2012 that directed scientists to focus vaccine development on the spike protein of the novel coronavirus. It was also learnt which modifications in the viral genome would stabilize the spike structure so that it could generate a robust and safe antibody response, making the mRNA safer and stabilizing it for the desired duration. To help the mRNA cross the cell membrane, the successful delivery vehicles deployed in therapeutic small interfering RNAs (siRNAs) were used in vaccine development (Pardi et al., 2018).

4.19 CONCLUSION

Any pandemic is highly dynamic and challenging, and so is the current one despite novel technologies playing a role. While writing this chapter the scenarios have been changing at the world level and the lessons learnt or being learnt are being updated at fast pace.

The lessons learned from the success of the SARS-CoV-2 testing strategies, the management of communities, and coordination of public health departments and governments have helped bring the pandemic under control. Molecular biology in terms of novel molecular testing technologies made a huge contribution. However, the novelty of the virus in itself was a challenge for molecular biology to understand its genome and functionality. Additionally, the emergence of many variants made things even more challenging. But key lessons have been learned along the way.

During the dominance of the delta variant, molecular biology of mRNA vaccine production finally did offer some respite. It was learned that a single dose or double dose of the vaccine and later boosters were needed to keep the virus under control. Soon after, the alarming spread of yet another variant omicron scared the whole world again with its unusual number of mutations (Callaway, 2021c; Hasteline, 2021a,b; Ingrid, 2021). In the main antigenic target of the virus where infection-induced antibodies and many vaccines target are hit by 32 mutations among which many mutations happen to be new (Callaway, 2021a).

The lessons learned in the development of the COVID-19 mRNA vaccine and molecular testing methods will go a long way beyond COVID-19. Vaccine production moved away from traditional strategies and a novel footpath was established to counter both emerging and established pathogens. Advances in thermostable mRNA vaccines (Zhang et al., 2020b) allowed them to be stored easily and transported to remote populations. Multiplex testing methods and microchip or microarray and DNA barcode-based test systems are in development too.

REFERENCES

Akyurek S, Demirci SNS, Bayrak Z, Isleyen A, and Akgoz M (2021). The production and characterization of SARS-CoV-2 RNA reference material. *Anal Bioanal Chem* 413: 3411–3419. https://doi.org/10.1007/s00216-021-03284-w

Alcoba-Florez J, González-Montelongo R, Íñigo-Campos A, Artola DG-M, Gil-Campesino H (2020). Fast SARS-CoV-2 detection by RT-qPCR in preheated nasopharyngeal swab samples. *Int J Infect Dis.* 97: 66–68. https://doi.org/10.1016/j.ijid.2020.05.099

Álvarez-Díaz DA, Franco-Muñoz C, Laiton-Donato K, Usme-Ciro JA, Franco-Sierra ND, Flórez-Sánchez, AC, Mercado-Reyes M (2020). Molecular analysis of several in-house rRT-PCR protocols for SARS-CoV-2 detection in the context of genetic variability of the virus in Colombia. *Infect Genet Evol* 84: 1–7. https://doi:10.1016/j.meegid.2020.104390

Andersen KG, Rambaut A, Lipkin WI, Holmes EC, Garry RF (2020). Correspondence: The proximal origin of SARS-CoV-2. *Nat Med* 26 (4): 450–452. https://doi:10.1038/s41591-020-0820-9

ANSES (2015). Guide de validation des méthodes d'analyses. www.anses.fr/fr/system/files/ANSES_GuideValidation.pdf

Belouafa S, Habti F, Benhar S, Belafkih B, Tayane S, Hamdouch S, Abourriche A (2017). Statistical tools and approaches to validate analytical methods: Methodology and practical examples. *Int J Metrol Qual Eng* 8 (9): 1–10. https://doi.org/10.1051/ijmqe/2016030

Benzigar MR, Bhattacharjee R, Baharfar M, Liu G (2021). Current methods for diagnosis of human coronaviruses: Pros and cons. *Anal Bioanal Chem* 413 (9): 2311–2330. https://doi.org/10.1007/s00216-020-03046-0

Bezier C, Anthoine G, and Charki A (2020). Reliability of real-time RT-PCR tests to detect SARS-Cov-2. *Int J Metrol Qual Eng* 11: 1–13. www.metrology-journal.org/articles/ijmqe/full_html/2020/01/ijmqe200027/ijmqe200027.html

Bosch A (1998). Human enteric viruses in the water environment: A minireview. *Int Microbiol* 1 (3): 191–196. https://pubmed.ncbi.nlm.nih.gov/10943359/

Bustin, S (2017). The continuing problem of poor transparency of reporting and use of inappropriate methods for RT-qPCR. *Biomol Detect Quantif* 12: 7–9. https://doi.org/10.1016/j.bdq.2017.05.001

Callaway E (2021a). Could new COVID variants undermine vaccines? *Nature* 589 (7841): 177–178. http://doi:10.1038/d41586-021-00031-0

Callaway E (2021b). The coronavirus is mutating—does it matter? *Nature* September 08, 2021. www.nature.com/articles/d41586-020-02544-6

Callaway, E (2021c). Heavily mutated coronavirus variant puts scientists on alert. *Nature* 600 (7887): 21. doi:10.1038/d41586-021-03552-w. PMID 34824381

Campbell C (2020). The Chinese Scientist Who Sequenced the First COVID-19 Genome Speaks Out About the Controversies Surrounding His Work. *The Time* August 24, 2020. https://time.com/5882918/zhang-yongzhen-interview-china-coronavirus-genome/

Canadian water network (2021). COVID-19 Wastewater Coalition. https://cwn-rce.ca/category/covid-19-wastewater-coalition/

Chandler-Brown D, Bueno AM, Atay O, Tsao DS (2020). A highly scalable and rapidly deployable RNA extraction-free COVID-19 assay by quantitative sanger sequencing. *bioRxiv* https://doi:10.1101/2020.04.07.029199

Charki A, and Pavese F (2019). Data comparisons and uncertainty: A roadmap for gaining in competence and improving the reliability of results. *Int J Metrol Qual Eng* 10 (1): 1–10. https://doi.org/10.1051/ijmqe/2018016

Chen CJ, Hsieh LL, Lin SK, Wang CF, Huang YH, Lin SY, Lu PL (2020). Optimization of the CDC protocol of molecular diagnosis of COVID-19 for timely diagnosis, *Diagnostics* 10: 333. https://doi:10.3390/diagnostics10050333

Cheng VCC, Lau SKP, Lau PCY, and Yuen KY (2007). Severe acute respiratory syndrome coronavirus as an agent of emerging and re-emerging infection. *Clin Microbiol Rev* 20 (4): 660–694. https://doi:10.1128/CMR.00023-07

Chik AHS, Glier MB, Servos M, Mangat CS, Pang X-L, Qiu X, D'Aoust PM, Burnet JB, Delatolla R, Dorner S, Geng Q, Giesy JP, McKay RM, Mulvey MR, Prystajecky N, Srikanthan N, Xie Y, Conant B, Hrudey SE (2021). Comparison of approaches to quantify SARS-CoV-2 in wastewater using RT-qPCR: Results and implications from a collaborative inter-laboratory study in Canada. *J Environ Sci* (China) 107: 218–229. https://doi.org/10.1016/j.jes.2021.01.029

CMS (2021). The Centers for Medicare & Medicaid Services (CMS). Clinical Laboratory Improvement Amendments (CLIA). www.cms.gov/Regulati ons-and-Guidance/Legislation/CLIA

Colton H, Ankcorn M, Yavuz M, Tovey L, Cope A, Raza M, Evans C (2020). Improved sensitivity using a dual target, E and RdRp assay for the diagnosis of SARS-CoV-2 infection: Experience at a large NHS Foundation Trust in the UK. J *Infect* 82 (1): 159–198. https://doi.org/10.1016/j.jinf.2020.05.061

Corman VM, Bleicker T, Brünink S, Drosten C (2020a). Diagnostic detection of 2019-nCoV by real-time RT-PCR-protocol and preliminary evaluation as of Jan 17 2020. www.who.int/docs/default-source/coronaviruse/protocol-v2-1.pdf?sfvrsn=a9ef618c_2

Corman VM, Landt O, Kaiser M, Molenkamp R, Meijer A, Chu DK, Drosten C (2020b). Detection of 2019 novel coronavirus (2019-nCoV) by real-time RT-PCR. *Euro Surveill* 25: 2000045 https://doi:10.2807/1560-7917.ES.2020.25.3.2000045

Coronaviridae (2020). Study Group of the International Committee on Taxonomy of Viruses. The species Severe acute respiratory syndrome-related coronavirus: classifying 2019-nCoV and naming it SARS-CoV-2. *Nature Microbiology* 5: 536–544 https://doi.org/10.1038/s41564-020-0695-z

Davies NG, Abbott S, Barnard RC, Jarvis CI, Kucharski AJ, Munday JD, Pearson CJB, Russell TW, Tully DC, Washburne AD, Wenseleers T, Waites W, Wong KLM, Zandvoort K, Silverman JD, Diaz-Ordaz K, Keogh R, Eggo RM, Funk E, Jit M, Atkins KE, Edmunds WJ (2021). Estimated transmissibility and impact of SARS-CoV-2 lineage B.1.1.7 in England. *Science* 372 (6538): eabg3055. https://doi:10.1126/science.abg3055

Dyer J (2021). Ready for the Next Pandemic? Infection Control Today, April 2021. 25 (3). www.infectioncontroltoday.com/view/ready-for-the-next-pandemic-spoiler-alert-it-s-coming

EU-CDC (2020). Rapid increase of a SARS-CoV-2 variant with multiple spike protein mutations observed in the United Kingdom. Dec 20, 2020. www.ecdc.europa.eu/sites/default/files/documents/SARS-CoV-2-variant-multiple-spike-protein-mutations-United-Kingdom.pdf

EVA (2021). SARS-CoV-2 collection. www.european-virus-archive.com/evag-news/sars-cov-2-collection.

Feldman J, Pavlovic MN, Gregory DJ, Poznansky MC, Sigal A, Schmidt AG, Lafrate AJ, and Naranbhai V, and Balazs AB (2021). Circulating SARS-CoV-2 variants escape neutralization by vaccine-induced humoral immunity. *medRxiv* 2021 Mar 2. https://doi.org/10.1101/2021.02.14.21251704

Fukumoto T, Iwasaki S, Fujisawa S, Hayasaka K, Sato K, Oguri S, Teshima T (2020). Efficacy of a novel SARS-CoV-2 detection kit without RNA extraction and purification. *Int J Infect Dis* 98: 16–17. https://doi.org/10.1101/2020.05.27.120410

Galloway SE, Paul P, MacCannell DR, Johansson MA, Brooks JT, MacNeil A (2021). Emergence of SARS-CoV-2 B.1.1.7 Lineage - United States, December 29, 2020-January 12, 2021. *MMWR Morb Mortal Wkly Rep* 70(3): 95–99. www.cdc.gov/mmwr/volumes/70/wr/mm7003e2.htm

Gill RK, Gill SS, Gelimson I, Slyadnev M, Martinez G, Nunley GR, Majoros T (2021). Color-coding of Microchip RT-PCR Test System for SARS-CoV-2 Detection. *J Biosc Medicine* 9 (5): 94–119. https://doi.org/10.4236/jbm.2021.95010

Gill S, Gill RK, Varankovich N, Gelimson I , Dancey C, Jackson K, Bailey S, Robinson E, Hui F, Brown A, Elliott R, Slyadnev M, and Eveleigh D. Molecular Controls for SARS-CoV-2 and Influenza A/B Testing in Microchip-based RT-PCR Test Systems. Poster AMP 2022, Phoenix AZ, USA

Giri B, Pandey S, Shrestha R, Pokharel K, Ligler FS, Neupane BB (2021). Review of analytical performance of COVID-19 detection methods. *Anal Bioanal Chem* 413 (1): 35–48. https://doi.org/10.1007/s00216-020-02889-x

Hagan A. (2021). SARS-CoV-2 Variants vs. Vaccines. American Society for Microbiology, March 3, 2021. https://asm.org/Articles/2021/February/SARS-CoV-2-Variants-vs-Vaccines

Hasan MR, Sundararaju S, Manickam C, Mirza F, Hal HA, Lorenz S, Tang P (2021). A novel point mutation in the N gene of SARS-CoV-2 may affect the detection of the virus2 by RT-qPCR. *J Clin Microbiol* 59(4): 1–3. (e03278-20). https://doi:10.1128/JCM.03278-20

Haseltine CE (2021a). Heavily mutated coronavirus variant puts scientists on alert. Nature. 600 (7887): 21. doi:10.1038/d41586-021-03552-w. PMID 34824381. S2CID 244660616.

Haseltine WA (2021b). Omicron Origins. Forbes. Archived from the original on 3 December 2021.

Hedman HD, Krawczyk E, Helmy YA, Zhang L, and Varga C (2021). Host Diversity and Potential Transmission Pathways of SARS-CoV-2 at the Human-Animal Interface. *Pathogens* 10 (2): 180–208. https://doi.org/10.3390/pathogens10020180

Holshue ML, DeBolt C, Lindquist S, Kathy MD, Lofy H, Wiesman J, Bruce H, Spitters C, Ericson K, Wilkerson S, Tural A, Diaz G, Cohn A, Fox L, Patel A, Gerber SI, Kim K, Tong S, Lu X, Lindstrom S, Pallansch MA, Weldon WC, Biggs HM, Uyeki TM, and Pillai SK (2020). First Case of 2019 Novel Coronavirus in the United States. *N Engl J Med* 382: 929–936. www.nejm.org/doi/full/10.1056/NEJMoa2001191

Iglói Z, Leven M, Abdel-Karem J, Weller B, and Matheeussen V, Coppens J, Molenkamp R (2020). Comparison of commercial realtime reverse transcription PCR assays for the detection of SARS-CoV-2. *J Clin Virol* 129: 1–3. https://doi.org/10.1016/j.jcv.2020.104510

Ingrid T 2021. Covid-19: Omicron may be more transmissible than other variants and partly resistant to existing vaccines, scientists fear. *BMJ*. 375: n2943. doi:10.1136/bmj.n2943. ISSN 1756-1833. PMID 34845008. S2CID 244715303.

Ishige T, Murata S, Taniguchi T, Miyab, A, Kitamur, Kawasak, Matsushit, K (2020). Highly sensitive detection of SARS-CoV-2 RNA by multiplex rRT-PCR for molecular diagnosis of COVID-19 by clinical laboratories. *Clinica Chimica Acta* 507: 139–142. https://doi.org/10.1016/j.cca.2020.04.023

Ivanova MS Yarmolskaya TP, Eremeeva GM, Babkina OY, Baykova LV, Akhmadishina AY, Krasota LIK, and Lukashev AN (2019). Environmental Surveillance for Poliovirus and Other Enteroviruses: Long-Term Experience in Moscow, Russian Federation, 2004–2017. *Viruses* 11 (5): 424–437. https://doi:doi:10.3390/v11050424.

Jarvis KF, and Kelley JB (2021). Temporal dynamics of viral load and false negative rate influence the levels of testing necessary to combat COVID-19 spread. *Scientific Reports* 11: 1–12. (9221). https://doi.org/10.1038/s41598-021-88498-9

Jasper F W, Chan S, Lau KP, Kelvin KW, Vincent C, Cheng C, Patrick C, Woo Y, Kwok-Yung Y (2015). Middle East respiratory syndrome coronavirus: another zoonotic betacoronavirus causing SARS-like disease. *Clin Microbiol Rev* 28 (2): 465–522. https://doi:10.1128/CMR.00102-14.

Jeffery K, Taubenberger MD, and Morens DM (2010). Influenza: The Once and Future Pandemic. *Public Health Rep* 125 (3): 16–26. https://doi.org/10.1177/00333549101250S305

JIMB (2021). Coronavirus standards working group. https://jimb.stanford.edu/

Jorgensen AU, Gamst J, Hansen LV, Knudsen IIH and Jensen SKS (2020). Eurofins Covid-19 sentinel TM wastewater test provide early warning of a potential COVID-19 outbreak. *MedRXiv*. https://doi.org/10.1101/2020.07.10.20150573

Kessler R (2021). Disease X: The Next Pandemic. *EcoHealth Alliance MISSION*. https://www.ecohealthalliance.org/2018/06/outbreak-next-pandemic

Koboldt DC (2020). Best practices for variant calling in clinical sequencing. *Genome Medicine* 12(91). https://doi.org/10.1186/s13073-020-00791-w

Kreier F (2021). The myriad ways sewage surveillance is helping fight COVID around the world. *Nature* 10 MAY 2021.

Kubina R, and Dziedzic A (2020). Molecular and Serological Tests for COVID-19. A Comparative Review of SARS-CoV-2 Coronavirus Laboratory and Point-of-Care Diagnostics. *Diagnostics* (Basel) 10 (6): 434.https://doi.org/10.3390/diagnostics10060434

Kucirka LM, Lauer SA, Laeyendecker O, Boon D, Lessler J (2020). Variation in false-negative rate of reverse transcriptase polymerase chain reaction–based SARS-CoV-2 tests by time since exposure. *Ann Intern Med* 173: 262–267.https://doi:10.7326/M20-1495

Kumar R, Verma H, Singhvi N, Sood U, Gupta V, Singh M, Kumari R, Hira P, Nagar S, Talwar C, Nayyar N, Anand S, Rawat CD, Verma M, Negi RK, Singh Y, Lal R (2020). Comparative Genomic Analysis of Rapidly Evolving SARS-CoV-2 Reveals Mosaic Pattern of Phylogeographical Distribution. *American Society for Microbiology* 5 (4): e00505–20. https://doi:10.1128/mSystems.00505-20

Lai C-C, Wang C-Y, Ko W-C, Hsueh P-R (2020). *In vitro* diagnostics of coronavirus disease 2019: Technologies and application. *J Microbiol Immunol Infect* 54 (2): 164–174. https://doi.org/10.1016/j.jmii.2020.05.016

Lamoril J, Bogard M, Ameziane, N, Deybach J-C, Bouizegarène P (2007). Biologie moléculaire et microbiologie clinique en, *Immuno-anal Biol Spé* 22 (2): 73–94. http://doi:10.1016/j.immbio.2006.11.002

Lapid N 2021 Omicron variant may have picked up a piece of common-cold virus. Reuters Healthcare & Pharmaceuticals. December 3, 2021.

Lauring AS, Hodcroft EB (2021). Genetic Variants of SARS-CoV-2—What Do They Mean? *JAMA* 325 (6): 529–531. https://doi:10.1001/jama.2020.27124

Lippi G, Simundic AM, Plebani M (2020). Potential preanalytical and analytical vulnerabilities in the laboratory diagnosis of coronavirus disease2019 (COVID-19), *Clin Chem Lab Med* 58: 1070–1076.https://doi:10.1515/cclm-2020-0285

Liu Y, Liu J, Xia H, Xianwen Z, Fontes-Garfias CR, Swanson KA, Weaver SC, Muik A, Jansen KU, Xi X, Jansen PR, Shi P-Y (2021). Neutralizing Activity of BNT162b2- Neutralizing Activity of BNT162b2-Elicited Serum. Elicited Serum - Preliminary Report. *N Engl J Med* 384 (15): 1–3. https://doi:10.1056/NEJMc2102017

Loeffelholz MJ, Tang Y-W (2020). Laboratory diagnosis of emerging human coronavirus infections – the state of the art. Emerg Microbes Infect 9: 747–756. https://doi:10.1080/22221751.2020.1745095.

Lu X, Wang L, Sakthivel SK, Whitaker B, Murray J, Kamili S, Lynch B, Malapati L, Burke SA, Harcourt J, Tamin A, Thornburg NJ, Villanueva JM, and Lindstrom S (2020). Synopsis- US CDC Real-Time Reverse Transcription PCR Panel for Detection of Severe Acute Respiratory Syndrome Coronavirus 2. *EID Journal* 26 (8): 1–4. https://wwwnc.cdc.gov/eid/article/26/8/20-1246_article

Made A, Dewantari AK, and Wiyatnoc A (2021). Molecular biology of coronaviruses: current knowledge. *Heliyon* 6(8): e04743. https://doi:10.1016/j.heliyon.2020.e04743.

Matic N, Christopher F Lowe Ritchie G, Stefanovic A, Lawson T, Jang W, Young M, Dong W, Brumme ZL, Brumme CJ, Leung V, Romney MG (2021). Rapid Detection of SARS-CoV-2 Variants of Concern, Including B.1.1.28/P.1, British Columbia, Canada. *Emerging Infectious Diseases* 27 (6): 1673–1676. http://doi:10.3201/eid2706.210532.

NCBI (2021). Severe acute respiratory syndrome coronavirus 2 isolate Wuhan-Hu-1, complete genome. *GenBank*: MN908947.3 https://www.ncbi.nlm.nih.gov/nuccore/MN908947

Nelson AC, Auch B, Schomaker M, Gohl DM, Grady P, Johnson D, Yohe S (2020). Analytical validation of a COVID-19 qRT-PCR detection assay using a 384-well format and three extraction methods. *bioRxiv* 2020.04.02.022186https://doi.org/10.1101/2020.04.02.022186

Nguyen LT, Smith BM, Jain PK (2020). Enhancement of trans-cleavage activity of Cas12a with engineered CrRNA enables amplified nucleic acid detection. *bioRxiv* 2020. https://doi.org/10.1101/2020.04.13.036079

Pan, Y., Long, L., Zhang, D., Yuan, T., Cui, S., Yang, P., Ren, S. 2020. Potential false-negative nucleic acid testing results for severe acute respiratory syndrome coronavirus 2 from thermal inactivation of samples with low viral loads. *Clin. Chem.*, 66: 794–801. https://doi:10.1093/clinchem/hvaa091.

Pandey S, Yadav B, Pandey A, Tripathi T, Khawary M, Kant S, and Tripathi S (2020). Lessons from SARS-CoV-2 Pandemic: Evolution, Disease Dynamics and Future. *Journals Biology* 9 (6): 141–153. https://doi.org/10.3390/biology9060141

Pardi N, Hogan MJ, Porter FW, Weissman D (2018). mRNA vaccines- a new era in vaccinology. *Nature Reviews Drug Discovery* 17: 261–279. http://doi:10.1038/nrd.2017.243

Pardo-Seco J, Gómez-Carballa F, Salas A (2021). Pitfalls of barcodes in the study of worldwide SARS-CoV-2 variation and phylodynamics. *Zool Res* 42 (1): 87–93. https://doi:10.24272/j.issn.2095-8137.2020.364

Patel V (2021). How Omicron, the New Covid-19 Variant, Got Its Name. The New York Times. ISSN 0362-4331. 28 November 2021.

Pennisi E (2020). Scientists discover virus with no recognizable genes. *Science* Feb. 7, 2020. https://doi:10.1126/science.abb2121.

Peterson S, Reynoso L, Downey GP, Frankel SK, Kapple J, Marrack P, Zhang G (2021. 501Y.V2 and 501Y.V3 variants of SARS-CoV-2 lose binding to Bamlanivimab *in vitro*. *bioRxiv* March 2, 2021. https://doi.org/10.1101/2021.02.16.431305

Poggiali E, Vercelli A, Vadacca GB, Schiavo R, Mazzoni G, Loannilli E, Demichele E, Magnacavallo A (2020). Negative nasopharyngeal swabs in COVID-19 pneumonia: the experience of an Italian Emergency Department (Piacenza) during the first month of the Italian epidemic. *Acta Biomed* 91(3): 1–4 (e2020024). https://doi.org/10.23750/abm.v91i3.9979

Posteraro B, Marchetti S, Romano L, Santangelo R, Morandotti GA, Sanguinetti M, Cattani P., Cacaci M, Carolis ED, Galuppi S, Giordano L, Graffeo R, Rosa ML, Sorda ML, Liotti FM, Martini C, Marturano C, Masucci L, Menchinelli G, Nicotra R, Palucci I, Rocchetti C, Sali M, Salvioni M, Torelli R (2020). Clinical microbiology laboratory adaptation to COVID-19 emergency: experience at a large teaching hospital in Rome, Italy. *Clin Microbiol Infect* 26: 1109–1111. https://doi.org/10.1016/j.cmi.2020.04.016

Public Health England (2021). Investigation of novel SARS-COV-2 variant: Variant of Concern 202012 / 01. https://assets.publishing.service.gov.uk/government/uploads/system/uploads/attachment_data/file/959426/Variant_of_Concern_VOC_202012_01_Technical_Briefing_5.pdf

Rahman H, Carter I, Basile K, Donovan L, Kumar S, Tran T, Kok J (2020). Interpret with caution: An evaluation of the commercial AusDiagnostics versus in-house developed assays for the detection of SARS-CoV-2 virus. *J Clin Virol* 127: 104374.https://doi.org/10.1016/j.jcv.2020.104374

Razvan C, Yaseen I, Unrau PJ, Lowe CF, Ritchie G, Romney MG, Sin DD, Gill S, Slyadnev M (2021). Microchip RT-PCR Detection of Nasopharyngeal SARS-CoV-2 Samples. *Journal of Molecular Diagnostics* 8: S1525–1578. https://www.jmdjournal.org/article/S1525-1578(21)00063-5/fulltext

Robin M (2020). "The vaccine miracle: how scientists waged the battle against Covid-19". *The Observer*. Retrieved 11 January 2021. https://www.theguardian.com/world/2020/dec/06/the-vaccine-miracle-how-scientists-waged-the-battle-against-covid-19

Robitzski D 20 Omicron Is WHO's Fifth Variant of Concern, Experts Urge Patience. The scientist. 2021

Saurabh S, Kuma, R, Gupta MK, Bhardwaj P, Nag VL, Garg MK, Misra S (2020). Prolonged persistence of SARS-CoV-2 in the upper respiratory tract of asymptomatic infected individuals, *QJM: Int J Med* 113: 556–560. https://doi.org/10.1093/qjmed/hcaa212

Shen Z, Xiao Y, Kang L, Ma W, Shi L, Zhang L, Li M (2020). Genomic diversity of severe acute respiratory syndrome–coronavirus 2 in patients with coronavirus disease 2019. *Clin Infect Dis* 71: 713–720.https://doi:10.1093/cid/ciaa203

Shental N, Levy S, Wuvshet V, Skorniakov S, Shalem B, Ottolenghi A, Greenshpan Y, Steinberg R, Edri A, Gillis R, Goldhirsh M, Moscovici K, Sachren S, Friedman LM, Nesher L, Shemer-Avni Porgador A, Hertz T (2021). Efficient high-throughput SARS-CoV-2 testing to detect asymptomatic carriers. *Science Advances* 6 (37): 1–8. (eabc5961). https://doiI:10.1126/sciadv.abc5961

Speers DJ (2006). Clinical Applications of Molecular Biology for Infectious Diseases *Clin Biochem Rev* 27(1): 39–51.https://www.ncbi.nlm.nih.gov/pmc/articles/PMC1390794/

Stanley, K. 2020. AusDiagnostics SARS-CoV-2 kits shown to be more sensitive than reference laboratory test. *J. Clin. Virol.*, 129: 104485.https://doi:10.1016/j.jcv.2020.104485

Toptan T, Hoehl S, Westhaus S, Bojkova D, Berger A, Rotter B, Widera M (2020). Optimized qRT-PCR approach for the detection of intra- and extra-cellular SARS-CoV-2 RNAs. *Int J Mol Sci* 21: 1–11. https://doi:10.3390/ijms21124396

Tsang NNY, So HC, Ng KY, Cowling BJ, Leung GM, Ming DK (2021). Diagnostic performance of different sampling approaches for SARS-CoV-2 RT-PCR testing: a systematic review and meta-analysis. *The Lancet* April 12, 2021:1–13. https://doi.org/10.1016/S1473-3099(21)00146-8

Udugama B, Kadhiresan P, Kozlowski HN, Malekjahani A, Osborne M, Li VYC, Chen H, Mubareka S, Gubbay JB, Warren CW (2020). Diagnosing COVID-19: The Disease and Tools for Detection. *ACS Nano* 14 (4): 3822–3835. https://doi.org/10.1021/acsnano.0c02624

Umair M, Ikram A, Salman M, Khurshid A, Alam M, Badar N, Suleman R, Tahir F, sharif S, Montgomery J, Whitmer J, Klena J (2021). Whole-genome sequencing of SARS-CoV-2 reveals the detection of G614 variant in Pakistan. *Plos One*https://doi.org/10.1371/journal.pone.0248371

US-CDC (2020a). Research Use Only 2019-Novel Coronavirus (2019-nCoV) Real-time RT-PCR Primers and Probes. https://www.cdc.gov/coronavirus/2019-ncov/lab/rt-pcr-panel-primer-probes.html.

US-CDC (2020b). Diagnostic Testing. Aug. 5, 2020. https://www.cdc.gov/coronavirus/2019-ncov/lab/testing.html

US-CDC (2021a). Emerging SARS-CoV-2 Variants. https://www.cdc.gov/coronavirus/2019-ncov/science/science-briefs/scientific-brief-emerging-variants.html

US-CDC (2021b). Genomic Surveillance for SARS-CoV-2. Apr. 2, 2021. https://www.cdc.gov/coronavirus/2019-ncov/cases-updates/variant-surveillance.html

US-CDC (2021c). Specimen Collection. https://www.cdc.gov/coronavirus/2019-ncov/lab/guidelines-clinical-specimens.html#:~:text=Store%20respiratory%20specimens%20at%202,70%C2%B0C%20or%20below

US-CDC (2021d). Test for COVID-19 Only. https://www.cdc.gov/coronavirus/2019-ncov/lab/virus-requests.html

US-FDA (2021). Coronavirus Disease 2019 Testing Basics. https://www.fda.gov/consumers/consumer-updates/coronavirus-disease-2019-testing-basics

Venkata KAJ, Anand P, Lenehan, PJ, Suratekar R, Raghunathan B, Michiel JM, Soundararajan VN (2021). Omicron variant of SARS-CoV-2 harbors a unique insertion mutation of putative viral or human genomic origin. OSF Preprints, 2021.

Vogels CBF, Watkins AE, Harden CA, Brackney DE, Shafer J, Wang J, Caraballo C, Kalinich CC, Ott IM, Fauver JR, Kudo E, Lu P, Venkataraman A, Tokuyama M, Moore AJ, Muenker MC, Casanovas-Massana A, Fournier J, Grubaugh ND (2021). SalivaDirect: A simplified and flexible platform to enhance SARS-CoV-2 testing capacity. *Med J* 2 (3): 263–280. https://doi.org/10.1016/j.medj.2020.12.010

Wang P (2020). Combination of serological total antibody and RT-PCR test for detection of SARS-COV-2 infections. *J Virol Methods* 283: 1–9. (113919). https://doi:10.1016/j.jviromet.2020.113919

Wang W, Xu Y, Gao R, Lu R, Han K, Wu G, Tan W (2020). Detection of SARS-CoV-2 in Different Types of Clinical Specimens. *JAMA* 323 (18): 1843–1844. https://doi:10.1001/jama.2020.3786

West R, Gronwall GK, Kobokovich A (2021). Variants, Vaccines and What They Mean For COVID-19 Testing. COVID-19 School of Public Health Experts Insights, John Hopkins School USA February, 2, 2021.https://www.jhsph.edu/covid-19/articles/variants-vaccines-and-what-they-mean-for-covid19-testing.html

WHO (2019). Novel Coronavirus (2019-nCoV). https://www.who.int/docs/default-source/coronaviruse/situation-reports/20200121-sitrep-1-2019-ncov.pdf

WHO (2020a). Director-General's opening remarks at the media briefing on COVID19 -March 2020 World Health Organization. Pneumonia of unknown cause–China. https://www.who.int/docs/default-source/coronaviruse/situation-reports/20200121-sitrep-1-2019-ncov.pdf

WHO (2020b). Laboratory testing for 2019 novel coronavirus (2019-nCoV) in suspected human cases. https://who.int/publications/i/item/10665-331501

WHO (2020c). 10 infectious diseases that could be the next pandemic. www.gavi.org/vaccineswork/10-infectious-diseases-could-be-next-pandemic

WHO (2020d). Protocol: Real-timeRT-PCR assays for the detectionof SARS-CoV-2. www.who.int/docs/default-source/coronaviruse/real-time-rt-pcr-assays-for-the-detection-of-sars-cov-2-institut-pasteur-paris.pdf?sfvrsn=3662fcb6_2

WHO (2021a). SARS-CoV-2 Variants. www.cdc.gov/coronavirus/2019-ncov/cases-updates/variant-surveillance/variant-info.html

WHO (2021b). Zoonotic disease: Emerging public health threats in the region. http://www.emro.who.int/fr/about-who/rc61/zoonotic-diseases.html

Wolff JA, Malone RW, Williams P, Chong W, Acsadi G, Jani A, Felgner PL (1990). Direct gene transfer into mouse muscle *in vivo*. *Science* 247: 1465–1468. http://doi:10.1126/science.1690918

Wong G, Bi YH, Wang QH, Chen XW, Zhang ZG, Yao YG (2020). "Zoonotic origins of human coronavirus 2019 (HCoV-19 / SARS-CoV-2): why is this work important?". *Zoological Research* 41 (3): 213–219. https:// doi:10.24272/j.issn.2095-8137.2020.031

Wu Y, Guo C, Tang L, Hong Z, Zhou J, Dong X, Yin H, Xiao Q, Tang Y, Qu X, Kuang L (2020). Prolonged presence of SARS-CoV-2 viral RNA in faecal samples. *The lancet Gastroenterology & hepatology* 5 (5): 434–435.https://doi.org/10.1016/S2468-1253(20)30083-2

Xie X, Liu Y, Liu J, Zhang X, Zou J, Fontes-Garfias CR, Xia H, Swanson KA, Cutler M, Cooper D, Menachery VD, Weaver SC, Dormitzer PR, Shi P-Y (2021). Neutralization of SARS-CoV-2 spike 69/70 deletion, E484K and N501Y variants by BNT162b2 vaccine-elicited sera. Nat Med 27: 620–621. https://doi.org/10.1038/s41591-021-01270-4

Yip CC-Y, Ho C-C, Chan JF-W, To KK-W, Chan HS-Y, Wong SC-Y, Yuen K-Y (2020). Development of a novel, genome subtraction-derived, SARS-CoV-2-specific COVID-19-nsp2 real-time RT-PCR assay and its evaluation using clinical specimens. *Int J Mol Sci* 21: 2574. https://doi:10.3390/ijms21072574.

Zhang N-N, Li X-F, Deng Y-Q, Zhao H, Huang YJ, Yang G, Huang W-J, Gao P, Zhou C, Zhang RR, Guo Y, Sun SH, Fan H, Zu S-L, Chen Q, He Q, Cao T-S, Huang X-Y, Qiu H-Y, Nie J-H, Jiang Y, Yan H-Y, Ye Q, Zhong X, Xue X-L, Zha Z-Y, Zhou D, Yang X, Wang YC, Ying B, Qin C-F (2020b). Thermostable mRNA Vaccine against COVID-19. *Cell* 182 (5): 1271–1283. http://doi:10.1016/j.cell.2020.07.024

Zhang X, Li M, Zhang B, Chen T, Pengfei DL, Sun XZ, Shentu X, Ligh HCL, Qiana X (2020a). The N gene of SARS-CoV-2 was the main positive component in repositive samples from a cohort of COVID-19 patients in Wuhan, China. *Clinica Chimica Acta* 511: 291–297.https://doi:10.1016/j.cca.2020.10.019

Zhang Y-Z (2020). Novel 2019 coronavirus genome. *Virological* 2020 Jan 10, 2020. http://virological.org/t/novel-2019-coronavirus-genome/319

Zhang Z, Bi Q, Fang S, Wei L, Wang X, He J, Wu J, Liu X, Gao W, Zhang R, Gong W, Su Q, Azman AS, Lessler J, Zou X (2021). Insight into the practical performance of RT-PCR testing for SARS-CoV-2 using serological data: a cohort study. *The Lancet* 2 (2): E79–E87.https://doi.org/10.1016/S2666-5247(20)30200-7

Zhu N, Zhang D, Wang W, Li X, Yang B, Song J, Zhao X, Huang B, Shi W, Lu R, Niu P, Zhan F, Ma X, Wang D, Xu W, Wu G, Gao GF, Tan F (2019). China Novel Coronavirus Investigating and Research Team. A novel coronavirus from patients with pneumonia in China, 2019. *N Engl J Med* 382: 727–733. https://doi:10.1056/NEJMoa2001017

5 The Mystery Virus, SARS-CoV-2
And Its Dynamics

Anita Yadav[1], Shivji Malviya,[2#] and Sandeep K. Malhotra[3]
[1]Department of Zoology, C.M.P. Post-Graduate Constituent College (A Constituent College of the University of Allahabad), India
[2]Department of Zoology, Government H.N.B. Post Graduate College, Naini (Prayagraj), India
[3]Parasitology Laboratory, Department of Zoology, University of Allahabad, India
#Corresponding author: Shivji Malviya, Email: philonym@gmail.com

CONTENTS

5.1 INTRODUCTION

The viruses of the Corona family are normally harboured in the respiratory tract of avians and mammals, particularly human beings. The animals usually act as reservoirs of these organisms that at times are transmitted to human beings, through intermediate host species that fulfill the requirement of a zoonotic invasion. The original UK variant of this virus, B.1.1.7, transformed in India by combining with the older variant, into double variant, B.1.617. Lancet journal brought to knowledge the prevalence of double variant in Delhi as well as Panjab and Maharashtra after its entry through Delhi, and the triple variant, B.1.617+S:V3821 emerged. These double and triple variants have spread through seventeen European, American, and other foreign countries.

5.2 HISTORICAL DEVELOPMENT

The origin and source of Severe Acute Respiratory Syndrome Coronavirus 2 (SARS-CoV-2) remains unknown, although initial cases have been associated with the Huanan South China Seafood Market where snakes, birds, and other animals such as bats were sold. Considering that many of the early patients worked in or visited the market in contrast to the exported cases, it was suggested that either a human-to-human transmission or a more widespread animal source was the cause of origin of the virus (Li et al., 2020). A suspected bat origin was suggested after 96% genome sequence identity was demonstrated between SARS-CoV-2 and another coronavirus named Bat-CoV-RaTG13 isolated from bat species that colonized a province nearly 2000 km away from Wuhan (Zhou et al., 2020; Andersen et al., 2020). Pangolins were also suggested as a natural

DOI: 10.1201/9781003358909-5

host of coronaviruses (Liu et al., 2020). However, evidence of human-to-human transmission became strongly supported on January 22, 2020 after a visit conducted by a WHO delegation to the city of Wuhan. Since the first outbreak recognized in February 2020, the disease spread rapidly around the world. According to the European Centre for Disease Prevention and Control, as of June 17, 2020 there were 8,142,129 cases of COVID-19 and 443,488 deaths. American countries were among the countries with the highest number of cases (3,987,543) with the United States and Brazil being the leading countries (2,137,731 and 923,189 respectively).

Several SARS-CoV-2 samples have been isolated from different people and genomic sequences have been made available aiming to better understand the virus and to provide information for the development of diagnostic tools and vaccines. To date more than 42,000 SARS-CoV-2 RNA genomes have been uploaded in the Global Initiative on Sharing All Influenza Data. SARS-CoV-2 belongs to the beta subgrouping of the *Coronaviridae* family and is an enveloped virus containing a positive-sense, single-stranded RNA with 29,891 bases of size (Chen et al., 2020; Paraskevis et al., 2020). The genome encodes for 29 proteins involved in the infection, replication, and virion assembly process. Like other coronaviruses it is characterized by the presence of crown-like spikes on its surface (Schoeman and Fielding, 2019). The spike S protein from SARS-CoV-2 contains a receptor binding domain (RBD) that binds the human angiotensin-converting enzyme 2 (ACE2) and thereby promotes membrane fusion and uptake of the virus into human cells by endocytosis (Yan et al., 2020). The RBD present in the spike protein is the most variable region of the coronavirus genome. Structural and biochemical studies have suggested that RBD from SARS-CoV-2 binds with high affinity to ACE2 compared to other SARS-CoV viruses. However, the human ACE2 protein variability may also be a factor for the high binding affinity (Wan et al., 2020).

Scientists from different countries around the world believe that the pattern of spread of the Coronavirus closely resembles that of the Spanish Flu in 1918. While the Spanish Flu broke out in 1918, the first vaccine to control it was not developed until 1940 in the United States. Thus, people were forced to fight the disease for many years by using masks and social distancing until 1940 when the first American vaccine was developed. Today, several vaccines have been developed by various countries to counter the ill effects of Covid-19 in record time. In addition to the double variant, triple variant, and UK variant of the Corona virus in India, the variant N 440 K that arrived in Andhra Pradesh has been duly replaced by the dominant UK strain of this virus, and the Bengal variant, B.1.618 has certainly coexisted. In the current context however, Omicron variant has finally replaced the latter two strains too.

5.3 CORONA VIRUS STRUCTURE AND SPREAD

The outbreak of the Corona virus (Figures 5.1 and 5.2) originated as a mysterious infection from a small local fish and wild animal market in the Chinese city Wuhan of the province Hubei (Lu et al., 2020) and later spread throughout all cities of China, and then across all nations of the world (Li, 2020; Chen, 2020). The spread was enhanced due to the virus attacking the ACE2 Receptor of the host (Hou et al., 2010) (Figure 5.1), resulting in virus mutations. Thus, most studies focused on finding the triggering alterations in the genome structure of SARS-CoV-2.

The causative organism of nCOV-19 is SARS-CoV-2 (Zhu et al., 2019). The heightened interest in determining (Paital et al., 2020) the genomes of different forms/strains of the coronavirus resulted in the submission of more than 17,000 genomes in GENBANK (GISAID, NCBI) by mid-May of 2020. It was indeed the expeditious sharing of these sequences in the international repositories that enabled comparison of the submitted sequences (*Nature news*; Nextstrain.org.). Since the virus had become an international emergency, saving time was an essential prerequisite, before the record of mutations were to be taken into account, and simultaneously, the designing followed by the production of vaccines and clinical trials. It was a most cumbersome task to segregate the sequences

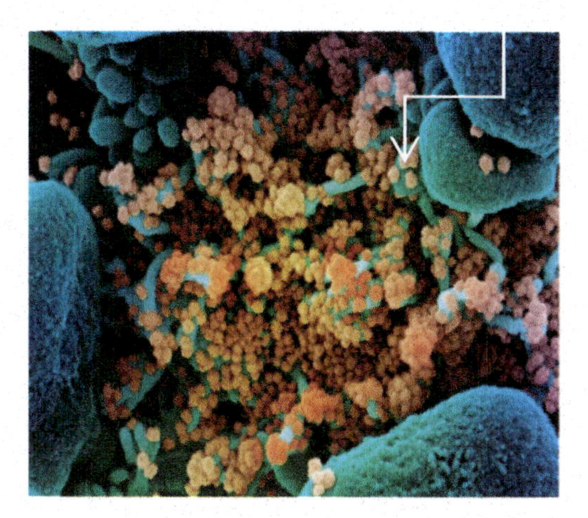

FIGURE 5.1 A Scanning electron microscope image of SARS-CoV-2 coronavirus particles (orange) on a cell (blue).

Source: After Laamarti et al., 2020.

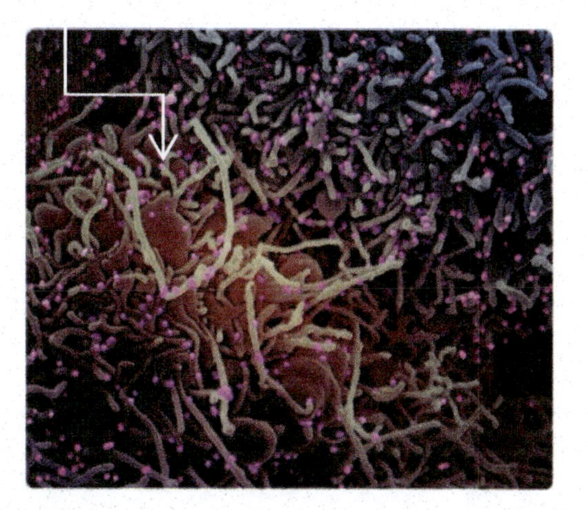

FIGURE 5.2 Novel Coronavirus SARS-CoV-2 colorized scanning electron micrograph of a VERO E6 cell (purple) exhibiting elongated cell projections and signs of apoptosis, after injection with SARS-CoV-2 virus particles (pink), which were isolated from a patient sample.

Source: After Laamarti, 2020.

that resulted because of mutations from the ones that were obtained from the genomes subjected to sequencing in the labs across the world (GISAID. NCBI, 2020).

This is a RNA-Dependent RNA Polymerase virus. Its genome of 29,903 base pair length comprises 11 genes coding for a variety of proteins, exemplified by spike proteins (Lu et al., 2020; NCBI, 2020).

A unique configuration of A, U, G, C that is the ribonucleotides, Adenine, Uracil, Guanine and Cytosine constituted the genome of SARS-CoV-2, whose simily could be drawn of to a lengthy paragraph. The inherent property of the genome of the virus, while it infected a cell, to replicate (i.e., making its copies), could be related to the copying of a paragraph. But the errors in such a process of copying resulted in the generation of mutations. Along with augmentation in the rate of production of copies, the chances of accumulating "mutations" was enhanced. Some such errors could result into an altered meaning of the paragraph, while a meaningless paragraph could result due to some such errors. On the other hand some errors (as typing mistakes) could have no effect on the meaning of this paragraph.

Seven sub-groups of human coronaviruses are known (scientific classification):

1. 229E
2. NL63
3. OC43
4. HKU1
5. MERS-CoV
6. SARS-CoV-1
7. SARS-CoV-2

5.3.1 Spike, Mutation, and Evolutionary Lineage

The mutations rechanging the genetic material of SARS-CoV-2 (nucleotide substitution rate of 8 × 10^-4 subs per site per year) at a fast speed. This speed is considerably slower than the influenza virus, that's why we are in a position to even think of having developed vaccine against it. There are thousands of complete genomes of SARS-CoV-2 available now on NCBI/GISAID. Phylogenetic analysis of these samples shows that currently at least 10–11 strains (based on sequence differences) are circulating the globe. After the outbreak in a particular geographic area, these 10–11 strains have formed hundreds of subvariants. This is the reason the Institute of Virology, Pune, India has claimed that there are 4,300 forms in India. The figure is more or less the same as in the United States.

The issues of a fastish spread as well as the suddenly enhanced virulence after every attack of the spike of Coronavirus on the ACE2 receptor (Figure 5.3) are explained by Hou et al. (2010).

The constituents of the SARS-CoV family comprise the nucleocapsid protein as its core component. During its early stages of infection, N protein is the most predominantly expressed protein,

FIGURE 5.3 The spike of 2019-nCoV virus in contact with host cell's angiotensin-converting enzyme 2 receptor.

and the pertinent immune response inhibition against it strengthens its position as an enticing diagnostic tool. As high as 89–91% alignment has been recorded between N protein and its counterparts in SARS-CoV and the Sl-CoV of bat. This provided the evidence of homology to establish evolutionary linkage. This provided implorative clues of cross-species transmission of SL-CoVs where suckling rodents were utilized as the experimental animals.

5.4 MODE OF ENTRY OF VIRUS

SARS-CoV-2 is always in an inactive state (Figures 5.1 and 5.2) until the time it comes into contact with a live cell (Figures 5.3, 5.4, and 5.5). As soon as the ACE2 receptor on the surface of a live cell is activated, it entangles itself to the spike of the virus (Figure 5.6) in an attempt to wade away the onslaught of spike onto the ACE2 receptor to enable the virus to successfully enter the cell. Initially, the ACE2 receptor is activated in contact with the spike protein of the spike of the virus. Later, it is necessary for the spike protein to neutralize the ACE2 receptor of its defense to make way for the RNA genetic material of the virus to enter the live cell, leaving the outer envelope of the virus outside the cell (Figure 5.3) (Shereen, 2020). The role of differential efficacy of ACE2 proteins

FIGURE 5.4 The genetic material of the virus enters into the live cell leaving the outer envelope of the virus outside the cell.

FIGURE 5.5 The genetic material of the virus interacting with ribosomes within the cell.

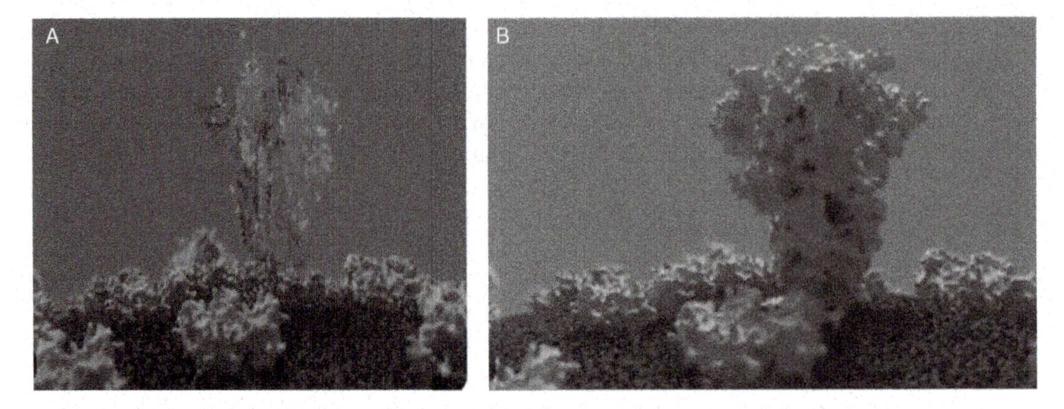

FIGURE 5.6 A, B ACE2 receptor entangles with the spike of virus on the surface of the cell.

Source: After Trucchi et al., 2020. https://doi.org/10.1101/2020.05.14.095620

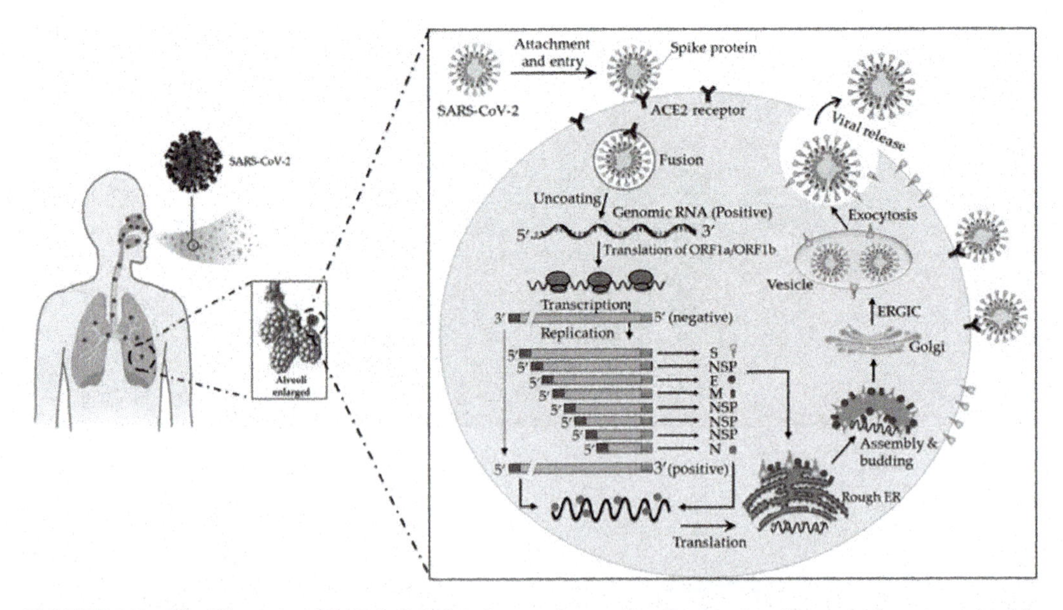

FIGURE 5.7 The life cycle of SARS-CoV-2 in human lung cells. Coronavirus is most often transmitted by droplets while sneezing and coughing and its journey begins in the first days after infiltration from the upper respiratory tract. The spike proteins of SARS-CoV-2 binds to ACE2 receptors. The virion then releases RNA genome into the cell and translation of structural and non-structural proteins follows. ORF1a and ORF1ab are translated to produce pp1a and pp1ab polyproteins, which are cleaved by the proteases that are encoded by ORF1a to yield non-structural proteins. This is followed by assembly and budding into the lumen of the ERGIC. Virions are then released from the infected cell through exocytosis (Shereen et al., 2000). NSP, non-structural proteins; ACE2, Angiotensin-Converting Enzyme 2; Rough ER, Rough Endoplasmic Reticulum; ERGIC, Endoplasmic Reticulum Golgi Intermediate Compartment.

Source: After Shereen et al., 2000.

from variable bat species was reported to affect the susceptibility of the mechanism of invasion by SARS-CoV by Hou et al. (2010).

The entry of the enveloped virus can occur directly at the cell surface, after binding to the receptor or after internalization via endocytosis. The fusion of viral membranes with host membranes is driven by large conformation changes of the spike protein. Coronaviruses contain a non-segmented, positive-sense RNA genome of 30 kb. Viral RNA synthesis follows the translation and assembly of the viral replicase complexes. Viral RNA synthesis produces both genomic and sub-genomic RNAs. Following replication and sub-genomic RNA synthesis, the viral structural proteins translate and insert into the endoplasmic reticulum (ER). These proteins move along the secretory pathway into the endoplasmic reticulum-Golgi intermediate compartment (ERGIC). Viral genomes enveloped by protein bud into membranes, forming mature virions.

5.5 THE ISSUE OF PHYLOGENY AND STRAINS IN INDIA

A recently published report included an analysis of 103 SARS-CoV-2 genomes, and it was noted that two distinct groups, S (Serine) and L (Leucine) types, tend to cluster (Tang et al., 2020). Thus, it was determined that prevalence is a reliable parameter to consider a type aggressive than the other one. The part of the genome that codes for the spike protein interacts with its receptor on human cells thus paving the way to entry of the virus into live cell as noted by Korber (2020) in their analysis of 4,535 GISAID genomes of SARS-CoV-2 genomes. The change of amino acid from Serine to Leucine was encountered in one of these (at position 84 in the ORF8 region). It was peculiar that a change from aspartic acid to glycine (D614G) was noted at position 614 of the protein; this was comparable inducing the formation of a variant spike protein than the original strain. Thus, its frequency was augmented all over the world. It has been hypothesized that because of the mutation on protein, the expeditious ability of the protein to bind to the ACE2 receptor hastens entry of the virus into the host cell.

Bio-segregated variants that might be generated due to the mutation triggering alterations in the characteristics of the SARS-CoV-2 are labelled as strains. The classification of SARS-CoV-2 genomes has been segregated into ten characteristic groups on the basis of sequencing data submitted to GENBANK. The phylogenetic tree thus constructed brings together the highly related sequences depicting as having emerged from a single common ancestor. Conclusively, several Chinese isolates were attributed to the Clade B, while the remaining ones from other parts of the world constituted Clade A. It was also found by Korber et al. (2020) that Clade A2a was similar to that ascribed to one of the mutations (i.e., D614G) on the spike protein analyzed by them. The latter has also been the most commonly sequenced clade. The variability of strains of CoV were taken into account to explore the possibility of medicinal candidates for treatment of this disease (Jin et al., 2020).

miR-27b, D614G, and A2a are believed to be three of the main strains prevalent in India as well as Asia on which investigations are currently in progress. The recent aspects of influence of environment on Covid-19 in India were illustrated by Malhotra (2020).

5.6 DEVELOPMENT OF COVAXIN IN INDIA

The process of the development of the vaccine Covaxin in India began with handing over of the genome extracted by the scientists of Indian Institute of Virology, Pune from the Indian strain of SARS-COV-2, to Bharat Biotech India (BBIL), at the high-containment facility in Hyderabad, under instructions of the Central Drugs Standard Control Organisation (CDSCO). This strain was subjected to genomic isolation from an inactivated virus than the attenuated virus, by NIV scientists from an asymptotic patient afflicted by Covid-19, and transferred to BBIL in May 2020. An antibody response was triggered against the inactivated virus on the injection of vaccine into a human subject during human trials. The animals were then subjected to pre-clinical trials involving small mammals such as experimental mice and guinea pigs, on which the efficacy of the vaccine candidate was assessed to determine safety levels. Later, the CDSCO was approached for permission to proceed with human trials. The head of CDSCO, the current Drug Controller General of India, granted permission to go ahead with phase I and II clinical trials, which is a significant step in the country for the attainment of an indigenously produced Covid-19 vaccine. The required phase I and II trials were to begin in July, and the ICMR had announced that if all the things and experiments were in order, the launch of first indigenously (domestically) produced vaccine of Covid-19, COVAXIN could see the light of the day on August 15, 2020, after its release by Hon'ble Prime Minister of India.

5.7 CONCLUSIONS

i. While controversially, the origin and prevalence of Covid-19 is attributed to the Corona virus that is thought to have emanated from the Chinese wild animal market in Wuhan. The virus supposedly spread through pangolin as intermediate host.
ii. The ferocity of attack of the mutated forms of this virus enhanced multiple times, after every strike of its spike, that succeeded by neutralizing ACE2 receptor on the cell surface.
iii. SARS-CoV-2 is an RNA-dependent RNA polymerase, and is always in an inactive state until coming into contact with live cell. It enters into the cell after binding to the receptor via endocytosis, where its genome wriggles through the internal environment of the cell.
iv. The virions are formed as soon as the genomes enveloped by protein bud into membranes.
v. The phylogeny of the virus was studied based on analysis of clustering of SARS-CoV-2 genomes. Every time the virus mutates, a new strain is expected.
vi. Covaxin, the indigenous Indian vaccine, was prepared via genome extraction of the Indian strain of SARS-COV-2, done by the Indian Institute of Virology, Pune by the scientists of Bharat Biotech India. Later, first, second, and third stage clinical trials were completed and the DCGI gave emergency approval to the vaccine.

REFERENCES

Adnan Shereen M (2020). COVID-19 infection: origin, transmission, and characteristics of human coronaviruses. *J Adv Res* 24: 91–98.

Andersen KG, Rambaut A, Lipkin WI, Holmes EC and Garry RF (2020). The proximal origin of SARS-CoV-2. *Nat Med* 26(4): 450–452. doi: 10.1038/s41591-020-0820-9

Callaway E, Ledford H and Mallapaty S (2020). Six months of coronavirus: the mysteries scientists are still racing to solve. *Nature* 583: 178–179. doi: 10.1038/d41586-020-01989-z

Chen N (2020). Epidemiological and clinical characteristics of 99 cases of 2019 novel coronavirus pneumonia in Wuhan, China: a descriptive study. *Lancet* 395 (10223): 507–513.

Chen Y, Liu Q and Guo D (2020). Coronaviruses: genome structure, replication, and pathogenesis. *J Med Virol* 92(4): 418–423. doi: 10.1002/jmv.25681

El-Aziz TMA and Stockand, JD (2020). Recent progress and challenges in drug development against COVID-19 coronavirus (SARS-CoV-2) – an update on the status. *Infect Genet Evol* 83: 104327. doi: 10.1016/j.meegid.2020.104327

Hou Y, Peng C, Yu M, Li Y, Han Z, Li F, Wang L-F and Shi Z (2010). Angiotensin-converting enzyme 2 (ACE2) proteins of different bat species confer variable susceptibility to SARS-CoV entry. *Arch Virology* 155: 1563–1569.

Hu D, Zhu C, Ai L, He T, Wang Y, Ye F, Yang L, Ding C, Zhu X, Lv R, Zhu J, Hassan B, Feng Y, Tan W and Wang C (2018). Genomic characterization and infectivity of a novel SARS-like coronavirus in Chinese bats. *Emerging Microbes & Inf* 7:154, doi: 10.1038/s41426-018-0155-5

Jin Y H, Cai L, Cheng, Z S, Cheng H, Deng T and Fan Y. (2020). (SARS CoV2) *Science.* doi: 10.1126/science.abb3221. pii: eabb3221

Korber B, Fischer W, Gnanakaran S G, Yoon H, Theiler J, Abfalterer W et al. (2020). Spike mutation pipeline reveals the emergence of a more transmissible form of SARS-CoV-2. *bioRxiv.* https://doi.org/10.1101/2020.04.29.069054

Laamarti M et al. (2020). Large scale genomic analysis 1 of 3067 SARS2 CoV-2 genomes reveals a clonal geo-distribution 3 and a rich genetic variations of hot spots mutations. *bioRxiv*. doi: www.biorxiv.org/content/10.1101/2020. 05.03. 074567v1

Lu R, Zhao X, Li J, Niu P, Yang B, Wu H et al. (2020). Genomic characterization and epidemiology of 2019 novel coronavirus: implications for virus origins and receptor binding. *Lancet*. https://doi.org/10.1016/S0140-6736(20)30251-8

Li Q (2020). Early transmission dynamics in Wuhan, China, of novel coronavirus–infected pneumonia. *N Engl J Med* 382(13): 1199–1207.

Li Q, Guan X, Wu P, Wang X, Zhou L, Tong Y, Ren R, Leung KSM, Lau EHY, Wong JY, Xing X, Xiang N, Wu Y, Li C, Chen Q, Li D, Liu T, Zhao J, Liu M, Tu W, Chen C, Jin L, Yang R, Wang Q, Zhou S, Wang R, Liu H, Luo Y, Liu Y, Shao G, Li H, Tao Z, Yang Y, Deng Z, Liu B, Ma Z, Zhang Y, Shi G, Lam TTY, Wu JT, Gao GF, Cowling BJ, Yang B, Leung GM, Feng Z (2020). Early transmission dynamics in Wuhan, China, of novel coronavirus-infected pneumonia. *N Engl J Med* 382(13): 1199–1207. doi: 10.1056/NEJMoa2001316

Liu P, Jiang JZ, Wan X, Hua Y, Li L, Zhou J, Wang X, Hou F, Chen J, Zou J, Chen J (2020). Are pangolins the intermediate host of the 2019 noval coronavirus (SARS-CoV-2)? *PLoSPathog* 16(5). doi: 10.1371/journal.ppat.1008421

Lu H, Stratton CW and Tang Y-W (2020). Outbreak of pneumonia of unknown etiology in Wuhan, China: the mystery and the miracle. *J Med Virol* 92(4): 401–402.

Malhotra Sandeep K (2020). 2019-nCoV: environmental aspects of the disease. *Indo-Egyptian Webposium on Covid Reality: Cost of Human Lives and Economy of the World*, 18th May 2020, CSJM Kanpur University, Kanpur. Keynote Address.

NCBI www.ncbi.nlm.nih.gov/

Nextstrain.org: https://nextstrain.org/ncov

Paital B, Das K and Parida SK (2020). Inter nation social lockdown versus medical care against COVID-19, a mild environmental insight with special reference to India. *Sci Total Environ* 728: 138914.

Paraskevis D, Kostaki EG, Magiorkinis G, Panayiotakopoulos G, Sourvinos G and Tsiodras S (2020). Full-genome evolutionary analysis of the novel corona virus (2019-nCoV) rejects the hypothesis of emergence as a result of a recent recombination event. *Infect Genet Evol* 79: 104212. doi: 10.1016/j.meegid.2020.104212

Schoeman D and Fielding B C (2019). Coronavirus envelope protein: current knowledge. *Virol J* 16(69): 1–22. doi: 10.1186/s12985-019-1182-0

Su S (2016). Epidemiology, genetic recombination, and pathogenesis of coronaviruses. *Trends Microbiol*, 24(6): 490–502.

Tang X, Wu C, Li X, Song Y, Yao X, Wu X et al. (2020). On the origin and continuing evolution of SARSCoV-2. *Natl Sci Rev.* 7(6): 1012–1023 https://doi.org/10.1093/nsr/ nwaa036

Trucchi E, Gratton P, Mafessoni F Motta S, Cicconardi F, Bertorelle GD'Annessa I and Marino D (2020). Unveilling diffusion pattern and structural impact of the most invasive SARS-CoV-2 spike mutation. https://doi.org/10.1101/2020.05.14.095620

Wan Y, Shang J, Graham R, Baric RS and Li F (2020). Receptor recognition by the novel coronavirus from Wuhan: analysis based on decade-long structural studies of SARS coronavirus. *J Virol:* 94(7). doi: 10.1128/JVI.00127-20. e00127-20

Yan R, Zhang Y, Li Y, Xia L, Guo Y and Zhou Q (2020). Structural basis for the recognition of SARS-CoV-2 by full-length human ACE2. *Science.* 367(6485): 1444–1448. doi: 10.1126/science.abb2762

Zhou P, Yang XL, Wang XG, Hu B, Zhang L, Zhang W, Si HR, Zhu Y, Li B, Huang CL, Chen HD, Chen J, Luo Y, Guo H, Jiang RD, Liu MQ, Chen Y, Shen XR, Wang X, Zheng XS, Zhao K, Chen QJ, Deng F, Liu L L, Yan B, Zhan FX, Wang YY, Xiao GF and Shi ZL (2020). A pneumonia outbreak associated with a new coronavirus of probable bat origin. *Nature* 579(7798): 270–273. doi: 10.1038/s41586-020-2012-7

Zhu N, Zhang D, Wang W, Li X, Yang B, Song J et al. (2019). "A novel coronavirus from patients with pneumoniain China, 2019." *N Eng J Med.* https://doi.org/10.1056/NEJMoa2001 017

Zumla A (2016). Coronaviruses – drug discovery and therapeutic options. *Nat Rev Drug Discov* 5(5): 327–347.

6 The Molecular Biology of SARS-CoV-2 and Its Evolving Variants

Zheng Yao Low,[1] *Ashley Jia Wen Yip,*[1] *and Sunil K. Lal*[1,2]
[1]School of science, Monash University, Selangor, Malaysia
[2]Tropical Medicine & Biology Platform, Monash University, Selangor, Malaysia
Correspondence to: sunil.lal@monash.edu

CONTENTS

6.1 INTRODUCTION

Coronavirus, a zoonotic virus that falls under the *Coronaviridae* family, is commonly found in avian species and mammals, such as masked palm civets, bats, and camels (Gong & Bao, 2018). There are four genera under the *Coronaviridae* family, namely, the alpha, beta, gamma, and delta-coronaviruses, in which the beta-coronavirus contribute to severe respiratory disease and numerous fatalities (Velavan & Meyer, 2020). Being a well-known source of respiratory-associated illness, the coronavirus is no stranger to the primary root of the common cold, accounting for up to 20% of all human common cold cases (Thiel et al., 2001). Before the emergence of the severe acute respiratory syndrome (SARS-CoV) in 2002 and Middle East Respiratory Syndrome (MERS-CoV) in 2012, the coronavirus was deemed to affect only immunocompromised individuals, causing mild illness (Zhong et al., 2003; Nature Research Custom Media, 2021). To date, seven strains of coronaviruses have been reported to cause disease in humans. Amongst these strains, the HCoV-OC43, HCoV-229E, HCoV-HKU1 and HCoV-NL63 give rise to the common cold and mild respiratory infections. In contrast, the SARS-CoV, MERS-CoV, and the recent SARS-CoV-2 (causative virus for the COVID-19 pandemic) have been more devastating and have contributed to significant worldwide mortality in pandemics (Ye et al., 2020). Accounting for more than 672,000,000 infected individuals and 6,740,000 associated deaths worldwide, the rapid spread of COVID-19 is a considerable burden and threat to public health and safety (Worldometer, 2023). The etiological agent responsible for this disastrous episode is known as SARS-CoV-2 (WHO, 2021). Akin to the previous SARS-CoV and MERS-CoV, the SARS-CoV-2 falls under the beta-coronavirus genera. Notably, the estimated fatality rate of SARS-CoV-2 is much lower at 3.4% compared to 9.6% and 40% for SARS-CoV and MERS-CoV, respectively (Peiris et al., 2003; Zumla, Hui & Perlman, 2015). Despite this, the

DOI: 10.1201/9781003358909-6

high infection rate of SARS-CoV-2 overwhelmingly surpasses both SARS-CoV and MERS-CoV, leading to the rapid progression of infection in individuals and the spatial range of epidemic regions (Zheng, 2020). What is even more alarming is that with rapid replication of the SARS-CoV-2 virus in humans, we are witnessing a vast number of new variants of this virus during this pandemic.

As a zoonotic virus, coronaviruses are often associated with animal reservoirs. Several animals, such as bats, can travel long distances and congregate within the community, resulting in the high occurrence of animal-human interspecies barrier crossing, increasing the occurrence of zoonotic transmission of coronaviruses and species spillover events (Wong et al., 2019). The previous SARS-CoV and MERS-CoV are examples of such circumstances. As such, the SARS-CoV and MERS-CoV are believed to be associated with bat and dromedary camels, respectively. In contrast, the origin of SARS-CoV-2 remains debatable, with evidence pointing towards bats as the natural reservoir (Andersen et al., 2020). Similar to its counterparts, the common symptoms of SARS-CoV-2 include fever, dry cough, headache, fatigue, pneumonia, and dyspnoea (Zhou et al., 2020). Less common symptoms such as loss of taste or smell, nasal congestion, diarrhoea, gastrointestinal distress, muscle weakness and rashes are also reported (WHO, 2021).

In this urgent time of need and the emergence of new variants, antiviral drug repositioning is a feasible approach in addition to vaccine development and the discovery of new antiviral compounds. In the quest for effective antiviral therapy against SARS-CoV-2, numerous drug candidates have been studied for drug repositioning potentials, such as remdesivir, molnupiravir, ivermectin and oseltamivir, to name a few (Yip et al., 2022). Moreover, many studies have revealed plausible targets for drug inhibitors, such as the RdRp protein complex, in hopes of inhibiting the viral replication of SARS-CoV-2 (Low, Yip & Lal, 2022). Given the rapid development of COVID-19 and its emerging variants, it is essential to revisit and understand the past and present coronavirus outbreaks in humans. We all understand that viral evolution towards new variants will undoubtedly decrease the efficacies of existing vaccines, but this will be a constant and ongoing task where antiviral drugs targeting internal proteins of the virus, rather than surface (spike) proteins may become a better alternative, in the long run. This chapter aims to provide consolidated information concerning SARS-CoV-2 and its emerging variants. We describe the molecular biology of transmission and replication, thereby highlighting the distinctions between the newly emerging variants.

6.2 GENERAL MORPHOLOGY OF SARS-COV-2

The most conspicuous structure found in coronavirus is the club-shaped spike projections on the virion surface, giving it a crown-like appearance under an electron microscope, thus the name, 'coronavirus'. A coronavirus is a group of enveloped, non-segmented positive-sense RNA viruses with sizes ranging from 65 to 125 nm in diameter (Shereen et al., 2020). Inside the envelope, the single-stranded RNA genome varies between 26 and 32 kilobases (kb) in length. Four main structural proteins can also be found within the envelope, namely, the spike (S), membrane (M), envelope (E) and nucleocapsid (N) proteins, all of which contribute to the host entry and assembly of coronaviruses (shown in Figure 6.1). Hemagglutinin-esterase, a fifth structural protein exclusive to a subset of beta-coronaviruses, is also identified to aid in viral entry and replication. Interestingly, coronaviruses possess helically symmetric nucleocapsids, uncommon amongst positive-sense RNA viruses (Fehr & Perlman, 2015). Falling under the same beta-coronavirus genera, the general morphology of SARS-CoV-2 closely resembles its predecessors, the SARS-CoV and MERS-CoV. The SARS-CoV-2 genome is approximately 29.9 kb in length with a 5' cap and poly-(A)-3' tail, allowing it to mimic an mRNA for translation. Taking up two-thirds of the RNA genome at the 5' end is the open reading frame (ORF) 1a and 1b replicase genes that further cleaves to yield the non-structural proteins (nsp1-16) that encode for many vital enzymes for viral replication, such as the RNA-dependent RNA polymerase (RdRp). The remaining one-third of the genome at the 3' end code for those above capsid-forming structural proteins (S, E, M, N). Generally, the typical layout

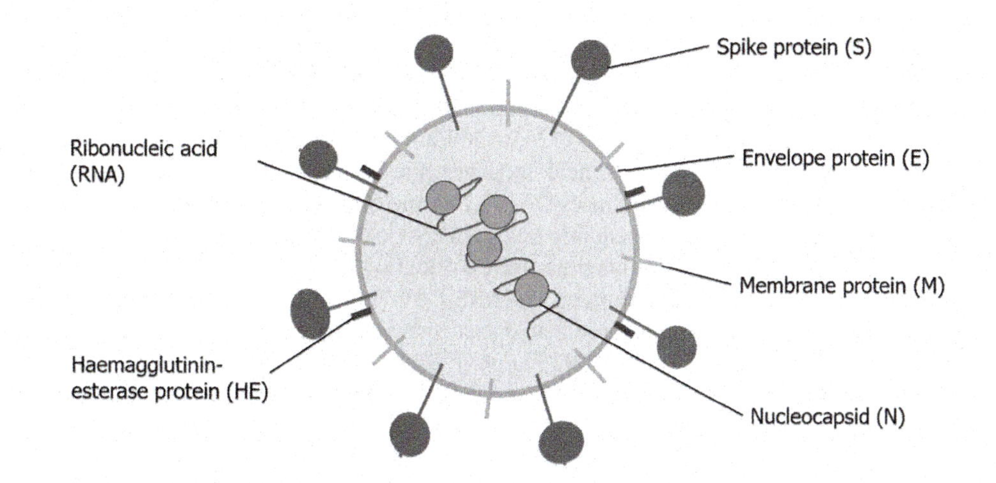

FIGURE 6.1 The morphology of SARS-CoV-2. Viral structure with its protein components and viral RNA.

of the SARS-CoV-2 genome denotes as follows, [5'-leader-UTR-replicase-S-E-M-N-3'-UTR-poly (A) tail], with seven putative ORFs encoding accessory genes scattered throughout the structural genes at the 3' end (Naqvi et al., 2020).

6.3 FEATURES AND DISTINCTION BETWEEN CORONAVIRUSES

As aforementioned, the beta-coronaviruses can be further classified into four lineages: A, B, C, and D. SARS-CoV and SARS-CoV-2 belong to lineage B. In contrast, MERS-CoV belongs to lineage C (Letko, Marzi & Munster, 2020). Amongst all, the genomic size of MERS-CoV is the largest (30.11 kb), followed by SARS-CoV-2 (29.9 kb) and SARS-CoV (29.75 kb) (N. Zhu et al., 2020). Phylogenetic analysis for the whole genome suggested that SARS-CoV-2 is closely associated with SARS-related coronaviruses and SARS-CoV, placing it in the subgenus *Sarbecovirus* of the beta-coronavirus genus (Hu et al., 2020). Indeed, SARS-CoV-2 shares a genome sequence similarity of 96.2% with the short region of RdRp of Bat CoVER aTG13, whereas an 86.9% nucleotide similarity was reported when compared with bat-derived SARS-like CoV (Z. Zhu et al., 2020; Sharma et al., 2022). When comparing the seven conserved replicase domains located in ORF 1a and b, an amino acid sequence similarity of 94.4% and 87.1% were revealed between SARS-CoV-2 with SARS-CoV and SARSr-CoV, respectively. Since SARS-CoV had descended from SARS-like bat CoVs, SARS-CoV-2 was postulated to be genetically closer to SARS-CoV rather than MERS-CoV. As such, SARS-CoV-2 shares a genetic similarity of 79% and 50% to SARS-CoV and MERS-CoV, respectively (Low et al., 2021).

The prominent difference between the genetic expression of SARS-CoV-2 and SARS-CoV lies in the absence of 8a protein and the difference in the number of amino acids in 8b and 3c proteins. Moreover, glutamine residue can be found in the amino acid position 50 (position 321 in NSP2) of SARS-CoV-2 instead of the threonine residue found in SARS-CoV, resulting in a side chain with higher polarity and a tendency to form hydrogen bonds, whilst increasing the protein stability. Another substitution at amino acid position 723 (position 543 in NSP3), where glycine was changed to a serine residue in SARS-CoV-2, had also resulted in a higher tendency of hydrogen bond formation along with an increment in local stiffness of the polypeptide chain, affecting the enzyme active sites. In addition, a change of isoleucine to proline in SARS-CoV-2 at amino acid position 1010 (position 192 in NSP3) had led to a steric bulge, increasing the stiffness to the overall molecular structure of SARS-CoV-2 (Angeletti et al., 2020).

The S protein in SARS-CoV-2 highly contributes to viral attachment and entry, in which the receptor-binding domain (RBD) located in the S1 subunit is responsible for the binding of the virus to host receptors (Ou et al., 2020). The S protein of SARS-CoV-2 is longer than SARS-CoV (1255 amino acids) and bat SARSr-CoVs (1245–1269 amino acids). Moreover, the S protein of SARS-CoV-2 shares the highest amino acid sequence similarities with pangolin coronaviruses (90.7–92.6%), followed by bat coronaviruses (75–97.7%) and SARS-CoVs from civets and humans (76.7–77%) (Hu et al., 2020). At the genomic level, SARS-CoV-2 shares a 93.1% nucleotide similarity in the RBD region with bat coronavirus (BatCoV RaTG13). In contrast, an 87.2% to 83.9% nucleotide similarity was reported compared to SARS-CoV (Lu et al., 2020). However, the RBD region of SARS-CoV-2 only shares an amino acid sequence similarity of 73% compared to SARS-CoV. Other than bat coronaviruses, the RBD region of SARS-CoV-2 also shares a high sequence similarity (97.4%) to pangolin-coronavirus. More notably, SARS-CoV-2 and pangolin-coronavirus share an identical set of the six key RBD residues, while bat RaTG13 only shares one amino acid amongst the six (Kumar et al., 2020). Based on the aforementioned, the animal reservoir for SARS-CoV-2 remains controversial and requires more investigation and research.

SARS-CoV and SARS-CoV-2 share the same host cell surface receptor, known as the angiotensin-converting enzyme 2 (ACE2) receptor, while MERS-CoV utilises dipeptidyl peptidase 4 (DPP4). Although SARS-CoV-2 and SARS-CoV utilise the same receptor, the ACE2 receptor possesses a higher affinity to SARS-CoV-2 than SARS-CoV due to the single mutation at N501T in the S protein (Wan et al., 2020). Besides that, the large interaction surface (18 interactions) between the ACE2 receptor and SARS-CoV-2 as compared to SARS-CoV (8 interactions) had contributed to a stronger binding affinity, up to 15-fold higher. The same study also demonstrated the ability of the S protein (amino acid position 471–486, 496–505, 404–416 and 446–456) of SARS-CoV-2 to form a robust multi-epitope synapse adhesion with the ACE2 receptor in which a stronger viral surface-host's epithelial adhesion was formed. This highlights the need for multi-epitope high-affinity antibodies targeting different adhesion synapse sites as antiviral treatment (Khatri, Staal & van Dongen, 2020). Distinct from SARS-CoV, three short insertions in the N-terminal domain were found in SARS-CoV-2. Such a region is similar to MERS-CoV, allowing sialic-acid binding activity (Milanetti et al., 2021).

Another distinctive genomic feature of SARS-CoV-2 is the insertion of 12 amino acid residues in the junction of the S1 and S2 region of the S protein, creating an additional polybasic cleavage site that was predicted to contribute to immune evasion and enabling cleavage by furin and related proteases (Coutard et al., 2020; Walls et al., 2020). The function of the polybasic cleavage site was later determined to contribute to viral entry and spread. The cleavage of the S protein is essential for viral entry, in which different cleavage sites require respective proteases. As such, the host cell tropism and virulence of RNA viruses highly depend on the abundance of proteases. As furin is ubiquitous in human tissues, especially lung tissues, the presence of furin cleavage sites will broaden the host cell tropism, subsequently facilitating the pathogenicity of SARS-CoV-2 (Xia, 2021). Generally, coronaviruses' S protein utilises four proteolytic activation modes: proprotein convertases, extracellular proteases, cell surface proteases and lysosomal proteases. It was reported that SARS-CoV-2 could utilise four modes of proteolytic activation, distinct from SARS-CoV and MERS-CoV. SARS-CoV was only reported to utilise extracellular protease (elastase), cell surface proteases (TMPRSS2) and lysosomal proteases (cathepsin L/B). In contrast, MERS-CoV utilises proprotein convertases (furin), cell surface proteases (TMPRSS2) and lysosomal proteases (cathepsin L) (Low et al., 2021).

6.4 VARIANTS OF SARS-COV-2

It is not surprising that coronavirus' evolve, including SARS-CoV-2, the cause of the COVID-19 pandemic. The changes in the virus' properties and their rapidly evolving antigenic epitopes pose a serious concern to public health and safety. The evolution of new variants may give rise to enhanced viral replication, increased transmission, greater disease severity, downgrade the performance of vaccines and antiviral therapeutics, and reduce the reliability of testing tool kits. While not all SARS-CoV-2 variations share similar significance and impact on human health, it is essential to investigate the different variants and their potential implications on human health. To accelerate the extermination of SARS-CoV-2, the World Health Organization (WHO) has established a standard nomenclature for naming SARS-CoV-2 variants to expedite the research progress. The currently designed nomenclature entails the use of the Greek alphabet to classify variants of concern (VOC), variants of interest (VOI) and variants under monitoring (WHO, 2022). Of these classifications, the VOC express an alarming level of transmissibility, and virulence, followed by a decrease in diagnostics, vaccines and the effectiveness of therapeutics in the international community. On the other hand, the VOI identify variants that cause significant community transmission in multiple countries or regions over time that could suggest an emerging risk to global public health (WHO, 2022). Lastly, the VUM suggests variants with genetic changes that may affect virus characteristics and indications of future risk but lacks clear phenotypic evidence accompanied by the uncertain epidemiological impact, which requires enhanced monitoring and repeat assessment (WHO, 2022). A similar approach was also conducted by the US government SARS-CoV-2 Interagency Group (SIG) for the variant classification, which entails the VOI, VOC and Variants Being Monitored (VBM) that includes data pointing towards potential increased transmission with authorised medical countermeasures or possible increase transmissions but are however no longer detected or at very low levels (CDC, 2022). With the alarming rate of SARS-CoV-2 infection, we aim to discuss some of the significant protein mutations in the more prevalent variants (summarised in Table 6.1).

The current circulating VOC remains the omicron variant entailing the sub-lineages BA.1 to BA.5 (Mahase, 2022; Tegally et al., 2022). The omicron variant, B.1.1.529, was first discovered in Bostwana and South Africa in November 2021 (Tian et al., 2022). Notably, the omicron variant possesses twice the number of spike protein mutations than that of the delta variant. It has the highest number of mutations, harbouring around 50, with more than 30 mutations originating from the aforementioned spike protein (Tian et al., 2022). Among the sublineages above, BA.1 to BA.2 are the most widely circulated omicron variant and shall be further discussed in this context (Kumar, Karuppanan & Subramaniam, 2022; Kurhade et al., 2022). The BA.1 to BA.3 omicron sublineages share 21 common mutations. In which, the G339D, N440K, S477N, and S371L mutations are the few well-known ones, atop the ones found in the previous circulating VOCs (Cao et al., 2021; Tian et al., 2022). The prevalence of the G339D mutation has been reported highest among these mutations (Li et al., 2022). It was postulated that the G339D mutation contributed to destabilising T-cell responses and enhancing binding affinity between T-cell epitopes and the major histocompatibility complex, subsequently increasing the infection capacity of the omicron variant (Li et al., 2022). The S477N mutation, on the other hand, has been reported to have a high binding free energy profile on the receptor binding domain (RBD) of the SARS-CoV-2 spike protein, indicating more robust transmission capacity via enhanced viral spike-host ACE2 receptor binding (Alkhatib et al., 2021; Chen et al., 2020). As for the S371L mutation, it was reported to enhance the stability of spike RBD-up conformation to resist the conformational change in RBD upon ACE2 binding, thereby enhancing the stability of spike-ACE2 binding (Zhao et al., 2022). The N440K mutation has been reported to confer additional resistance towards class 3 RBD monoclonal antibodies and is substantial resistance to neutralisation by currently available antibodies, including those generated from vaccines (J. Liu et al. 2021; L. Liu et al., 2021). Apart from the BA subvariants mentioned above, many more subvariants are emerging worldwide at a rapid pace due to their capacity for recombination, such as the XBB and BQ.1 variant that has been reported to confer additional mutations at R346T, K444N/T and F486S/P/V/I that confers greater immune evasions and enhances ACE2 binding (Cao et al., 2022). Notably, the most recently emerged XBB 1.5 variant carrying an additional S486P

TABLE 6.1
The Confirmed Key Mutations and Their Significance in SARS-CoV-2 Variants

Key Mutations	Mutation Significance	VOC/VOI
N501Y	• Increases the binding affinity for human ACE2 receptors.	Alpha*, beta*, gamma*, mu*
P681H	• Affects the furin cleavage site between S1 and S2 subunits in S protein, promoting viral entry into cells.	Alpha*, gamma*
E484K	• Eliciting considerable loss of neutralising activity of convalescent sera and monoclonal antibodies. • Interacts with the K31 residues in the human ACE2 receptor, enhancing the binding affinity for SARS-CoV-2.	Alpha*, beta*, gamma*, mu*
K417N	• Interacts with the D30 ACE2 protein residue and enhances the binding affinity to human ACE2 receptors. • Alters the key interactions associated with class 1 neutralising antibodies, contributing to immune evasion.	Beta*, gamma*, delta*
L18F	• Associated with the escape in numerous N-terminal domain (NTD) binding monoclonal antibodies and reduces neutralisation by several antibodies.	Beta*, gamma*
T478K	• Increasing the electrostatic potential and interfering with the spike RBD interaction with convalescent sera or vaccine-elicited antibodies, contributing to immune evasion.	Delta*
P681R	• Enhances furin-mediated spike cleavage, accelerating SARS-CoV-2 to cell fusion.	Delta*
L452R	• Enhances the binding affinity for human ACE2 receptors and promotes spike stability. • Reduces cell-mediated cellular immunity, causing rapid viral replication.	Delta*
L542Q	• Immune escapes and enhances the binding affinity for human ACE2 receptors.	Lambda*
F490S	• Reduces susceptibility to antibody neutralisation.	Lambda*
D614G	• Increase the binding affinity for ACE2 receptors. • Reduces S1 shedding and increases the total S protein incorporated into the virion, enhancing virus infection.	Alpha*, beta*, gamma*, delta*, omicron, mu* lambda*
S371L	• Enhances the stability of spike RBD-up conformation to resist the conformational change in RBD upon ACE2 binding. • Enhances the stability of spike-ACE2 binding.	Omicron
146N ins	• Affects the spike S1 conformation, leading to an enhanced affinity for the ACE2 receptor.	Mu*
G339D	• Destabilising T-cell responses. • Enhancing binding affinity between T-cell epitopes and the major histocompatibility complex, subsequently increasing the infection capacity.	Omicron
S477N	• More robust transmission capacity via enhanced viral spike-host ACE2 receptor binding	Omicron
N440K	• Confer additional resistance towards class 3 RBD monoclonal antibodies and substantial resistance to neutralisation by currently available antibodies.	Omicron

Note: *denoted as previous circulating VOI/VOCs.

substitution has outcompeted the other XBB and BQ variants by exhibiting both enhanced ACE2 binding and substantial immune evasion capabilities (Yue et al., 2023).

On the other hand, the alpha, beta, gamma, and delta variants are now identified as previously circulating VOCs (WHO, 2022). The alpha (B.1.1.7) variant was first identified in Denmark, the United Kingdom and Northern Ireland back in December 2020 (Galloway et al., 2021). This variant differs from the original variant discovered during the initial COVID-19 outbreak; the alpha variant S protein carries a deletion of 2 amino acids at positions 69 and 70 (del 69–70), allosterically changes the S1 subunit conformation in the S protein, confers an evasion point associated with S-gene target failure (SGTF) in the RT-PCR assay and leads to increasing false-negative results for asymptomatic spread (Brown et al., 2021; Xie et al., 2021). In addition, the alpha variant carries an amino acid change at position 501 (N501Y), trading asparagine for tyrosine in the RBD of the spike protein, concomitantly increases the binding affinity (~2x) to the human ACE2 receptor (Ramanathan et al., 2021). Besides that, the P681H mutation in the alpha variant trades proline for histidine at position 681 and was postulated to provide an additional basic residue and affect the furin cleavage site between S1 and S2 subunits in the spike protein, promoting viral entry into the respiratory epithelial cells (Lubinski et al., 2021; Munitz et al., 2021). More recently, the E484K substitution has also emerged in the B.1.1.7 variant, disrupting the N-terminal domain of spike protein, eliciting considerable loss of the neutralising activity of the convalescent sera and monoclonal antibodies (Collier et al., 2021; Wang et al., 2021). This was also reported in mRNA-1273 (Moderna) vaccinated individuals, where there was a 3- to 6-fold reduction in neutralisation by sera against the alpha (B.1.1.7) variant carrying the E484K mutation (Wu et al., 2021). Apart from that, the E484K mutation in the S protein has been shown to interact with the K31 residues in the human ACE2 receptor, concomitantly enhancing the binding affinity for SARS-CoV-2 (Lim et al., 2020).

The following variant, the beta (B.1351/B.1351.2/B.1351.3), was first discovered in Nelson Mandela Bay, South Africa, in December 2020 (Zhou et al., 2021). The beta variant carries four significant protein mutations (N501Y, E484K, K417N and, more recently, the L18F). Apart from the N501Y and E484K mutations found in the alpha variant mentioned above, the K417N mutation that substitutes lysine for asparagine at position 417 in the RBD of the S protein in the beta variant interacts with the D30 ACE2 protein residue, contributing to the significant enhancement of the binding affinity to human ACE2 receptors (Han, Penn-Nicholson & Cho, 2006). Furthermore, reports describing the K417N mutation have also contributed to the alteration of key interactions associated with class 1 neutralising antibodies, contributing towards immune evasion (Tegally et al., 2020). The K417N mutation is also postulated to increase the S1 RBD binding to the ACE2 receptor, which is further enhanced by the presence of both the N501Y and K417N mutations in the RBD of the S protein (Harvey et al., 2021). More recently, the L18F mutation has also emerged in the beta (B.1.351) variant that substitutes leucine for phenylalanine, conferring greater spreading capacity measured via replicative advantage (Focosi & Maggi, 2021; Grabowski, Kochańczyk & Lipniacki, 2021). The L18F mutation is also associated with escape in numerous N-terminal domain (NTD) monoclonal antibodies and reduces neutralisation by several antibodies (McCallum et al., 2021).

The gamma (P.1/P.1.1/P.1.2) variant was first discovered by the National Institute of Infectious Disease (NIID) in Tokyo, Japan, in January 2021 (Fujino et al., 2021). Akin to the beta variant, the gamma variant carries the K417N, E484K, N501Y and L18F mutations, in addition to the recently discovered P681H mutation. As mentioned above, the P681H mutation was first discovered in the B.1.1.7 alpha variant, conferring an advantage of enhancing viral properties and entry.

The delta variant (B.1.617.2/AY.1/AY.2) was first discovered in India in October 2020 (Duong, 2021). It carries the spike protein's T478K, P681R, L452R and K417N mutation (in the delta-plus variant) (Ferreira et al., 2021). The T478K mutation exchange uncharged threonine for charged lysine at position 478, increasing the electrostatic potential on the spike RBD, affecting the interaction between the spike RBD and convalescent sera or vaccine-elicited antibodies interaction, allowing for immune evasion (Arora et al., 2021; Di Giacomo et al., 2021; Farinholt et al., 2021). On the other hand, the P681R mutation exchanges proline for arginine at position 681, enhancing furin-mediated spike cleavage, accelerating SARS-CoV-2 to undergo cell fusion (C. Liu et al., 2021; Saito et al., 2021). Moving on to the L452R mutation, in spike protein, it exchanges leucine for arginine at position 452, resulting in significantly enhanced binding affinity for the human ACE2 receptor and promoting spike stability, thereby increasing viral infectivity and viral replication (Motozono et al., 2021). It was also postulated that the L452R mutation reduces cell-mediated cellular immunity, thereby causing rapid viral replication (Motozono et al., 2021). The K417N mutation was found in the beta and gamma variants (Davis et al., 2021). As aforementioned, the K417N mutation contributes to immune evasion and enhanced ACE2 binding affinity, thereby conferring the delta plus variant a much higher infection capacity as compared to its counterpart (delta/B.1.617.2) (Han, Penn-Nicholson & Cho, 2006; Tegally et al., 2020; Harvey et al., 2021).

Conversely, there are currently no circulating VOIs. Previously circulating VOIs include the epsilon, zeta, eta, iota, theta, kappa, lambda and mu variants, with the latter two representing the latest identified with greater prevalence, and have garnered significant attention (WHO, 2022). The lambda variant (C.37) was first reported in Peru in December 2020 (ECDC, 2022; Kimura et al., 2022). It possesses a deletion in positions 246 to 252, and G75V, T76I, L452Q, F490S and T859N nonsynonymous mutations in the spike protein. Of which, only the 246 to 252 deletion and L542Q and F490S mutation have been studied. The 246 to 252 deletion in the N-terminal domain of the spike protein was postulated to contribute to immune escape (Acevedo et al., 2021). Moreover, the L452Q mutation in the lambda spike protein exchanging leucine for glutamine was proposed to show similar immune escape and enhanced binding affinity for human ACE2 receptors found in L452R mutation in the delta variant (Tada et al., 2021). Studies have shown that the L452Q mutation in lambda spike protein has caused a 3-fold increase in ACE2 binding accompanied by a 2-fold increase in viral infectivity (Starr et al., 2020; Tada et al., 2021). On the other hand, the F490S mutation in the lambda spike protein exchanges phenylalanine for serine. This leads to reduced susceptibility to antibody neutralisation (Wink et al., 2021). This was pronounced in a study where the F490S mutation showed a 2- to 3-fold resistance to neutralisation by mRNA vaccine-elicited antibodies and convalescent serum (Tada et al., 2021).

The mu variant (B.1.621) was first reported in Columbia in January 2021 (Halfmann et al., 2022). It carries a slew of common mutations found in other variants, such as the E484K and N501Y mutations found in the alpha, beta and gamma variants. Notably, the mu variant possesses several unique mutations, the YY144-145TS and 146N insertion mutation in the N-terminal domain, where the latter has been reported to affect the spike S1 conformation, leading to an enhanced affinity for the ACE2 receptor (Chatterjee et al., 2022; Hossain et al., 2022).

Interestingly, the D614G mutation in the S protein, which exchanges aspartic acid for glycine at position 614, is present in all the variants mentioned above, accounting for the increase of binding affinity for ACE2 receptors (Zhang et al., 2020; ECDC, 2022). The D614G mutation also reduces S1 shedding and increases the total S protein incorporated into the virion, thereby contributing to enhanced virus infection (Zhang et al., 2020; ECDC, 2022).

6.5 VIRAL TRANSMISSION AND REPLICATION OF SARS-COV-2

The transmission of SARS-CoV-2 was postulated to be via animal-to-human and human-to-human. For animal-to-human transmission, an intermediate host must be present. A study indicated that a possible viral recombination of SARS-CoV-2 happened between the bat and the intermediate host of the pangolin before transmitting the virus to humans (Xiao et al., 2020; Sharma, Farouk & Lal, 2021). Moving to human-to-human transmission, the common mode for SARS-CoV-2 transmission entails aerosols. Generally, the transmission is airborne, where respiratory droplets from an infected individual undergo expulsion, such as coughing and sneezing (Carlos et al., 2020). In addition, the SARS-CoV-2 also exhibited a significant survival duration of up to 72 hours on surfaces such as plastic, stainless steel, copper and cardboard, suggesting possible fomite transmission (van Doremalen et al., 2020). Besides that, other factors, such as nosocomial transmission, were also reported in SARS-CoV-2. Common hospital items, such as toiletries, were found to possess a significant amount of SARS-CoV-2 viral RNA on the surface. It is no stranger that hospital rooms are used to treat COVID-19 patients, thereby spreading the virus in the air. This is more pronounced in patients having a nasal oxygen cannula, where a study noted the highest airborne viral concentrations in the hospital (19.17 and 48.22 gene copies/L) (Santarpia et al., 2020).

The replication cycle of SARS-CoV-2 begins with receptor recognition, where the human angiotensin-converting enzyme-2 (ACE2) receptor was reported to be recognised by the spike (S) surface glycoprotein of SARS-CoV-2 for attachment (Khatri, Staal & van Dongen, 2020). Following the attachment of spike glycoprotein, the host cell proteolytic enzyme is then activated. Generally, there are four different proteolytic modes in coronaviruses: proprotein convertases (e.g., furin), extracellular proteases (e.g., elastase), cell surface proteases (e.g., TMPRSS2) and lysosomal proteases (e.g., cathepsin L/B) (Walls et al., 2020; Zheng et al., 2020). Notably, the SARS-CoV-2 has a unique insert at the S1/S2 site of the S protein, which is absent in the SARS-CoV that allows cleavage by the furin, and enhances the viral attachment and replication (Li, 2016; Tang et al., 2021). Upon proteolytic activation, the S protein will be cleaved into two domains, the S1 receptor binding domain and the S2 fusion domain (Belouzard, Chu & Whittaker, 2009). This process facilitates the membrane fusion of the virus with the host cell membrane via endocytosis which subsequently releases the viral RNA genomes in the host cell cytoplasm. In the viral genome, polyproteins pp1a and pp1b are produced via translation of ORF1a and 1b, which are then further cleaved to yield non-structural proteins (nsp1–16) that encode for many vital enzymes, such as the RdRp (Naqvi et al., 2020). The RdRp will synthesise negative-sense RNA copies during viral genome replication from the positive-strand template (shown in Figure 6.2). The negative strand now serves as the new template for the replications of positive-sense RNA genomes to facilitate virus replication. This is followed by a series of mRNA translations, which are assembled and packaged. Finally, the resulting new virion will undergo budding and release from the host cell via exocytosis, ready to

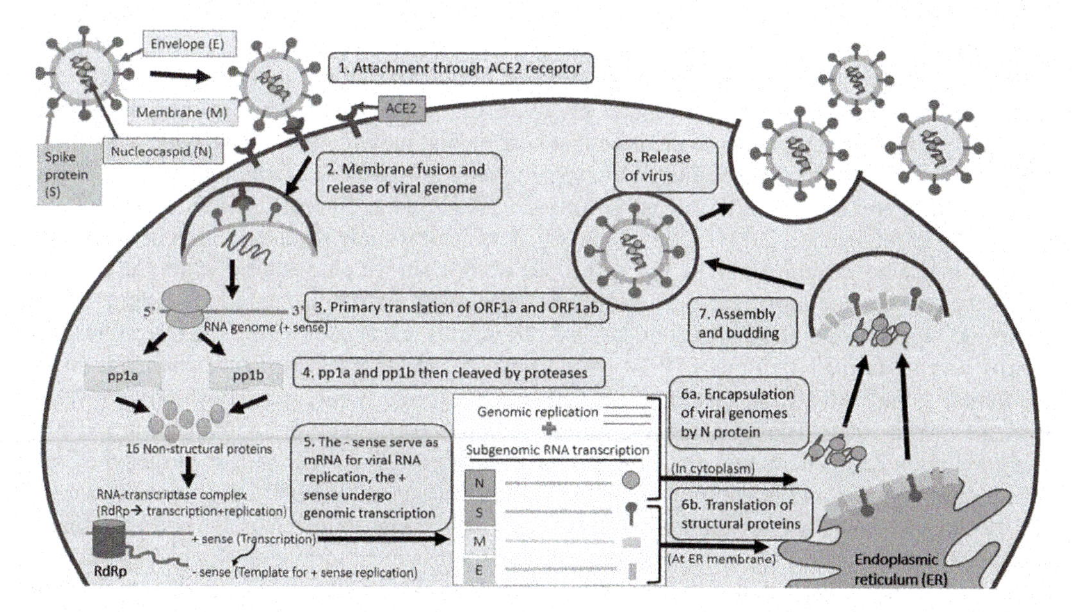

FIGURE 6.2 The general life cycle layout for the SARS-CoV-2, beginning from the *[1-]* attachment through the ACE2 receptor to the *[-7]* assembly and budding of new virus particles.

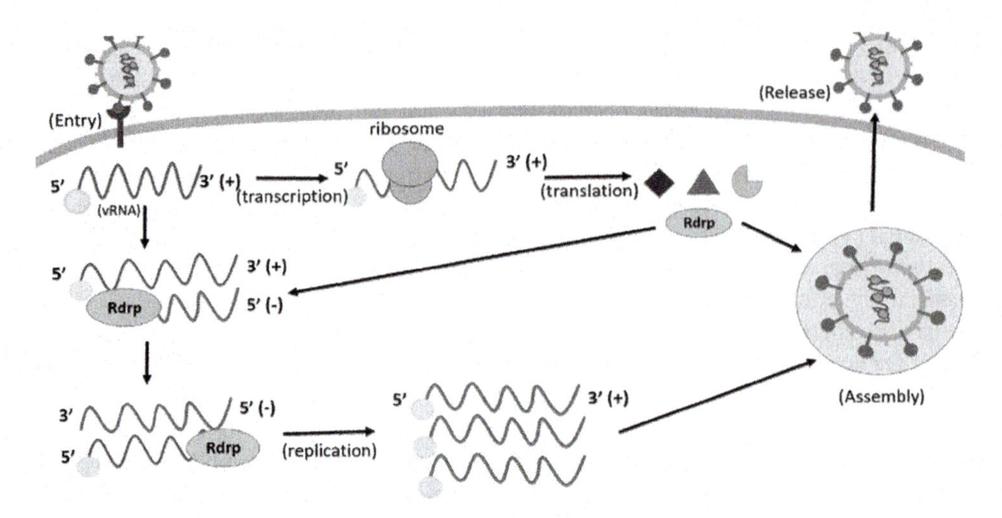

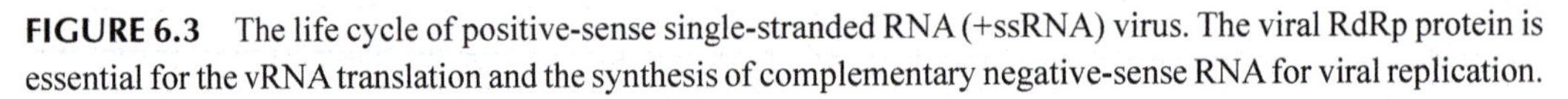

FIGURE 6.3 The life cycle of positive-sense single-stranded RNA (+ssRNA) virus. The viral RdRp protein is essential for the vRNA translation and the synthesis of complementary negative-sense RNA for viral replication.

Source: Adapted from Low et al., 2022.

infect adjacent or other host cells. Having explained this intricate mechanism of virus binding and entry in this section, we wish to emphasise the high fidelity of receptor-virus interaction and how mutations in this region can affect this process. The evolution of variants described a clear advantage that these new variants gain by changing single amino acids, thereby altering receptor binding dynamics.

6.5.1 RdRp in Virus Evolution

As aforementioned, the RdRp is a vital component for viral RNA replication and translation (shown in Figure 6.3). RdRp is required in many RNA viruses for replication and transcription of the viral genome, a crucial enzyme to the survival of the virus (Wu, Liu & Gong, 2015). In the past, several positive-sense single-stranded RNA viruses (+ssRNA) have been studied concerning their RdRp, and it is noteworthy to mention that single point mutations in this enzyme have long-term implications on the adaptability of the virus to its new host. For example, the nsp12-RdRp in SARS-CoV and MERS-CoV, the nsp5-RdRp in Zika virus and hepatitis C virus (HCV) and many more (Afdhal et al., 2014; Min et al., 2020). A structural study revealed a high homology in RdRp domains between the SARS-CoV and SARS-CoV-2. In which, 96.35%, 98.8% and 97.5% similarity are reported in nsp12, nsp7 and nsp8, respectively (Vicenti, Zazzi & Saladini, 2021). Given the highly conserved RdRp domains in RNA viruses, especially within the *Coronaviridae* family, it is no surprise that the SARS-CoV-2 carries a similar RdRp structure compared to its counterparts, the SARS-CoV, making it a suitable target for antiviral drug repositioning (Weber et al., 1999; Xu, 2003). The SARS-CoV-2 RdRp is composed of three nsps, with the central nsp12 being the main component supported by nsp7 and nsp8 co-factors to stabilise further the conformation of nsp12 for efficient binding and processivity (Gao et al., 2020). The general process of RdRp begins from the nucleotide triphosphate (NTP) binding, followed by the conformational change in the active site and phosphatidyl transfer with subsequent formation of a phosphodiester bond with the existing nucleotide chain (Wu & Gong, 2018). The Mg^{2+} ions aid this process to ensure smooth translocation, creating ways for new NTP binding vital for chain elongation (Wu & Gong, 2018).

Nucleotide and nucleoside analogue has been widely used for antiviral and anticancer treatment. It works by replacing the usual nucleoside/nucleotide for incorporation in the viral RdRp. Upon transporting the analogues into the cells, the nucleoside analogue is then phosphorylated by a cascade of kinases such as nucleoside kinase, nucleoside monophosphate kinase and nucleoside diphosphate kinase (Jordheim et al., 2013). The production of the nucleoside analogue phosphates inhibits nucleic acid synthesis, terminating the growing chain in virus replication (Ahmad & Dealwis, 2013). The nucleoside analogues can also induce early chain termination due to the absence of the 3′ hydroxyl group in nature, preventing essential 3′–5′ phosphodiester bond formation, terminating the growing chain once again (Jockusch et al., 2020). Apart from that, incorporating nucleotide analogues, such as remdesivir, carries modifications on the sugar or base that have been shown to delay chain termination irreversibly, exhibiting antiviral properties (Tchesnokov et al., 2019). Aside from direct binding of NTP sites for RdRp inhibition, the replacement of two nsp8 subunit residues (P183 and R190) that is essential for interaction and binding with nsp12, followed by a third residue (K58) can be lethal to the SARS-CoV RdRp complex, terminating the RdRp function (Subissi et al., 2014). Given the similarity in genomic structure of the SARS-CoV-2 RdRp complex, this could be another feasible route for RdRp inhibition. It may be too early, but there may be point mutations in the new variants that may have helped the SARS-CoV-2 evolve from its original bat reservoir to its intermediate host to humans. Mutations in the RdRp help and contribute subtly to virus evolution and increased host specificity.

6.6 CONCLUSION

Being a beta-coronavirus, the SARS-CoV-2 shows many resemblances to its counterpart, the SARS-CoV. As such, the general morphology and structure resemblance entails the ORFs for nsps that encode vital viral enzymes and the capsid-forming structural proteins (S, E, M, N). Akin to the predecessors, common symptoms such as fever, dry cough, headache, fatigue, dyspnoea and pneumonia are also present in SARS-CoV-2. Given the genomic similarities between SARS-CoV-2 and SARS-CoV, such as a 94.4% amino-acid sequence similarity and 87.2% to 83.9% nucleotide similarity in the RBD, it is no surprise that they share the same mode of transmission (e.g., ACE2 host cell-surface receptor recognition). It was postulated that SARS-CoV-2 has a stronger binding affinity for ACE2 than SARS-CoV due to factors like the larger interaction surface and the ability to utilise furin for proteolytic activation. In light of the above, the origin of SARS-CoV-2 is suggested to be the bats-pangolin reservoir. Notably, critical mutations such as the D614G in the spike protein of SARS-CoV-2 variants (B.1.1.7, B.1.351, P.1, B.1.617.2, C.37, B.1.621) contribute to enhanced binding affinity towards ACE2 receptors. Besides, there are also key genetic differences between the SARS-CoV and SARS-CoV-2, such as the exchange of glutamine residue for threonine in nsp2 of SARS-CoV-2, thereby increasing the viral protein stability. The typical replication cycle of SARS-CoV-2 entails the RdRp, an essential enzyme for viral translation and replication and may contribute to virus and variant evolution and may play a significant role in new host adaptation. On the other hand, the similarity in structure of the highly conserved RdRp in coronaviruses opens up many opportunities for antiviral drug repositioning, which include the typical nucleotide and nucleoside analogues such as remdesivir.

REFERENCES

Acevedo, M., Alonso-Palomares, L., Bustamante, A., et al. (2021) Infectivity and immune escape of the new SARS-CoV-2 variant of interest Lambda. *medRxiv*. doi:10.1101/2021.06.28.21259673

Afdhal, N., Zeuzem, S., Kwo, P., et al. (2014) Ledipasvir and sofosbuvir for untreated HCV genotype 1 infection. *New England Journal of Medicine* 370:1889–1898. doi:10.1056/nejmoa1402454

Ahmad, M., Dealwis, C. (2013) The structural basis for the allosteric regulation of ribonucleotide Reductase. *Progress in Molecular Biology and Translational Science* 389–410. doi:10.1016/b978-0-12-386931-9.00014-3

Alkhatib, M., Svicher, V., Salpini, R., et al. (2021) SARS-CoV-2 variants and their relevant mutational profiles: Update Summer 2021. *Microbiology Spectrum* 9. doi:10.1128/spectrum.01096-21

Andersen, K., Rambaut, A., Lipkin, W., et al. (2020) The proximal origin of SARS-CoV-2. *Nature Medicine* 26:450–452. doi:10.1038/s41591-020-0820-9

Angeletti, S., Benvenuto, D., Bianchi, M., et al. (2020) COVID-2019: The role of the nsp2 and nsp3 in its pathogenesis. *Journal of Medical Virology* 92:584–588. doi:10.1002/jmv.25719

Arora, P., Kempf, A., Nehlmeier, I., et al. (2021) Increased lung cell entry of B.1.617.2 and evasion of antibodies induced by infection and BNT162b2 vaccination. *bioRxiv*. doi:10.1101/2021.06.23.449568

Belouzard, S., Chu, V., Whittaker, G. (2009) Activation of the SARS coronavirus spike protein via sequential proteolytic cleavage at two distinct sites. *Proceedings of the National Academy of Sciences* 106:5871–5876. doi:10.1073/pnas.0809524106

Brown, K., Gubbay, J., Hopkins, J. et al. (2021) S-Gene target failure as a marker of variant B.1.1.7 among SARS-CoV-2 isolates in the greater toronto area, December 2020 to March 2021. *JAMA* 325:2115. doi:10.1001/jama.2021.5607

Cao, Y., Jian, F., Wang, J., et al. (2022) Imprinted SARS-CoV-2 humoral immunity induces convergent Omicron RBD evolution. *bioRxiV*. doi:10.1101/2022.09.15.507787

Cao, Y., Wang, J., Jian, F., et al. (2021) Omicron escapes the majority of existing SARS-CoV-2 neutralising antibodies. *Nature* 602:657–663. doi:10.1038/s41586-021-04385-3

Carlos, W., Dela Cruz, C., Cao, B., et al. (2020) Novel Wuhan (2019-nCoV) Coronavirus. *American Journal of Respiratory and Critical Care Medicine* 201:P7–P8. doi:10.1164/rccm.2014p7

Centers for Disease Control and Prevention (CDC) (2022) SARS-CoV-2 Variant Classifications and Definitions. www.cdc.gov/coronavirus/2019-ncov/variants/variant-classifications.html?CDC_AA_refVal=https%3A%2F%2Fwww.cdc.gov%2Fcoronavirus%2F2019-ncov%2Fvariants%2Fvariant-info.html. Accessed 29 October 2022.

Chatterjee, D., Tauzin, A., Laumaea, A., et al. (2022) Antigenicity of the mu (b.1.621) and A.2.5 SARS-CoV-2 spikes. *Viruses* 14:144. doi:10.3390/v14010144

Chen, J., Wang, R., Wang, M., Wei, G-W. (2020) Mutations strengthened SARS-CoV-2 infectivity. *Journal of Molecular Biology* 432:5212–5226. doi:10.1016/j.jmb.2020.07.009

Collier, D., De Marco, A., Ferreira, I., et al. (2021) Sensitivity of SARS-CoV-2 B.1.1.7 to mRNA vaccine-elicited antibodies. *Nature* 593:136–141. doi:10.1038/s41586-021-03412-7

Coutard, B., Valle, C., de Lamballerie, X., et al. (2020) The spike glycoprotein of the new coronavirus 2019-nCoV contains a furin-like cleavage site absent in CoV of the same clade. *Antiviral Research* 176:104742. doi:10.1016/j.antiviral.2020.104742

Davis, C., Logan, N., Tyson, G., et al. (2021) Reduced neutralisation of the Delta (B.1.617.2) SARS-CoV-2 variant of concern following vaccination. *medRxiv*. doi:10.1101/2021.06.23.21259327

Di Giacomo, S., Mercatelli, D., Rakhimov, A., Giorgi, F. (2021) Preliminary report on severe acute respiratory syndrome coronavirus 2 (SARS-CoV-2) Spike mutation T478K. *Journal of Medical Virology* 93:5638–5643. doi:10.1002/jmv.27062

Duong, D. (2021) Alpha, Beta, Delta, Gamma: What's important to know about SARS-CoV-2 variants of concern?. *Canadian Medical Association Journal* 193:E1059–E1060. doi:10.1503/cmaj.1095949

European Centre for Disease Prevention and Control (ECDC) (2022) SARS-COV-2 variants of concern as of 27 October 2022. www.ecdc.europa.eu/en/covid-19/variants-concern. Accessed 29 October 2022.

Farinholt, T., Doddapaneni, H., Qin, X., et al. (2021) Transmission event of SARS-CoV-2 Delta variant reveals multiple vaccine breakthrough infections. *medRxiv*. doi:10.1101/2021.06.28.21258780

Fehr, A., Perlman, S. (2015) Coronaviruses: An overview of their replication and pathogenesis. *Coronaviruses* 1–23. doi:10.1007/978-1-4939-2438-7_1

Ferreira, I., Kemp, S., Datir, R., et al. (2021) SARS-CoV-2 B.1.617 mutations L452 and E484Q are not synergistic for antibody evasion. *Journal of Infectious Diseases*. doi:10.1093/infdis/jiab368

Focosi, D., Maggi, F. (2021) Neutralising antibody escape of SARS-CoV-2 spike protein: Risk assessment for antibody-based Covid-19 therapeutics and vaccines. *Reviews in Medical Virology*. doi:10.1002/rmv.2231

Fujino, T., Nomoto, H., Kutsuna, S., et al. (2021) Novel SARS-CoV-2 variant in travelers from Brazil to Japan. *Emerging Infectious Diseases*. doi:10.3201/eid2704.210138

Galloway, S., Paul, P., MacCannell, D., et al. (2021) Emergence of SARS-CoV-2 B.1.1.7 lineage – United States, December 29, 2020–January 12, 2021. *MMWR Morbidity and Mortality Weekly Report* 70:95–99. doi:10.15585/mmwr.mm7003e2

Gao, Y., Yan, L., Huang, Y., et al. (2020) Structure of the RNA-dependent RNA polymerase from COVID-19 virus. *Science* 368:779–782. doi:10.1126/science.abb7498

Gong, S., Bao, L. (2018) The battle against SARS and MERS coronaviruses: Reservoirs and animal models. *Animal Models and Experimental Medicine* 1:125–133. doi:10.1002/ame2.12017

Grabowski, F., Kochańczyk, M., Lipniacki, T. (2021) L18F substrain of SARS-CoV-2 VOC-202012/01 is rapidly spreading in England. *medRxiv*. doi:10.1101/2021.02.07.21251262

Halfmann, P., Kuroda, M., Armbrust, T., et al. (2022) Characterisation of the SARS-CoV-2 B.1.621 (Mu) Variant. *Science Translational Medicine* 14. doi:10.1126/scitranslmed.abm4908

Han, D., Penn-Nicholson, A., Cho, M. (2006) Identification of critical determinants on ACE2 for SARS-CoV entry and development of a potent entry inhibitor. *Virology* 350:15–25. doi:10.1016/j.virol.2006.01.029

Harvey, W., Carabelli, A., Jackson, B., et al. (2021) SARS-CoV-2 variants, spike mutations and immune escape. *Nature Reviews Microbiology* 19:409–424. doi:10.1038/s41579-021-00573-0

Hossain, M., Rabaan, A., Mutair, A., et al. (2022) Strategies to tackle SARS-CoV-2 Mu, a newly classified variant of interest likely to resist currently available COVID-19 vaccines. *Human Vaccines & Immunotherapeutics* 18. doi:10.1080/21645515.2022.2027197

Hu, B., Guo, H., Zhou, P., Shi, Z. (2020) Characteristics of SARS-CoV-2 and COVID-19. *Nature Reviews Microbiology* 19:141–154. doi:10.1038/s41579-020-00459-7

Jockusch, S., Tao, C., Li, X., et al. (2020) A library of nucleotide analogues terminate RNA synthesis catalysed by polymerases of coronaviruses that cause SARS and COVID-19. *Antiviral Research* 180:104857. doi:10.1016/j.antiviral.2020.104857

Jordheim, L., Durantel, D., Zoulim, F., Dumontet, C. (2013) Advances in the development of nucleoside and nucleotide analogues for cancer and viral diseases. *Nature Reviews Drug Discovery* 12:447–464. doi:10.1038/nrd4010

Khatri, I., Staal, F., van Dongen, J. (2020) Blocking of the high-affinity interaction-synapse between SARS-CoV-2 spike and human ACE2 proteins likely requires multiple high-affinity antibodies: An immune perspective. *Frontiers in Immunology*. doi:10.3389/fimmu.2020.570018

Kimura, I., Kosugi, Y., Wu, J., et al. (2022). The SARS-CoV-2 lambda variant exhibits enhanced infectivity and immune resistance. *Cell Reports* 38:110218. doi:10.1016/j.celrep.2021.110218

Kumar, S., Karuppanan, K., Subramaniam, G. (2022) Omicron (BA.1) and sub-variants (BA.1, BA.2 and BA.3) of SARS-CoV-2 spike infectivity and pathogenicity: A comparative sequence and structural-based computational assessment. *bioRxiv*. doi:10.1101/2022.02.11.480029

Kumar, S., Maurya, V., Prasad, A., et al. (2020) Structural, glycosylation and antigenic variation between 2019 novel coronavirus (2019-nCoV) and SARS coronavirus (SARS-CoV). *VirusDisease* 31:13–21. doi:10.1007/s13337-020-00571-5

Kurhade, C., Zou, J., Xia, H., et al. (2022) Neutralisation of Omicron Ba.1, Ba.2, and Ba.3 SARS-CoV-2 by 3 doses of BNT162b2 vaccine. *Nature Communications* 13. doi:10.1038/s41467-022-30681-1

Letko, M., Marzi, A., Munster, V. (2020) Functional assessment of cell entry and receptor usage for SARS-CoV-2 and other lineage B betacoronaviruses. *Nature Microbiology* 5:562–569. doi:10.1038/s41564-020-0688-y

Li ,Y., Wang, X., Jin, J., et al. (2022) T-cell responses to Sars-Cov-2 omicron spike epitopes with mutations after the third booster dose of an inactivated vaccine. *Journal of Medical Virology* 94:3998–4004. doi:10.1002/jmv.27814

Li, F. (2016) Structure, function, and evolution of coronavirus spike proteins. *Annual Review of Virology* 3:237–261. doi:10.1146/annurev-virology-110615-042301

Lim, H., Baek, A., Kim, J., et al. (2020) Hot spot profiles of SARS-CoV-2 and human ACE2 receptor protein protein interaction obtained by density functional tight binding fragment molecular orbital method. *Scientific Reports*. doi:10.1038/s41598-020-73820-8

Liu, C., Ginn, H., Dejnirattisai, W., et al. (2021) Reduced neutralisation of SARS-CoV-2 B.1.617 by vaccine and convalescent serum. *Cell*. doi:10.1016/j.cell.2021.06.020

Liu, J., Liu, Y., Xia, H., et al. (2021) BNT162b2-elicited neutralisation of B.1.617 and other SARS-CoV-2 variants. *Nature*. doi:10.1038/s41586-021-03693-y

Liu, L., Iketani, S., Guo, Y., et al. (2021) Striking antibody evasion manifested by the Omicron variant of SARS-CoV-2. *Nature* 602:676–681. doi:10.1038/s41586-021-04388-0

Low, Z., Yip, A., Lal, S. (2022) Repositioning anticancer drugs as novel COVID-19 antivirals: Targeting structural and functional similarities between viral proteins and cancer. *Expert Reviews in Molecular Medicine* 24. doi:10.3390/v14091991

Low, Z., Yip, A., Sharma, A., Lal, S. (2021) SARS coronavirus outbreaks past and present – a comparative analysis of SARS-CoV-2 and its predecessors. *Virus Genes*. doi:10.1007/s11262-021-01846-9

Lu, R., Zhao, X., Li, J., et al. (2020) Genomic characterisation and epidemiology of 2019 novel coronavirus: implications for virus origins and receptor binding. *The Lancet* 395:565–574. doi:10.1016/s0140-6736(20)30251-8

Lubinski, B., Tang, T., Daniel, S., et al. (2021) Functional evaluation of proteolytic activation for the SARS-CoV-2 variant B.1.1.7: role of the P681H mutation. *bioRxiv*. doi:10.1101/2021.04.06.438731

Mahase, E. (2022) Covid-19: What do we know about Omicron sublineages? *British Medical Journal* o358. doi:10.1136/bmj.o358

McCallum, M., De Marco, A., Lempp, F., et al. (2021) N-terminal domain antigenic mapping reveals a site of vulnerability for SARS-CoV-2. *Cell* 184:2332–2347.e16. doi:10.1016/j.cell.2021.03.028

Milanetti, E., Miotto, M., Di Rienzo, L., et al. (2021) In-silico evidence for a two receptor based strategy of SARS-CoV-2. *Frontiers in Molecular Biosciences*. doi:10.3389/fmolb.2021.690655

Min, J., Kim, G., Kwon, S., Jin, Y. (2020) A cell-based reporter assay for screening inhibitors of MERS coronavirus RNA-dependent RNA polymerase activity. *Journal of Clinical Medicine* 9:2399. doi:10.3390/jcm9082399

Motozono, C., Toyoda, M., Zahradnik, J., et al. (2021) SARS-CoV-2 spike L452R variant evades cellular immunity and increases infectivity. *Cell Host & Microbe* 29:1124–1136.e11. doi:10.1016/j.chom.2021.06.006

Munitz, A., Yechezkel, M., Dickstein, Y., et al. (2021) BNT162b2 vaccination effectively prevents the rapid rise of SARS-CoV-2 variant B.1.1.7 in high-risk populations in Israel. *Cell Reports Medicine* 2:100264. doi:10.1016/j.xcrm.2021.100264

Naqvi, A., Fatima, K., Mohammad, T., et al. (2020) Insights into SARS-CoV-2 genome, structure, evolution, pathogenesis and therapies: Structural genomics approach. *Biochimica et Biophysica Acta (BBA) – Molecular Basis of Disease* 1866:165878. doi:10.1016/j.bbadis.2020.165878

Nature Research Custom Media (2021) The rapid journey of a deadly MERS outbreak. www.nature.com/articles/d42473-019-00422-y. Accessed 18 July 2021.

Ou, X., Liu, Y., Lei, X., et al. (2020) Characterisation of spike glycoprotein of SARS-CoV-2 on virus entry and its immune cross-reactivity with SARS-CoV. *Nature Communications*. doi:10.1038/s41467-020-15562-9

Peiris, J., Yuen, K., Osterhaus, A., Stöhr, K. (2003) The Severe Acute Respiratory Syndrome. *New England Journal of Medicine* 349:2431–2441. doi:10.1056/nejmra032498

Ramanathan, M., Ferguson, I., Miao, W., Khavari, P. (2021) SARS-CoV-2 B.1.1.7 and B.1.351 spike variants bind human ACE2 with increased affinity. *The Lancet Infectious Diseases*. doi:10.1016/s1473-3099(21)00262-0

Saito, A., Nasser, H., Uriu, K., et al. (2021) SARS-CoV-2 spike P681R mutation enhances and accelerates viral fusion. *bioRxiv*. doi:10.1101/2021.06.17.448820

Santarpia, J., Rivera, D., Herrera, V., et al. (2020) Aerosol and surface contamination of SARS-CoV-2 observed in quarantine and isolation care. *Scientific Reports*. doi:10.1038/s41598-020-69286-3

Sharma, A., Farouk, I. A., Lal, S. (2021) COVID-19: A Review on the novel coronavirus disease evolution, transmission, detection, control and prevention. *Viruses* 13:202. doi:10.3390/v13020202

Sharma, A., Low, Z., Lal, S., Chow, V. (2022) The Novel SARS-CoV-2: An overview on its detection, prevention and control. In *Uncovering the Science of Covid-19*, 1st ed. Singapore: *World Scientific*, 1–29. doi:10.1142/9789811254338_0001

Shereen, M., Khan, S., Kazmi, A., et al. (2020) COVID-19 infection: Emergence, transmission, and characteristics of human coronaviruses. *Journal of Advanced Research* 24:91–98. doi:10.1016/j.jare.2020.03.005

Starr, T., Greaney, A., Hilton, S., et al. (2020) Deep mutational scanning of SARS-CoV-2 receptor binding domain reveals constraints on folding and ACE2 binding. *Cell* 182:1295–1310.e20. doi:10.1016/j.cell.2020.08.012

Subissi, L., Posthuma, C., Collet, A., et al. (2014) One severe acute respiratory syndrome coronavirus protein complex integrates processive RNA polymerase and exonuclease activities. *Proceedings of the National Academy of Sciences* 111:E3900–E3909. doi:10.1073/pnas.1323705111

Tada, T., Zhou, H., Dcosta, B., et al. (2021) SARS-CoV-2 lambda variant remains susceptible to neutralisation by mRNA vaccine-elicited antibodies and convalescent serum. *bioRxiv*. doi:10.1101/2021.07.02.450959

Tang, T., Jaimes, J., Bidon, M., et al. (2021) Proteolytic activation of SARS-CoV-2 spike at the S1/S2 boundary: Potential role of proteases beyond furin. *ACS Infectious Diseases* 7:264–272. doi:10.1021/acsinfecdis.0c00701

Tchesnokov, E., Feng, J., Porter, D., Götte, M. (2019) Mechanism of inhibition of Ebola virus RNA-dependent RNA polymerase by Remdesivir. *Viruses* 11:326. doi:10.3390/v11040326

Tegally, H., Moir, M., Everatt, J., et al. (2022). Emergence of SARS-CoV-2 Omicron lineages BA.4 and BA.5 in South Africa. *Nature Medicine* 28:1785–1790. doi:10.1038/s41591-022-01911-2

Tegally, H., Wilkinson, E., Giovanetti, M., et al. (2020) Emergence and rapid spread of a new severe acute respiratory syndrome-related coronavirus 2 (SARS-CoV-2) lineage with multiple spike mutations in South Africa. *medRxiv*. doi:10.1101/2020.12.21.20248640

Thiel, V., Herold, J., Schelle, B., Siddell, S. (2001) Infectious RNA transcribed in vitro from a cDNA copy of the human coronavirus genome cloned in vaccinia virus. *Journal of General Virology* 82:1273–1281. doi:10.1099/0022-1317-82-6-1273

Tian, D., Sun, Y., Xu, H., Ye, Q. (2022) The emergence and epidemic characteristics of the highly mutated Sars-Cov-2 omicron variant. *Journal of Medical Virology* 94:2376–2383. doi:10.1002/jmv.27643

van Doremalen, N., Bushmaker, T., Morris, D., et al. (2020) Aerosol and surface stability of SARS-CoV-2 as compared with SARS-CoV-1. *New England Journal of Medicine* 382:1564–1567. doi:10.1056/nejmc2004973

Velavan, T., Meyer, C. (2020) The COVID-19 epidemic. *Tropical Medicine & International Health* 25:278–280. doi:10.1111/tmi.13383

Vicenti, I., Zazzi, M., Saladini, F. (2021) SARS-CoV-2 RNA-dependent RNA polymerase as a therapeutic target for COVID-19. *Expert Opinion on Therapeutic Patents* 31:325–337. doi:10.1080/13543776.2021.1880568

Walls, A., Park, Y., Tortorici, M., et al. (2020) Structure, function, and antigenicity of the SARS-CoV-2 spike glycoprotein. *Cell* 181:281–292.e6. doi:10.1016/j.cell.2020.02.058

Wan, Y., Shang, J., Graham, R., et al. (2020) Receptor recognition by the novel coronavirus from Wuhan: An analysis based on decade-long structural studies of SARS coronavirus. *Journal of Virology*. doi:10.1128/jvi.00127-20

Wang, P., Nair, M., Liu, L., et al. (2021) Antibody resistance of SARS-CoV-2 variants B.1.351 and B.1.1.7. *bioRxiv*. doi:10.1101/2021.01.25.428137

Weber, P., Lesburg, C., Cable, M., et al. (1999) Crystal structure of the RNA-dependent RNA polymerase from hepatitis C virus reveals a fully encircled active site. *Nature Structural Biology* 6:937–943. doi:10.1038/13305

Wink, P., Zempulski Volpato, F., Monteiro, F., et al. (2021) First identification of SARS-CoV-2 Lambda (C.37) variant in Southern Brazil. *medRxiv*. doi:10.1101/2021.06.21.21259241

Wong, A., Li, X., Lau, S., Woo, P. (2019) Global Epidemiology of Bat Coronaviruses. *Viruses* 11:174. doi:10.3390/v11020174

World Health Organization (WHO) (2021) Naming the Coronavirus disease (COVID-19) and the virus that causes it. www.who.int/emergencies/diseases/novel-coronavirus-2019/question-and-answers-hub/q-a-detail/coronavirus-disease-covid-19. Accessed 18 July 2021.

World Health Organization (WHO) (2022) Tracking SARS-CoV-2 variants. www.who.int/en/activities/tracking-SARS-CoV-2-variants/. Accessed 29 October 2022.

Worldometer (2023) COVID Live Update: Worldometer. www.worldometers.info/coronavirus/. Accessed 17 January 2023.

Wu, J., Gong, P. (2018) Visualising the nucleotide addition cycle of viral RNA-dependent RNA polymerase. *Viruses* 10:24. doi:10.3390/v10010024

Wu, J., Liu, W., Gong, P. (2015) A structural overview of RNA-dependent RNA polymerases from the flaviviridae family. *International Journal of Molecular Sciences* 16:12943–12957. doi:10.3390/ijms160612943

Wu, K., Werner, A., Moliva, J., et al. (2021) mRNA-1273 vaccine induces neutralising antibodies against spike mutants from global SARS-CoV-2 variants. *bioRxiv*. doi:10.1101/2021.01.25.427948

Xia, X. (2021) Domains and functions of spike protein in SARS-Cov-2 in the context of vaccine Design. *Viruses* 13:109. doi:10.3390/v13010109

Xiao, K, Zhai, J., Feng, Y., et al. (2020) Isolation and characterisation of 2019-nCoV-like Coronavirus from Malayan Pangolins. *bioRxiv*. doi:10.1101/2020.02.17.951335

Xie, X., Liu, Y., Liu, J., et al. (2021) Neutralisation of SARS-CoV-2 spike 69/70 deletion, E484K and N501Y variants by BNT162b2 vaccine-elicited sera. *Nature Medicine* 27:620–621. doi:10.1038/s41591-021-01270-4

Xu, X. (2003) Molecular model of SARS coronavirus polymerase: Implications for biochemical functions and drug design. *Nucleic Acids Research* 31:7117–7130. doi:10.1093/nar/gkg916

Ye, Z., Yuan, S., Yuen, K., et al. (2020) Zoonotic origins of human coronaviruses. *International Journal of Biological Sciences* 16:1686–1697. doi:10.7150/ijbs.45472

Yip, A., Low, Z., Chow, V., Lal, S. (2022) Repurposing Molnupiravir for Covid-19: The mechanisms of anti-viral activity. *Viruses* 14:1345. doi:10.3390/v14061345

Yue, C., Song, W., Wang, L., et al. (2023) Enhanced transmissibility of XBB.1.5 is contributed by both strong ACE2 binding and antibody evasion. *BioRxiV*. doi:10.1101/2023.01.03.522427

Zhang, L., Jackson, C., Mou, H., et al. (2020) The D614G mutation in the SARS-CoV-2 spike protein reduces S1 shedding and increases infectivity. *bioRxiv*. doi:10.1101/2020.06.12.148726

Zhao, Z., Zhou, J., Tian, M., et al. (2022) Omicron SARS-CoV-2 mutations stabilise spike up-RBD conformation and lead to a non-rbm-binding monoclonal antibody escape. *Nature Communications* 13. doi:10.1038/s41467-022-32665-7

Zheng, J. (2020) SARS-CoV-2: An emerging coronavirus that causes a global threat. *International Journal of Biological Sciences* 16:1678–1685. doi:10.7150/ijbs.45053

Zheng, Y., Ma, Y., Zhang, J., Xie, X. (2020) COVID-19 and the cardiovascular system. *Nature Reviews Cardiology* 17:259–260. doi:10.1038/s41569-020-0360-5

Zhong, N., Zheng, B., Li, Y., et al. (2003) Epidemiology and cause of severe acute respiratory syndrome (SARS) in Guangdong, People's Republic of China, in February, 2003. *The Lancet* 362:1353–1358. doi:10.1016/s0140-6736(03)14630-2

Zhou, D., Dejnirattisai, W., Supasa, P., et al. (2021) Evidence of escape of SARS-CoV-2 variant B.1.351 from natural and vaccine-induced sera. *Cell* 184:2348–2361.e6. doi:10.1016/j.cell.2021.02.037

Zhou, P., Yang, X., Wang, X., et al. (2020) A pneumonia outbreak associated with a new coronavirus of probable bat origin. *Nature* 579:270–273. doi:10.1038/s41586-020-2012-7

Zhu, N., Zhang, D., Wang, W., et al. (2020) A novel coronavirus from patients with pneumonia in China, 2019. *New England Journal of Medicine* 382:727–733. doi:10.1056/nejmoa2001017

Zhu, Z., Lian, X., Su, X., et al. (2020) From SARS and MERS to COVID-19: A brief summary and comparison of severe acute respiratory infections caused by three highly pathogenic human coronaviruses. *Respiratory Research*. doi:10.1186/s12931-020-01479-w

Zumla, A., Hui, D., Perlman, S. (2015) Middle East respiratory syndrome. *The Lancet* 386:995–1007. doi:10.1016/s0140-6736(15)60454-8

7 Covid-19

Pathology, Clinical Implications, and Prevention[1]

Varidmala Jain[1] *and Jitendra Kumar Jain*[2]
[1]Secretary, Trishla Foundation, Prayagraj, Uttar Pradesh, India
[2]President, Trishla Foundation, Prayagraj, Uttar Pradesh, India

CONTENTS

7.1 INTRODUCTION

Covid-19 is a coronavirus disease that started in 2019 in Wuhan, China. This pandemic is different from previous large pandemics in two aspects. First, this is the era of fast tourism and travel like never before, which has led to very fast spread of the disease. Second, social media and information technology and fast communication in advancement of medical sciences has led to fast development of awareness among masses and development of methods of prevention and treatment. While healthcare institutions are able to handle the situation more efficiently it is we are still learning about pathogenesis, natural history, prevention, and treatment of this disease.

Covid-19 is a viral infection caused by the coronavirus SARS-CoV-2 that can progress to severe acute respiratory syndrome with pneumonia and acute respiratory distress syndrome (Shanmugam et al., 2020). SARS-CoV-2 is a single-stranded RNA virus from the coronavirus family that uses human ACE2 receptors (angiotensin-converting enzyme 2) for its attachment to cell. It has very high affinity to these receptors (Wan et al., 2020).

According to current evidence, SARS-CoV-2 is primarily transmitted between people through respiratory droplets and contact routes. Transmission of SARS-CoV-2 can occur by **direct contact** with infected people through coughing, sneezing, etc., and through **indirect contact** with surfaces in the immediate environment or with objects used on the infected person (fomite) (WHO a, 2020).

7.2 CLINICO-PATHOLOGICAL CHARACTERISTICS

A large number of people may be asymptomatic or may present with fever or chills, cough, sore throat, congestion or runny nose, shortness of breath or difficulty breathing, fatigue, muscle or body

DOI: 10.1201/9781003358909-7

aches, headache, new loss of taste or smell, nausea or vomiting, and diarrhoea. Symptoms may appear 2–14 days after exposure to the virus (average time is 5 days) (CDC a, 2020).

7.2.1 Severity of Disease

The World Health Organization (WHO) classified Covid-19 into five categories according to severity for the purposes of proper management (WHO b, 2020):

Mild disease: Symptomatic patients without evidence of viral pneumonia or hypoxia

Moderate disease (Pneumonia): Clinical signs of pneumonia (fever, cough, dyspnoea, fast breathing) but no signs of sever pneumonia including $SpO_2 \geq 90\%$

Severe disease (Severe pneumonia): Clinical signs of pneumonia (fever, cough, dyspnoea, fast breathing) along with one of the following: respiratory rate 30 breaths per minute, severe respiratory distress, or $SpO_2 < 90\%$

Critical disease (Acute respiratory distress syndrome): Onset within 1 week of pneumonia or new or worsening respiratory symptoms, on chest imaging bilateral opacities, lung collapse, or nodules, where respiratory failure not fully explained by cardiac failure or fluid overload.

Oxygenation impairment in adults: • Mild ARDS: 200 mmHg $< PaO_2/FiO_2$ a $\leq$ 300 mmHg • Moderate ARDS: 100 mmHg $< PaO_2/FiO_2 \leq$ 200 mmHg • Severe ARDS: $PaO_2/FiO_2 \leq$ 100 mmHg (Force ADT et al., 2012).

Critical disease including either sepsis (acute life-threatening condition including multi-organ dysfunction) or **septic shock** (persistent hypotension despite volume resuscitation).

It has been observed that there are various complications among hospitalized patients especially in late stages of the disease. Pneumonia is the most common complication (75%) followed by acute respiratory distress syndrome (15%), acute liver injury (19%), cardiac injury (7–17%), prothrombotic coagulopathy resulting in venous and arterial thrombo-embolic event (10–15%), acute kidney injury (9%), acute cerebro-vascular disease (3%), and shock in (6%) patients (Wiersinga, 2020).

7.2.2 Pathologenesis

Research on Covid-19 has become the first priority among all the concerned fields but especially in pathology and virology. There is now of a vast body of knowledge about the processes and mechanisms involved in what happens when SARS-CoV-2 enters the human body.

Lungs are the primary site of infection by SARS-CoV-2. Alveoli are the smallest units of the lungs. There are two type of cells in the alveoli: type I and type II. The ACE II receptors are more abundant in type II cells and SARS-CoV-2 has high affinity for ACE II receptors.

The spike glycoprotein present on the surface of the virus binds to the host via receptor-binding domains of ACE II, which are more abundant in alveolar cell type II (Wan et al., 2020). Then virion releases RNA into the cell initiating replication of the virus, which further disseminates to infect more cells (Hoffmann, 2020).

As severity of disease increases, the virus induces macrophage activation syndrome (MAS), cytokine storm, and secondary hemophagocyticlymphohistocytosis (sHLH). This state involves activation of immune systems of the body. The inflammatory mediators start damaging alveolar epithelium, which ultimately leads to diffuse alveolar damage (DAD). Covid-19-induced DAD is characterized by damage to alveolar capillary endothelium and type II pneumocytes. There is alveolar septal oedema, formation of hyaline membranes, and accumulation of immune-mediating cells megakaryocytes, platelets, and neutrophils in alveolar capillaries. These lead to precipitation of fibrin inside and outside alveolar capillaries. Also there is mild accumulation of lymphocytes and macrophages within alveoli (Barth et al, 2020) (Figure 7.1). There is severe endothelial injury and

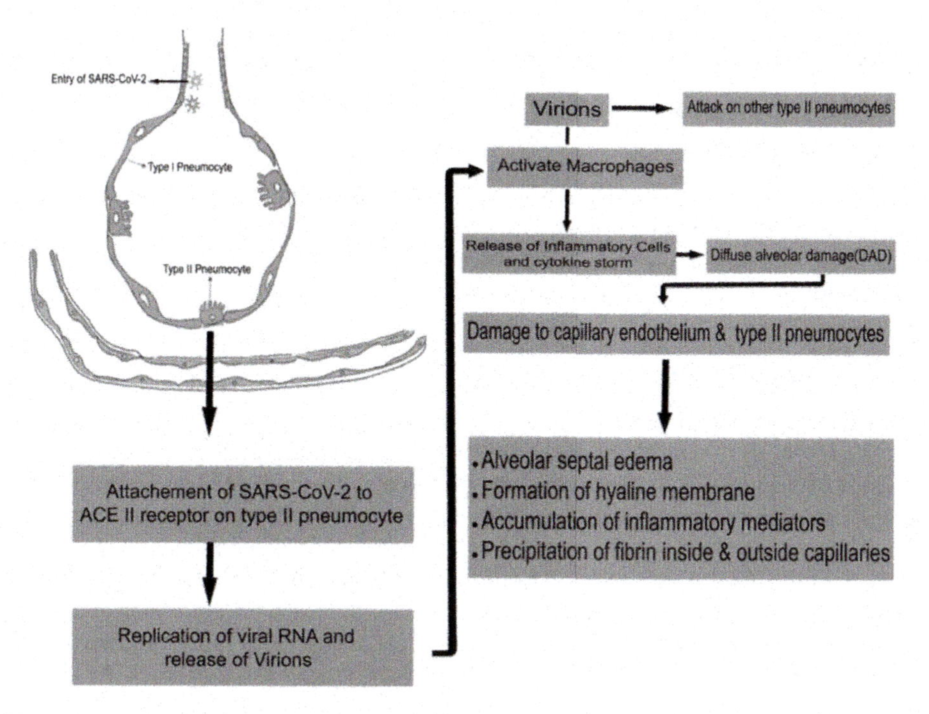

FIGURE 7.1 Pathogenesis of SARS-CoV-2 in lungs alveoli.

pulmonary thrombotic microangiopathy. This thrombus formation and angiogenesis in lung alveoli is many times higher and severe in SARS-CoV-2-induced injury than in other similar viral infections such as influenza (Ackermann et al, 2020). There is involvement of multiple organs in severe cases of Covid-19. The underlying mechanism may be due to the presence of ACE II receptors in these organs and activation of mediators of inflammation and cytokine storm that start affecting the body's own organs (Mokhtari et al., 2020).

7.3 DIAGNOSIS

Two types of lab **tests** are available for Covid-19: viral tests and antibody tests (CDC b, 2020).

7.3.1 Viral Tests

These are done for current infection and are done either for detection of viral antigen or viral genome. Sample is taken by swab either from upper or lower respiratory tract.

- **Antigen tests** are a type of rapid diagnostic test done to detect specific proteins on the surface of the virus in a sample from the respiratory tract. This test gives very quick results usually within 30 minutes. However, the sensitivity of these tests is highly variable (WHO c, April 2020). The government of India suggests the use of Rapid Antigen Test (RAT) as routine surveillance test in containment zones (MOHFW-ICMR, Sept 2020).
- Confirmation of Covid-19 infection is done by using real-time **RT-PCR test**. In this test sequencing of RNA present in the SARS-CoV-2 virus is done. A sample from the respiratory tract is taken for this test. In the case of negative result in a highly suspicious case a sample

from the lower respiratory tract must be obtained (WHO d, March 2020). A single positive test of RT-PCR is considered confirmatory. No repeat test is required (MOHFW-ICMR, 2020).

7.3.2 Antibody Tests

These tests are performed for detection of past infection since antibodies are formed only after 1 week of infection. These tests can determine if the person had the virus, but it will not spot active infection. These serological tests are used to detect the extent of spread of infection in a population (Hopkins, 2020).

7.3.3 Radiological Tests

Radiological tests are not done as screening tests, but are done either in symptomatic patients with negative lab report or in confirmed cases for review of treatment protocol. Chest X-rays are of limited value but can be used to determine lung involvement in moderate-to-severe cases. CT scan is a highly sensitive tool if lungs are involved and is very effective at detecting lung abnormalities. CT is considered ideal for assessing severity of disease in confirmed cases. Serial CT tells the evolution of the disease over a period of time. The CT severity score specifies severity of lung involvement and has prognostic value. Total score is 25, with 5 each for every lobe of lung. A score of >18 indicates severe involvement of lungs and is associated with higher mortality (Francone et al., 2020).

7.4 PREVENTION

According to the WHO the following preventive measures are very effective in preventing the spread of Covid-19 (WHO e, 2020):

- Social distancing: It is advisable to practice social distancing of at least 6 feet since respiratory droplets travel upto this distance.
- Wearing a mask: Mask wearing is a new normal when around people. It is also important to wear it properly. We must also wash hands before wearing the mask and also after removing it. It should cover nose, mouth, and chin properly.
- Regular cleaning of hands with either soap and water or alcohol-based hand rub is very important.
- Avoid touching eyes, nose, and mouth without properly cleaning hands.
- Coughing into a bent elbow or tissue: Make it a habit of coughing or sneezing in a tissue and washing your hands after safely disposing that tissue or if not possible because of some reason then in bent elbow.
- Regularly clean and disinfect frequently touched surfaces such as door handles, phone screens, etc.
- Keeping rooms well ventilated: Ventilation is important as it is a natural way of washing away impurities in the environment. It is also important to keep car windows open, especially if sharing the vehicle.
- Avoiding crowds: Avoid the 3Cs: spaces that are closed, crowded, or involve close contact. Try to meet people outside, since outdoor gatherings are safer. Avoid crowded or indoor settings, but when not possible wear a mask and open a window.
- Check local advice where you live and work: Always keep current with advice of local authorities and follow it fully.
- If you are feeling unwell then isolate yourself immediately and get yourself checked if have any symptoms and follow all directions.

- Keep up-to-date on the latest information from trusted sources like the WHO or government authorities. Do not believe in rumours.

7.4.1 Vaccine for SARS-CoV-2

Now after almost three years after start of a pandemic that shook the world there are many safe and effective vaccines available that give promising protection from serious illness and death from COVID-19. Thanks to the advances in modern sciences billions of people have been vaccinated against the deadly disease across the world so far in one of the largest vaccination drives to date. It is recommended that all the doses be taken as per guiding authorities. According to the WHO the vaccines that are safe and effective are AstraZenecea/Oxford vaccine, Johnson and Johnson, Moderna, Pfizer/BionTech, Sinovac, COVAXIN, Covovax, Nuvaxovid, and CanSino (WHO). Although the vaccines prevent serious illness, they cannot completely prevent transmission so we need to continue taking other precautions as well.

7.5 CONCLUSIONS

History is witnessing another large pandemic after a gap of around 100 years. With every pandemic comes large changes in lifestyle, functioning, new discoveries, etc. Changes may be positive and negative. As witnesses we carry the responsibility to give a better world to future generations. We have to reconsider current lifestyle patterns, population overload, and overuse of natural resources, which lead to so many health issues.

NOTE

1 No permission is required from third party for publication of this chapter and I permit CRC to publish this chapter.

REFERENCES

Ackermann M, Verleden SE, Kuehnel M, Haverich A, Welte T, Laenger F et al. (Jul 2020) Pulmonary Vascular Edothelialitis, Thrombosis, and Angiogenesis in COVID-19. *New England Journal of Medicine*. https://doi:10.1056/NEJMoa2015 432. Epub 2020 May 21.

Barth RF, Buja LM, Parwani AV. (Jul 2020) The spectrum of pathological findings in coronavirus disease (COVID-19) and the pathogenesis of SARS-CoV-2. *Diagnostic Pathology*. https://doi:10.1186/s13000-020-00999-9

Bharat Biotech. (Dec 2020) COVAXIN-India's First Indigenous COVID-19 Vaccine. www.bharatbiotech.com/covaxin.html

CDC a (May 2020) Your Health, Symptoms of Coronavirus. www.cdc.gov/coronavirus/2019-ncov/symptoms-testing/symptoms.html

CDC b (Oct 2020) COVID-19 Testing Overview. www.cdc.gov/coronavirus/2019-ncov/symptoms-testing/testing.html

Force ADT, Ranieri VM, Rubenfeld GD, Thompson BT, Ferguson ND, Caldwell E et al. (2012) Acute respiratory distress syndrome: the Berlin definition. doi: 10.1001/jama. 2012.5669

Francone M, Iafrate F, Masci GM, Coco S, Cilia F, Manganaro L. (Jul 2020) Chest CT score in COVID-19 patients: correlation with disease severity and short term prognosis, *European Radiology*. https://doi: 10.1007/s00330-020-07033-y. Epub 2020 Jul 4.

Hoffmann M, Kleine-Weber H, Schroeder S, Kruger N, Herrler T, Erichsen S et al. (Mar 2020) SARS-CoV-2 entry depends on ACE2 and TMPRSS2 and is blocked by a clinically proven Protease Inhibitor, https://doi:10.1016/j.cell.2020.02.052. Epub 2020 Mar 5

John Hopkins, Centre for Health Security (June 2020) Serology ttesting for COVID-19. www.centerforhealthsecurity.org/resources/COVID-19/COVID-19-fact-sheets/200228-Serology-testing-COVID.pdf

MOHFW-ICMR (Sept 2020) Advisory on strategy for COVID-19 testing in India. www.mohfw.gov.in/pdf/AdvisoryonstrategyforCOVID19TestinginIndia.pdf

Mokhtari T, Hasani F, Ghaffari N, Ebrahimi B, Yarahmadi A, Hassanzadeh G. (Oct 2020) COVID-19 and multiorgan failure: A narrative review on potential mechanisms. *Journal of Molecular Histology*. https://doi:10.1007/s10735-020-09915-3 (Epub ahead of print).

PTI (17 December 2020) Serum Institute of India applies for emergency use authorization for COVID-19 vaccine. http://ptinews.com/news/12016707_Serum-Institute-applies-for-emergency-use-authorisation- for-COVID-19-vaccine.html

Shanmugam C, Mohammed AR, Ravuri S, Luthra V. Rajagopal N, Karre S. (Oct 2020) COVID-19 A Comprehensive Pathology Insight. https://doi:10.1016/j.prp.2020.153222. Epub 18 Sep 2020.

Wan Y, Shang J, Graham R, Baric RS, Li F. (Mar 2020) Receptor recognition by the novel coronavirus from Wuhan: An analysis based on decade long structural studies of SARS coronavirus. *Journal of Virology*. https://doi:10.1128/ JVI.00127-20

WHO a (Mar 2020) Scientific brief, modes of transmission of virus causing COVID-19: Implications for IPC precaution recommendations. www.who.int/news-room/commentaries/detail/modes-of-transmission-of-virus-causing-covid-19-implications-for-ipc-precaution-recommendations

WHO b (May 2020) Clinical management of COVID-19, interim Guidance. www.who.int/publications/i/item/clinical-management-of-covid-19

WHO c (April 2020) www.who.int/news-room/commentaries/ detail/advice-on-the-use-of-point-of- care-immunodiagnostic-tests-for-covid-19

WHO d (March 2020) Laboratory testing for corona virus disease (COVID-19) in suspected human cases. https://apps.who.int/ iris/bitstream/handle/10665/331501/WHO-COVID-19-laboratory-2020.

WHO e (Dec 2020) Coronavirus disease (COVID-19) advice for the public. www.who.int/emergencies/diseases/novel-coronavirus-2019/advice-for-public

WHO f (Dec 2020) COVID-19 Vaccine. www.who.int/emergencies/diseases/novel-coronavirus-2019/covid-19-vaccines

Wiersinga WJ, Rhodes A, Cheng AC, Peacock SJ, Prescott HC. (Aug 2020) Pathophysiology, transmission, diagnosis and treatment of Coronavirus Disease 2019 (COVID-19): A Review. *Journal of the American Medical Association*. 25;324(8):782–793. https://doi:10.1001/jama.2020.12839.

8 COVID-19 and Comorbidities

Sanjay K. Bhadada[*] *and Rimesh Pal*
Department of Endocrinology, Post Graduate Institute of Medical Education and Research, Chandigarh, India
* Corresponding author: Sanjay K. Bhadada,
Email: bhadadask@rediffmail.com

CONTENTS

8.1 INTRODUCTION

The novel coronavirus disease (COVID-19) has been on the rampage ever since its inception in December 2019 in Wuhan, China. COVID-19 has affected more than 190 million people and claimed over 4 million lives globally (1). The clinical manifestations of individuals diagnosed with COVID-19 are predominantly characterized by a cluster of flu-like symptoms; however, asymptomatic cases not infrequent. The majority of cases are mild and do not require hospitalization; only about 5% of patients develop severe to critical disease often complicated by life-threatening complications like acute respiratory distress syndrome (ARDS), acute renal injury (AKI), and multi-organ dysfunction (2). Overall, the global mortality rate is around 2.1% (1).

Although the overall mortality rate does not seem to be too alarming, COVID-19 tends to be more severe and fatal in patients who are elderly (>60 years) and/or with one or more comorbidities (3–6). Initial data from Wuhan suggested that 32% of COVID-19 patients had underlying diseases, namely, cardiovascular disease (CVD), hypertension (HTN), diabetes mellitus, and chronic obstructive pulmonary disease (COPD) (7). Subsequently, data from the rest of the world also confirmed that individuals with comorbidities including CVD, HTN, diabetes mellitus, COPD, CKD, and malignancy are at a higher risk for severe disease and mortality with COVID-19 (8). Hospitalization and intensive care unit (ICU) admissions with COVID-19 have been observed in about 20% of cases with polymorbidity, with mortality rates as high as 14% (9). Overall, composite data suggests that individuals with chronic underlying illness may have severe outcome risks as high as 10-fold as compared to individuals without any comorbidity (10).

The present chapter aims to summarize the impact of various comorbidities on COVID-19 (Figure 8.1). Table 8.1 summarizes the mortality rates in COVID-19 patients with diverse comorbid medical conditions.

DOI: 10.1201/9781003358909-8

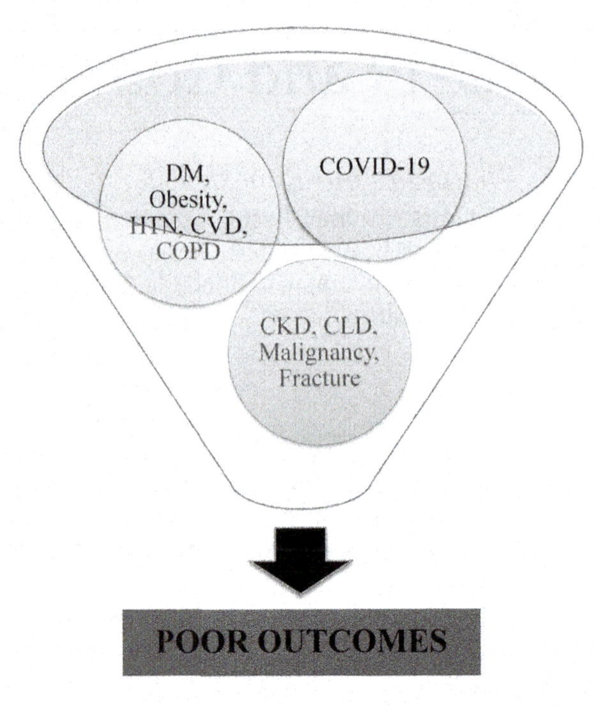

FIGURE 8.1 Schematic diagram representing poor outcomes associated with various comorbidities in patients with Covid-19.

TABLE 8.1
Mortality Rates in COVID-19 Patients with Various Comorbidities

Comorbidity	Mortality (Reference)
Diabetes mellitus	7.3% (14)
Hypertension	6.0% (14)
Obesity	50%* (36)
Cardiovascular disease	10.5% (14)
Renal disease	~ 50% (43)
Chronic liver disease	10.3% (49)
COPD	19.2% (53)
Malignancy	13% (58)
Hip fracture	40% (62)

In COVID-19 patients with severe obesity.

8.2 DIABETES MELLITUS AND COVID-19

Diabetes mellitus has consistently been shown to be associated with severe disease, ICU admissions, need for mechanical ventilation, and mortality in COVID-19 (4–6,11–14). Both type 1 diabetes (T1D) and type 2 diabetes (T2D) portends a poor prognosis in COVID-19 (15). A large systematic review and meta-analysis revealed that after adjustment for confounders, *diabetes mellitus was the best predictor of mortality rate* in an age- and sex-dependent manner [odds ratio (OR) 1.85, 95% confidence interval (CI): 1.35, 2.51, $p<0.001$], followed by COPD and malignancies (16).

Multiple mechanisms have been implicated in the association between diabetes mellitus and COVID-19. Impaired innate immunity in patients with diabetes mellitus might be contributory. The dysregulated adaptive immune system in patients with diabetes mellitus is also believed to increase the risk of cytokine storm when infected by the severe acute respiratory syndrome coronavirus 2 (SARS-CoV-2). Cytokine storm is responsible for ARDS and multiorgan dysfunction in patients with severe/critical COVID-19 (6,12,13).

In addition, the expression of angiotensin converting enzyme 2 (ACE2), the host protein that acts as a receptor for SARS-CoV-2, is altered in diabetes mellitus. Diabetes mellitus, *per se*, reduces the expression of ACE2; pulmonary ACE2 exerts an antiinflammatory and antioxidant effect and protects the lungs against acute lung injury. Reduced expression of ACE2 would increase the chances of lung damage with COVID-19 (6,11).

There seems to be a two-way interaction between diabetes mellitus and COVID-19 wherein diabetes mellitus worsens COVID-19 severity while COVID-19 worsens dysglycemia, thereby setting up a vicious cycle (4). The SARS-CoV-2 has been shown to damage the β-cell of the pancreas (17), thereby leading to new-onset diabetes mellitus or worsening of dysglycemia. Furthermore, multiple drugs used in the management of COVID-19, notably glucocorticoids, remdesivir, and lopinavir-ritonavir aggravates hyperglycemia (4). COVID-19 and diabetes mellitus also provides a perfect milieu for the breakthrough of opportunistic infections like mucormycosis (18,19).

Of all the factors, in-hospital hyperglycemia has consistently been shown to be associated with poor prognosis in COVID-19 patients with diabetes mellitus. Hence, control of hyperglycemia in COVID-19 patients with diabetes mellitus is of utmost importance (12). Insulin remains the mainstay of therapy in critically ill COVID-19 patients with diabetes mellitus (13). In patients with T2D, certain oral antidiabetic drugs like metformin and dipeptidyl peptidase-4 inhibitors have been shown to improve clinical outcomes in COVID-19, and hence may be prescribed if not contraindicated (20). Moreover, hydroxychloroquine, a drug that had initially shown promising results in COVID-19, can also be used for the control of hyperglycemia in patients with COVID-19 (21,22).

Apart from severe and fatal clinical outcomes, it has been hypothesized that diabetes mellitus might increase the risk of re-infections with COVID-19 (23,24). Viral clearance of SARS-CoV-2 is also delayed in patients with diabetes mellitus (25). Thus, considering the plethora of complications observed in patients with diabetes mellitus, it is imperative that they get vaccinated against COVID-19 at the earliest (26).

8.3 HYPERTENSION AND COVID-19

Hypertension represents the most common comorbidity seen in patients with COVID-19 (10,16,27,28). Systematic reviews and meta-analyses have shown that the prevalence of hypertension in patients with COVID-19 has been reported to range from 20.96% to 59.6% (16,28). Like diabetes mellitus, hypertension is associated with severe/critical disease, disease progression, ARDS, ICU care, and mortality (14,27,29,30). Patients with hypertension had a 1.82-fold higher risk for critical COVID-19 (adjusted OR 1.82, 95% CI: 1.19, 2.77, $p=0.005$) and a 2.17-fold higher risk for COVID-19 mortality (adjusted OR 2.17, 95% CI: 1.67, 2.82, $p<0.001$) (30).

The underlying pathophysiological link between COVID-19 and hypertension is probably ACE2. As already mentioned, ACE2 acts as a receptor for SARS-CoV-2 that facilitates entry of the virus into host cells. Certain antihypertensive medications, commonly prescribed in clinical practice, namely angiotensin-converting enzyme inhibitors (ACEi) and angiotensin receptor blockers (ARBs), lead to upregulation of ACE2, thus ultimately supplying the SARS-CoV-2 with a larger number of "anchors" for the infecting host cells. Overexpression of ACE2 by ACEi/ARBs might be responsible for the increased disease severity and poor outcomes seen in patients with hypertension (6,11).

The role of ACE2 in the context of COVID-19 is dichotomous. While overexpression of ACE2 is thought to be counterproductive, ACE2 expression in the lungs is also believed to protect lungs against acute lung injury, as has already been described. However, once SARS-CoV-2 uses ACE2 to gain entry into host cells, the enzyme activity of ACE2 is diminished. Downregulation of ACE2 following viral intrusion results in decreased degradation of angiotensin-II, increased aldosterone secretion, and subsequent increased urinary potassium loss. In fact, early normalization of serum potassium has been proposed to be a predictor of good prognosis in COVID-19 (31). Thus, ACE2 overexpression, while facilitating entry of SARS-CoV-2, is unable to protect against lung injury as the enzyme gets degraded by the virus (6).

Despite the putative causative role of ACEi/ARBs, receipt of ACEIs or ARBs has not been shown to be associated with a higher risk of multivariable-adjusted mortality and severe adverse events among patients with COVID-19 who had either hypertension or multiple comorbidities, supporting the recommendations of medical societies. On the contrary, ACEi and ARBs may be associated with protective benefits in patients with hypertension (32,33).

8.4 OBESITY AND COVID-19

Obesity has emerged as a novel risk factor for severe disease and hospitalization due to COVID-19 (34). A quantitative analyses of 22 studies from 7 countries in North America, Europe, and Asia found that obesity is associated with an increased likelihood of presenting with more severe COVID-19 symptoms (OR 3.03, 95% CI: 1.45, 6.28, p=0.003), developing ARDS (OR 2.89, 95% CI: 1.14, 7.34, p=0.025), requiring hospitalization (OR 1.68, 95% CI: 1.14, 1.59, p=0.001), being admitted to an ICU (OR 1.35, 95% CI: 1.15, 1.65, p=0.001), and undergoing invasive mechanical ventilation (OR 1.76, 95% CI: 1.29, 2.40, p<0.001) compared to non-obese patients (35).

The association between obesity and COVID-19 is inconsistent. The aforementioned systematic review and meta-analysis reported that obese patients had similar likelihoods of death from COVID-19 as non-obese patients (OR 0.96, 95% CI: 0.74, 1.25, p=0.750). On the contrary, another study showed that patients with moderate obesity had a lower mortality rate (13.8%) than patients with normal weight, overweight, or severe obesity (17.6%, 21.7%, and 50%, respectively; p=0.011), suggesting a possible obesity paradox. Logistic regression showed that patients with a body mass index (BMI) ≤29 kg/m^2 (OR 3.64, 95% CI: 1.38, 9.60) and those with a BMI >39 kg/m^2 (OR 10.04, 95% CI: 2.45, 41.09) had a higher risk of mortality than those with a BMI from 29 to 39 kg/m^2 (36).

In another prospective, community-based, cohort study that had included 6,910,695 eligible individuals, the authors reported non-linear associations between BMI and hospital admission and death due to COVID-19, and a linear association between BMI and ICU admission due to COVID-19. Each excess BMI unit above a BMI of 23 kg/m^2 was associated with increased risk of hospital admission [adjusted hazard ratio (HR) 1.05, 95% CI: 1.05, 1.05], ICU admission (adjusted HR 1.10, 95% CI: 1.09, 1.10), and death (adjusted HR 1.04, 95% CI: 1.04, 1.05) (37).

The association between COVID-19 and obesity is multifactorial. Adipose tissue express ACE2; the greater the adipose tissue, the greater the overall ACE2 expression that would act as receptors for SARS-CoV-2 (38). As in diabetes mellitus, even in basal state, obese patients have a higher concentration of several pro-inflammatory cytokines such as TNFα, IL-6 and MCP-1, produced by visceral and subcutaneous adipose tissue. This could again predispose an obese individual to an exaggerated cytokine response in the presence of SARS-CoV-2, manifesting as severe disease and ARDS (39).

8.5 CARDIOVASCULAR DISEASE AND COVID-19

Clinical studies have also reported an association between COVID-19 and cardiovascular disease (CVD). The prevalence of cardiovascular comorbidities in patients with COVID-19 ranges from 4.2% to 15% (2,7). On the contrary, the prevalence of CVD in critically ill COVID-19 patients is as high as 25% (40), emphasizing the fact that pre-existing CVD seems to be linked with worse outcomes in COVID-19 (3,14,41).

Underlying CVD also increases the risk of death with COVID-19. The overall case fatality rate of COVID-19 reported by the Chinese Center for Disease Control and Prevention as of February 11, 2020 was 2.3% (1,023 deaths among 44,672 confirmed cases). The individual case fatality rate of patients with CVD was 10.5% (highest among those with other comorbidities, including COPD, cancer, diabetes mellitus, and hypertension) (2,41).

The precise mechanism behind the association between CVD and COVID-19 is not clear. As in patients with hypertension, the confounding role of ACEi/ARBs might play a role in patients with CVD by altering ACE2 expression (3,42). COVID-19 can also cause cardiovascular disorders, including myocardial injury, arrhythmias, acute coronary syndrome, and venous thromboembolism. In patients with pre-existing CVD, a new cardiovascular event precipitated by COVID-19 can be fatal, hence partly explaining the high mortality rate and poor outcomes in COVD-19 patients with CVD. Moreover, certain drugs that have been used for the management of COVID-19, notably, azithromycin and hydroxychloroquine, can also contribute to cardiovascular morbidity and mortality (41).

8.6 RENAL DISEASES AND COVID-19

Individuals with underlying kidney disease are particularly vulnerable to develop severe COVID-19, characterized by multisystem organ failure, thrombosis, and a heightened inflammatory response. Among 4,264 critically ill adults with COVID-19 admitted to 68 intensive care units across the United States, Flythe et al. found that both non-dialysis-dependent chronic kidney disease (CKD) patients and maintenance dialysis patients had a 28-day in-hospital mortality rate of ~50%. Patients with underlying kidney disease had higher in-hospital mortality than patients without pre-existing kidney disease, with patients receiving maintenance dialysis having the highest risk (43).

The recently published OpenSAFELY project analyzed factors associated with COVID-19 death in 17 million patients. Interestingly, hypertension did not emerge as an independent risk factor for COVID-19 death (adjusted HR 0.89). On the contrary, dialysis (adjusted HR 3.69), organ transplantation (adjusted HR 3.53), and CKD (adjusted HR 2.52 for patients with eGFR <30 mL/min/1.73 m^2) represented the three of the four comorbidities associated with the highest mortality risk from COVID-19. The risk associated with CKD Stages 4 and 5 was higher than the risk associated with diabetes mellitus (adjusted HR range 1.31–1.95, depending upon glycemic control) or chronic heart disease (adjusted HR 1.17) (44). In another recent publication, the Global Burden of Disease collaboration identified that worldwide, CKD is the most prevalent risk factor for severe COVID-19 (45).

The association between underlying renal disease and COVID-19 can partly be explained on the basis of uremia-induced innate immune dysfunction marked by impaired neutrophil, monocyte, and B- and T-cell functions (46,47). Moreover, the plethora of associated comorbidities already prevalent in patients with renal disease might augment the risk of severe disease and mortality from COVID-19.

8.7 LIVER DISEASE AND COVID-19

Patients with chronic liver disease (CLD) and cirrhosis have high rates of hepatic decompensation, acute-on-chronic liver failure, and death from respiratory failure following COVID-19 (48,49). An analysis of 745 patients with CLD and cirrhosis infected with SARS-CoV-2 from 29 countries showed that once infected with COVID-19, there is a stepwise increase in morbidity and mortality with increasing severity of cirrhosis as measured by Child–Pugh (CP) class. There was a stepwise increase in frequency of ICU admission, renal replacement therapy, invasive ventilation, and death across CP classes A, B, and C. There was also an increase in mortality for all patients as they required more intense medical support, with patients classed as CP-C having only a 10% survival once undergoing mechanical ventilation. After adjusting for baseline characteristics, COVID-19-related mortality was significantly associated with the severity of pre-existing liver cirrhosis and the odds ratio for death increased across the stages of cirrhosis: CP-A 1.90 (95% CI: 1.03, 3.52), CP-B 4.14 (95% CI: 2.24, 7.65), and CP-C 9.32 (95% CI: 4.8, 18.08) (50).

Both CLD and cirrhosis are characterized by immune dysregulation. CLD causes impairment of the liver's homeostatic role in the systemic immune response. The molecular patterns from damaged liver cells prompt circulating immune cells to induce systemic inflammation in the form of activated circulating immune cells and increased serum levels of pro-inflammatory cytokines (e.g., TNF and IL-6). Importantly, liver-associated immune dysfunction can increase susceptibility to infections. In this context, it is unsurprising that patients with CLD, especially those with decompensated cirrhosis, are at higher risk of COVID-19-related morbidity and mortality. The combination of SARS-CoV-2 infection and cirrhosis appears to be a lethal duo, likely due to the combination of biological processes characterized by immune dysregulation. Importantly, liver transplantation restores hepatic function in patients with decompensated cirrhosis, thus reducing the risk of COVID-19 mortality to that of the general population (51).

8.8 LUNG DISEASE AND COVID-19

Chronic respiratory diseases like chronic obstructive pulmonary disease (COPD) and asthma are common in the general population. Guan et al. evaluated the risk for serious adverse outcomes in COVID-19 patients in China by stratifying them according to the number of comorbidities. A greater number of comorbidities was correlated with worse clinical outcomes, and COPD patients had the highest HR (2.68) for admission to the ICU, invasive ventilation, or death among patients with various types of chronic underlying diseases (52). A nationwide population-based study in South Korea reported that compared to non-COPD patients relatively greater proportions of patients with COPD received intensive critical care (3.7% vs. 7.1%, p=0.041) and mechanical ventilation (2.4% vs. 5.7%, p= 0.015). Multivariate analyses showed that COPD was a significant independent risk factor for all-cause mortality (OR 1.80, 95% CI: 1.11, 2.93) after adjustment for age, sex, and Charlson Comorbidity Index score (53).

The relationship between COPD and COVID-19 has not been fully elucidated, but several studies have suggested pathways related to ACE2. The expression level of ACE2 is significantly higher in COPD patients than in controls and in current smokers than in former or never smokers (54,55). Moreover, inactivation of the innate immune system and underexpression of pulmonary interferon-β, a cytokine involved in the defense against coronavirus, are observed in patients with COPD (3). Furthermore, patients with COPD also have poor pulmonary reserve that further portends a poor prognosis in COVID-19.

The data regarding bronchial asthma and COVID-19 is contrasting. Most of the available data do not suggest an association between asthma and severe COVID-19. Wang et al. conducted a meta-analysis with data from four studies, including 744 people with asthma and 8151 people without asthma. They found no significant association between asthma and mortality with COVID-19 (56). In fact, one study detected a statistically significant reduced odds of death with COVID-19 in people with asthma (57).

8.9 MALIGNANCY AND COVID-19

Patients with underlying malignancies tend to have poor COVID-19 outcomes. Out of 928 COVID-19 patients with underlying malignancies, 121 had died equating to a 30-day mortality rate of 13% (58). Among cancer patients, some appear to be more vulnerable than others when it comes to COVID-19. A multicenter study showed that patients with hematologic, lung, or other metastatic malignancies and those who had undergone surgical procedures were more vulnerable to serious COVID-19 illness (59). In another study, patients with hematological malignancies (leukemia, lymphoma, and myeloma) were found to have more severe COVID-19 trajectory compared with patients with solid organ tumors (OR 1.57, 95% CI: 1.15, 215, p<0.0043) (60).

One particular area of concern is cardiovascular complications observed in cancer patients with COVID-19. Cardiovascular complications are a common feature of COVID-19 infection. Likewise, it is well known that malignancy and its treatment promote various cardiovascular complications and coagulopathies; indeed, both cancer and COVID-19 infection in isolation fulfill Virchow's triad for thrombosis by promoting blood stasis, vascular wall damage, and hypercoagulation; so it follows that COVID-19 infection further exacerbates these complications in cancer patients. One prospective cohort study provides strong evidence in favor of this hypothesis, indicating that COVID-positive cancer patients with cardiovascular disease have an odds ratio for mortality of 2.32 (95% CI: 1.47, 3.64) compared to COVID-positive cancer patients without comorbidities (61).

8.10 FRACTURES AND COVID-19

The co-existence of hip fracture and COVID-19 is associated with an exceptionally high mortality rate. Hitherto available studies suggest that almost 30–40% of COVID-19 patients with hip fractures die (62–64). This raises special concerns among the geriatric population, since they are among individuals at high-risk of contracting COVID-19 or experiencing osteoporotic fractures (65). It has also been found that the 30-day mortality of fracture patients during the pandemic was up to two times higher than that observed in the pre-pandemic situation (66).

REFERENCES

1. Weekly Operational Update on COVID-19 20 July 2021 [Internet] [cited 2021 Jul 20]. Available from: www.who.int/publications/m/item/weekly-operational-update-on-covid-19---20-july-2021
2. Wu Z, McGoogan JM. Characteristics of and Important Lessons from the Coronavirus Disease 2019 (COVID-19) outbreak in China: Summary of a Report of 72 314 Cases from the Chinese center for disease control and prevention. J Am Med Assoc. 2020;323:1239.
3. Pal R, Bhadada SK. COVID-19 and non-communicable diseases. Postgrad Med J. 2020;96:429–30.
4. Pal R, Bhadada SK. COVID-19 and diabetes mellitus: An unholy interaction of two pandemics. Diabetes Metab Syndr Clin Res Rev. 2020;14:513–17.
5. Pal R, Bhadada SK. Managing common endocrine disorders amid COVID-19 pandemic. Diabetes Metab Syndr Clin Res Rev. 2020;14:767–71.
6. Pal R, Bhansali A. COVID-19, diabetes mellitus and ACE2: The conundrum. Diabetes Res Clin Pract. 2020;162:108132.
7. Huang C, Wang Y, Li X, Ren L, Zhao J, Hu Y, et al. Clinical features of patients infected with 2019 novel coronavirus in Wuhan, China. The Lancet. 2020;395:497–506.
8. CDC COVID-19 Response Team, CDC COVID-19 Response Team, Chow N, Fleming-Dutra K, Gierke R, Hall A, et al. Preliminary estimates of the prevalence of selected underlying health conditions among patients with coronavirus disease 2019—United States, February 12–March 28, 2020. MMWR Morb Mortal Wkly Rep. 2020;69:382–6.
9. Rodriguez-Morales AJ, Cardona-Ospina JA, Gutiérrez-Ocampo E, Villamizar-Peña R, Holguin-Rivera Y, Escalera-Antezana JP, et al. Clinical, laboratory and imaging features of COVID-19: A systematic review and meta-analysis. Travel Med Infect Dis . 2020;34:101623.
10. Bajgain KT, Badal S, Bajgain BB, Santana MJ. Prevalence of comorbidities among individuals with COVID-19: A rapid review of current literature. Am J Infect Control. 2021;49:238–46.
11. Pal R, Bhadada SK. Should anti-diabetic medications be reconsidered amid COVID-19 pandemic? Diabetes Res Clin Pract. 2020;163:108146.
12. Apicella M, Campopiano MC, Mantuano M, Mazoni L, Coppelli A, Del Prato S. COVID-19 in people with diabetes: Understanding the reasons for worse outcomes. Lancet Diabetes Endocrinol [Internet]. 2020 [cited 2020 Jul 28]; Available from: https://linkinghub.elsevier.com/retrieve/pii/S2213858720302382

13. Lim S, Bae JH, Kwon H-S, Nauck MA. COVID-19 and diabetes mellitus: From pathophysiology to clinical management. Nat Rev Endocrinol. 2021;17:11–30.
14. Riddle MC, Buse JB, Franks PW, Knowler WC, Ratner RE, Selvin E, et al. COVID-19 in people with diabetes: Urgently needed lessons from early reports. Diabetes Care. 2020;43:1378–81.
15. Cariou B, Hadjadj S, Wargny M, Pichelin M, Al-Salameh A, Allix I, et al. Phenotypic characteristics and prognosis of inpatients with COVID-19 and diabetes: The CORONADO study. Diabetologia. 2020;63:1500–15.
16. Corona G, Pizzocaro A, Vena W, Rastrelli G, Semeraro F, Isidori AM, et al. Diabetes is most important cause for mortality in COVID-19 hospitalized patients: Systematic review and meta-analysis. Rev Endocr Metab Disord [Internet]. 2021 [cited 2021 Apr 23]; Available from: http://link.springer.com/10.1007/s11154-021-09630-8
17. Müller JA, Groß R, Conzelmann C, Krüger J, Merle U, Steinhart J, et al. SARS-CoV-2 infects and replicates in cells of the human endocrine and exocrine pancreas. Nat Metab. 2021;3:149–65.
18. Pal R, Singh B, Bhadada SK, Banerjee M, Bhogal RS, Hage N, et al. COVID-19-associated mucormycosis: An updated systematic review of literature. Mycoses. 2021;myc.13338.
19. Banerjee M, Pal R, Bhadada SK. Intercepting the deadly trinity of mucormycosis, diabetes and COVID-19 in India. Postgrad Med J. 2021 Jun 8;postgradmedj-2021-140537.
20. Pal R, Banerjee M, Mukherjee S, Bhogal RS, Kaur A, Bhadada SK. Dipeptidyl peptidase-4 inhibitor use and mortality in COVID-19 patients with diabetes mellitus: An updated systematic review and meta-analysis. Ther Adv Endocrinol Metab. 2021;12:204201882199648.
21. Pal R, Banerjee M, Kumar A, Bhadada S. Glycemic efficacy and safety of hydroxychloroquine in type 2 diabetes mellitus: A systematic review and meta-analysis of relevance amid the COVID-19 pandemic. Int J Noncommunicable Dis. 2020;5:184.
22. Pal R, Bhasin MK, Bhadada SK. COVID-19 and Type 2 Diabetes Mellitus: Hydroxychloroquine May Be The Holy Grail. Infect Disord - Drug Targets [Internet]. 2020 [cited 2020 Oct 27];20. Available from: www.eurekaselect.com/186842/article
23. Pal R, Banerjee M. Are People With Uncontrolled Diabetes Mellitus At High Risk of Reinfections with COVID-19? Prim Care Diabetes [Internet]. 2020 [cited 2020 Aug 11]; Available from: https://linkinghub.elsevier.com/retrieve/pii/S1751991820302382
24. Pal R, Sachdeva N, Mukherjee S, Suri V, Zohmangaihi D, Ram S, et al. Impaired anti-SARS-CoV-2 antibody response in non-severe COVID-19 patients with diabetes mellitus: A preliminary report. Diabetes Metab Syndr Clin Res Rev. 2021;15:193–6.
25. Chen X, Hu W, Ling J, Mo P, Zhang Y, Jiang Q, et al. Hypertension and Diabetes Delay the Viral Clearance in COVID-19 Patients [Internet]. 2020 [cited 2020 Aug 1]. Available from: http://medrxiv.org/lookup/doi/10.1101/2020.03.22.20040774
26. Pal R, Bhadada SK, Misra A. COVID-19 vaccination in patients with diabetes mellitus: Current concepts, uncertainties and challenges. Diabetes Metab Syndr Clin Res Rev. 2021;15:505–8.
27. Honardoost M, Janani L, Aghili R, Emami Z, Khamseh ME. The association between presence of comorbidities and COVID-19 severity: A systematic review and meta-analysis. Cerebrovasc Dis. 2021;50:132–40.
28. Atkins JL, Masoli JAH, Delgado J, Pilling LC, Kuo C-L, Kuchel GA, et al. Preexisting comorbidities predicting COVID-19 and mortality in the UK biobank community cohort. Newman AB, editor. J Gerontol Ser A. 2020;75:2224–30.
29. Pranata R, Lim MA, Huang I, Raharjo SB, Lukito AA. Hypertension is associated with increased mortality and severity of disease in COVID-19 pneumonia: A systematic review, meta-analysis and meta-regression. J Renin Angiotensin Aldosterone Syst. 2020;21:147032032092689.
30. Du Y, Zhou N, Zha W, Lv Y. Hypertension is a clinically important risk factor for critical illness and mortality in COVID-19: A meta-analysis. Nutr Metab Cardiovasc Dis. 2021;31:745–55.
31. Chen D, Li X, Song Q, Hu C, Su F, Dai J, et al. Assessment of hypokalemia and clinical characteristics in patients with coronavirus disease 2019 in Wenzhou, China. JAMA Netw Open. 2020;3:e2011122.
32. Baral R, Tsampasian V, Debski M, Moran B, Garg P, Clark A, et al. Association between renin-angiotensin-aldosterone system inhibitors and clinical outcomes in patients with COVID-19: A systematic review and meta-analysis. JAMA Netw Open. 2021;4:e213594.

33. Lee HW, Yoon C-H, Jang EJ, Lee C-H. Renin-angiotensin system blocker and outcomes of COVID-19: A systematic review and meta-analysis. Thorax. 2021;76:479–86.
34. Mohammad S, Aziz R, Al Mahri S, Malik SS, Haji E, Khan AH, et al. Obesity and COVID-19: What makes obese host so vulnerable? Immun Ageing. 2021;18:1.
35. Zhang X, Lewis AM, Moley JR, Brestoff JR. A systematic review and meta-analysis of obesity and COVID-19 outcomes. Sci Rep. 2021;11:7193.
36. Dana R, Bannay A, Bourst P, Ziegler C, Losser M-R, Gibot S, et al. Obesity and mortality in critically ill COVID-19 patients with respiratory failure. Int J Obes [Internet]. 2021 [cited 2021 Jul 22]; Available from: www.nature.com/articles/s41366-021-00872-9
37. Gao M, Piernas C, Astbury NM, Hippisley-Cox J, O'Rahilly S, Aveyard P, et al. Associations between body-mass index and COVID-19 severity in 6·9 million people in England: A prospective, community-based, cohort study. Lancet Diabetes Endocrinol. 2021;9:350–9.
38. Kassir R. Risk of COVID-19 for patients with obesity. Obes Rev [Internet]. 2020 [cited 2020 Apr 22]; Available from: http://doi.wiley.com/10.1111/obr.13034
39. Pal R, Banerjee M. COVID-19 and the endocrine system: Exploring the unexplored. J Endocrinol Invest. 2020;43:1027–31.
40. Wang D, Hu B, Hu C, Zhu F, Liu X, Zhang J, et al. Clinical characteristics of 138 hospitalized patients with 2019 novel coronavirus–infected pneumonia in Wuhan, China. J Am Med Assoc. 2020;323:1061.
41. Nishiga M, Wang DW, Han Y, Lewis DB, Wu JC. COVID-19 and cardiovascular disease: From basic mechanisms to clinical perspectives. Nat Rev Cardiol. 2020;17:543–58.
42. Ejaz H, Alsrhani A, Zafar A, Javed H, Junaid K, Abdalla AE, et al. COVID-19 and comorbidities: Deleterious impact on infected patients. J Infect Public Health. 2020;13:1833–9.
43. Flythe JE, Assimon MM, Tugman MJ, Chang EH, Gupta S, Shah J, et al. Characteristics and outcomes of individuals with pre-existing kidney disease and COVID-19 admitted to intensive care units in the United States. Am J Kidney Dis. 2021;77:190–203.e1.
44. Williamson EJ, Walker AJ, Bhaskaran K, Bacon S, Bates C, Morton CE, et al. Factors associated with COVID-19-related death using OpenSAFELY. Nature. 2020;584:430–6.
45. Clark A, Jit M, Warren-Gash C, Guthrie B, Wang HHX, Mercer SW, et al. Global, regional, and national estimates of the population at increased risk of severe COVID-19 due to underlying health conditions in 2020: A modelling study. Lancet Glob Health. 2020;8:e1003–17.
46. Ando M, Shibuya A, Tsuchiya K, Akiba T, Nitta K. Reduced expression of Toll-like receptor 4 contributes to impaired cytokine response of monocytes in uremic patients. Kidney Int. 2006;70:358–62.
47. Syed-Ahmed M, Narayanan M. Immune dysfunction and risk of infection in chronic kidney disease. Adv Chronic Kidney Dis. 2019;26:8–15.
48. Marjot T, Webb GJ, Barritt AS, Moon AM, Stamataki Z, Wong VW, et al. COVID-19 and liver disease: Mechanistic and clinical perspectives. Nat Rev Gastroenterol Hepatol. 2021;18:348–64.
49. Wang Q, Davis PB, Xu R. COVID-19 risk, disparities and outcomes in patients with chronic liver disease in the United States. E Clinical Medicine. 2021;31:100688.
50. Marjot T, Moon AM, Cook JA, Abd-Elsalam S, Aloman C, Armstrong MJ, et al. Outcomes following SARS-CoV-2 infection in patients with chronic liver disease: An international registry study. J Hepatol. 2021;74:567–77.
51. Martinez MA, Franco S. Impact of COVID-19 in Liver Disease Progression. Hepatol Commun. 2021;5:1138–50.
52. Guan W, Liang W, Zhao Y, Liang H, Chen Z, Li Y, et al. Comorbidity and its impact on 1590 patients with Covid-19 in China: A Nationwide Analysis. Eur Respir J. 2020;2000547.
53. Lee SC, Son KJ, Han CH, Park SC, Jung JY. Impact of COPD on COVID-19 prognosis: A nationwide population-based study in South Korea. Sci Rep. 2021;11:3735.
54. Cai G, Bossé Y, Xiao F, Kheradmand F, Amos CI. Tobacco smoking increases the lung gene expression of ACE2, the receptor of SARS-CoV-2. Am J Respir Crit Care Med. 2020;201:1557–9.
55. Li M, Li L, Zhang Y, Wang X. An investigation of the expression of 2019 novel coronavirus cell receptor gene ACE2 in a wide variety of human tissues [internet]. In Review; 2020 [cited 2020 Apr 1]. Available from: www.researchsquare.com/article/rs-15309/v2
56. Wang Y, Chen J, Chen W, Liu L, Dong M, Ji J, et al. Does asthma increase the mortality of patients with COVID-19?: A systematic review and meta-analysis. Int Arch Allergy Immunol. 2021;182:76–82.

57. Santos MM, Lucena EES, Lima KC, Brito AAC, Bay MB, Bonfada D. Survival and predictors of deaths of patients hospitalised due to COVID-19 from a retrospective and multicentre cohort study in Brazil. Epidemiol Infect. 2020;148:e198.
58. Kuderer NM, Choueiri TK, Shah DP, Shyr Y, Rubinstein SM, Rivera DR, et al. Clinical impact of COVID-19 on patients with cancer (CCC19): A cohort study. The Lancet. 2020;395:1907–18.
59. Dai M, Liu D, Liu M, Zhou F, Li G, Chen Z, et al. Patients with cancer appear more vulnerable to SARS-CoV-2: A multicenter study during the COVID-19 Outbreak. Cancer Discov. 2020;10:783–91.
60. Lee LYW, Cazier J-B, Starkey T, Briggs SEW, Arnold R, Bisht V, et al. COVID-19 prevalence and mortality in patients with cancer and the effect of primary tumour subtype and patient demographics: A prospective cohort study. Lancet Oncol. 2020;21:1309–16.
61. Lee LY, Cazier J-B, Angelis V, Arnold R, Bisht V, Campton NA, et al. COVID-19 mortality in patients with cancer on chemotherapy or other anticancer treatments: A prospective cohort study. Lancet Lond Engl. 2020;395:1919–26.
62. Biarnés-Suñé A, Solà-Enríquez B, González Posada MÁ, Teixidor-Serra J, García-Sánchez Y, Manrique Muñóz S. Impacto de la pandemia COVID-19 en la mortalidad del paciente anciano con fractura de cadera. Rev Esp Anestesiol Reanim. 2021;68:65–72.
63. De C, Wignall A, Giannoudis V, Jimenez A, Sturdee S, Aderinto J, et al. Peri-operative outcomes and predictors of mortality in COVID-19 positive patients with hip fractures: A multicentre study in the UK. Indian J Orthop. 2020;54:386–96.
64. Lim MA, Pranata R. Coronavirus disease 2019 (COVID-19) markedly increased mortality in patients with hip fracture—A systematic review and meta-analysis. J Clin Orthop Trauma. 2021;12:187–93.
65. Lim MA, Kurniawan AA. Dreadful consequences of sarcopenia and osteoporosis due to COVID-19 containment. Geriatr Orthop Surg Rehabil. 2021;12:215145932199274.
66. Lim MA, Mulyadi Ridia KG, Pranata R. Epidemiological pattern of orthopaedic fracture during the COVID-19 pandemic: A systematic review and meta-analysis. J Clin Orthop Trauma. 2021;16:16–23.

9 Immunology of COVID-19

Prasenjit Mitra[1], Taru Goyal[1], Malavika Lingeswaran[1], Praveen Sharma[1], and Sanjeev Misra[2]

[1]Department of Biochemistry, All India Institute of Medical Science, Jodhpur, Rajasthan, India

[2]Department of Surgical Oncology, All India Institute of Medical Science, Jodhpur, Rajasthan, India

CONTENTS

9.1 INTRODUCTION

On December 31, 2019, an epidemic of severe pneumonia of unknown cause that started in Wuhan, China has thrashed the whole globe today. The causative agent was a newly identified coronavirus of zoonotic origin. The disease was initially named novel coronavirus disease (nCoV 2019). The earlier conditions implicated by coronavirus include severe acute respiratory syndrome (SARS) (caused by SARS-CoV-1) during 2002–2003 and the Middle East Respiratory Syndrome (MERS) (caused by MERS-CoV) in 2011. Later, the name nCOV 2019 was replaced with the name Severe Acute Respiratory Syndrome Coronavirus 2 (SARS-CoV-2) and coronavirus disease 2019 (COVID-19), respectively (Chang et al., 2020). Owing to its highly contagious nature and rapid spread across the world, leading to overwhelming morbidity and mortality rate, the World Health Organization (WHO) declared it a pandemic in March 2020 (WHO, 2020).

Like SARS and MERS, patients with COVID-19 suffer from cough and fever with 8–19% of patients showing progression to respiratory distress and Acute Respiratory Distress Syndrome (ARDS) ultimately leading to multiorgan failure (Mitra, Misra, and Sharma, 2020). Older adults and individuals with underlying comorbidities like diabetes mellitus, hypertension, and cardiovascular diseases are more vulnerable to COVID-19 infection, which is the cause of varied mortality rates across various countries (Mitra et al., 2020). However, a proportion of COVID-19 patients may develop respiratory symptoms that require supplemental oxygen and mechanical ventilation (Mitra, Misra, and Sharma, 2021). The clinical presentation of COVID-19 patients requiring intensive care support is severe pneumonia characterized with fever, lymphopenia, increased C-reactive protein, serum ferritin, pro-inflammatory cytokines, and D-dimers (J. Zhang et al., 2020; W. Zhang et al., 2020). Based on the clinical picture, disease pathology, and homology to SARS and MERS, it could

DOI: 10.1201/9781003358909-9

be speculated that the hyperinflammatory response in COVID-19 patients is associated with the increased mortality and morbidity.

9.2 TYPES OF CORONAVIRUSES/THE CORONAVIRUS FAMILY

Coronaviruses (CoVs) are respiratory illness-causing viruses belonging to the *Coronaviridae* family, which includes many species that may infect various wild animals, some of which may also affect humans. CoVs are pleomorphic, enveloped, single-stranded RNA viruses with a crown-shaped peplomer that forms the name coronavirus (Mitra, Misra, and Sharma, 2020). Four categories of coronaviruses, alpha, beta, gamma, and delta, are known. However, human coronaviruses (i.e., HCoV-229E and NL63) belong to the alpha genera while HCoV-OC43, SARS-CoV, HCoV-HKU1, MERS-CoV, and SARS-CoV-2 belong to beta genera of CoVs. In contrast to SARS-CoV and MERS-CoV, which also belong to the beta family of coronaviruses and are associated with severe respiratory syndromes, SARS-CoV-2 has an exceptionally higher transmission rate and lethality (García, 2020).

The genome of SARS-CoV-2 contains at least ten open reading frames. About 75% of the RNA constitutes the ORF-1 a/b that is translated to a multitude of non-structural proteins such as pp1a and pp1b. These proteins are the viral replicase transcriptase complex that plays an important role in viral replication and transcription. Another 25% of the genome at the 3' end codes for the virus's main structural proteins that do not have any role in viral replication. The membrane of the virus is composed of four main structural components: the spike (S), envelope (E), membrane (M), and nucleocapsid (N) protein inside the virion that covers the RNA. The presence of the receptor-binding domain (RBD) is essential for the S protein-mediated fusion of the virus into the host cell membrane. The presence of epitopes for T and B cells helps to produce neutralizing antibodies (Fung and Liu, 2019). Because of 75% sequence homology of the RBD between SARS and SARS-CoV-2, both the viruses utilize similar mechanisms to enter the host cell membrane.

Angiotensin-converting enzyme 2 (ACE2) receptor, which encodes a homolog of ACE is responsible for cleavage of C terminal residues, facilitates the binding site of viral anchoring or spike (S) proteins of SARS CoV and SARS-CoV-2 (Letko, Marzi and Munster, 2020). The primary location of these receptors is in lung and gastrointestinal epithelia. Binding of RBD and ACE2 initiates conformational alterations on the S protein leading to cleavage of S1 and S2, facilitating the virus's fusion to the host cell membrane and thus permitting entry of the viral RNA to the target cell (Shang et al., 2020). Unlike SARS-CoV, the entry of SARS-CoV-2 is facilitated by an enzyme Furin present in the host cell (Walls et al., 2020). The binding affinity of S protein and ACE2 defines the pathogenesis of the infection. Although some studies suggest that the binding of SARS-CoV-2 is not as strong as SARS-CoV (Lu et al., 2020; Wan et al., 2020), other studies show a significantly higher binding affinity of SARS-CoV-2 RBD with ACE2 than of SARS-CoV (Tai et al., 2020). While SARS-CoV and SARS-CoV-2 utilize ACE2 as the receptor for host entry, MERS-CoV utilizes dipeptidyl peptidase-4 (DPP4) as a receptor.

9.3 PATHOGENESIS OF SARS-CoV-2

Among all the coronaviruses, SARS-CoV, MERS, and SARS-CoV-2 infect the upper respiratory tract and may also replicate to lower respiratory tract triggering severe pneumonia. Individuals with COVID-19 infection exhibit a wide range of clinical manifestations with heterogeneous symptoms varying from asymptomatic infections to mild, moderate, and severe infections requiring hospitalization and respiratory support. Infection may result in death in some cases. Risk factors like gender, aging, and underlying comorbidities like diabetes, cardiovascular disease, and hypertension may further implicate the disease's severity. Similar to other respiratory coronaviruses, the primary mode of transmission of SARS-CoV-2 is via respiratory droplets (from coughing and sneezing) or by contaminated fomites, whereby it enters the nasal system by inhalation and starts replicating. However, there is a possibility of fecal and oral transmission, although unproven as of now. Once encountered with the virus, it takes 4–5 days of incubation for symptom onset with an average of symptom onset within 11.5 days. Unlike SARS-CoV where viral load peaks at 10 days after symptom onset, the viral load of SARS-CoV-2 peaks within 5–6 days of symptom onset and may progress to ARDS (average of 8–10 severe cases). The sequence homology of SARS-CoV-2 with SARS-CoV and MERS implicates that the pathogenesis of COVID-19 may utilize a similar immune evasion strategy. As the virus enters the host cell via ACE2 receptor binding to the S protein, it triggers the endocytosis of SARS-CoV-2 exposing it to the endosomal proteases, which further processes the viral package release to host cytoplasm. The main targets of the SARS-CoV-2 virus are the bronchial epithelial cells, alveolar epithelial cells, vascular endothelial cells, and macrophages in the lung and pneumocytes of the alveolar epithelium, all of which express ACE2 receptor (Sungnak et al., 2020; Hamming et al., 2004). After binding to the receptor, it induces an aggressive inflammatory response, autophagy, basal membrane detachment, and inhibition of ACE2 expression, which in turn cause binding of angiotensin II to the AT1aR receptor, resulting in acute lung injury (Sims et al., 2005). Inhibition of ACE2 expression results in renin-angiotensin system dysfunction, further influencing the blood pressure, fluid/electrolyte balance, and augment inflammation and vascular permeability in the airways (Kuba et al., 2005). The pathogenesis of SARS-CoV-2 involves abnormal host response and overreactive immune system resulting in increased production of inflammatory cytokines, along with increased secretion of chemokines and free radicals responsible for severe insults to lungs and other organs resulting in multiorgan failure and finally death in severe cases (Channappanavar and Perlman, 2017; Collange et al., 2020). The disease severity thus depends on both the viral infection as well as the host response.

9.4 IMMUNE RESPONSE AGAINST CORONAVIRUS

SARS-CoV-2 infection, initially in non-severe cases, activates the local immune response, thus recruiting macrophages, monocytes at the site of infection, inducing secretion of cytokines, and priming the T and B lymphocytes for infection. However, in a proportion of cases, it may result in immune dysregulation and hyperinflammation, resulting in severe lung injury and systemic effects.

9.4.1 Innate Immune Response in COVID-19

Generally, viruses provoke several host immune responses via increased production of inflammatory cytokines, maturation of the antigen presenting cells (macrophages and dendritic cells) as well as by upregulation of type I interferons (IFNs), which plays an essential role in curbing viral spread.

An innate immune response is mediated via a set of cellular and chemical mediators that helps in recognition and restraining the invading pathogen following stimulation of the adaptive immune response. Innate immunity is activated by association of the pathogen-associated molecular pattern (PAMPs) with the respective pattern recognition receptors (PRRs). This leads to transcription factor activation and subsequent production of IFNs and cytokines via IFN-α/β receptor (IFNAR). These molecules further stimulate other interferon-stimulated genes (ISGs), thereby restricting viral replication and helping in viral clearance from the host (Taefehshokr et al., 2020).

The major endosomal RNA PRRs for the recognition of RNA viruses like SARS-CoV, SARS-CoV-2, and MERS include toll-like receptors (TLR-) 3 and 7, a retinoic acid-inducible gene I (RIG-I), and melanoma differentiation-associated protein 5 (MDA5). Activation of TLR3/7 and RIG-I/MDA5 brings about translocation and expression of NFκβ and IRF3, respectively, which induces increased expression of type 1 IFN and other pro-inflammatory cytokines (Alunno et al., 2019, de Wit et al., 2016).

SARS-CoV-2 in a similar fashion as SARS-CoV and MERS-CoV utilizes multiple mechanisms to alter the host's antiviral response. The three primary mechanisms by which SARS-CoV-2 abandons the host innate response include:

1) Replication in a double-membrane vesicle, which avoids the recognition by respective PRRs.
2) Inhibition of the IFN-mediated activation of the transcription factor required for cytokine production.
3) Downregulation of the IFNAR-mediated signaling cascade.

Since SARS-CoV-2 virus replicates in a double membranous vesicle, it quickly escapes recognition via PRRs. Furthermore, the structural proteins (i.e., N proteins) activates RIG-I's ubiquitination, consequently leading to its degradation. NSP1 protein in SARS-CoV which degrades the IFN- β mRNA and ORF6 which stops the transportation of IRF3 and STAT1 to the nucleus resulting in inhibition of IFN signalling, M protein of SARS-CoV-2 in similar fashion inhibits the production of type I and III IFNs via RIG-I/MDA-5-MAVS signalling pathway (Spiegel et al., 2005; Frieman et al., 2007). Further, they induce downregulation of the downstream cascade and signalling pathways, thereby inactivating cytokine production. The ubiquitination of RIG-I and MDA-5 results in inactivation of mitochondrial antiviral signalling protein (MAVS). Activated MAVS plays a crucial role in the activation and nuclear translocation of IRF3. Inactivated IRF3 and TRAF 3/6 leads to inhibition of the central NFκBand STAT3 signalling pathways, resulting in the inactivation of type-1IFN signalling, thereby delaying the pathogen clearance and progressing to fatal outcomes. A significant delay in the secretion of IFNs has also been detected in patients with COVID-19 (Kindler, Thiel, and Weber, 2016; Lu et al., 2010).

Macrophages, the primary managers of innate immunity, play a pivotal role in inflammation and the pathology of COVID-19. Studies have reported the existence of distinct pro-fibrotic ($SPP1^{high}$) and inflammatory ($FAPB4^{+}$) macrophage population in the lungs of COVID-19 patients (Liao et al., 2020). The primary mechanism of how SARS-CoV-2 affects the macrophages is still unclear. Still, preliminary findings from other CoVs suggest that inhibition of cytokine-induced MHC-II expression in macrophages could be a potential player. Further, epigenetic alteration of histone deacetylase 2 (HDAC2), which plays an essential role in MHC-II expression and NSP-5 mediated production, has been attributed to the inhibition of antigen-presentation property of macrophages seen in COVID-19. Studies have reported interaction between Nsp13 and ORF8 of SARS-CoV-2 with Golgi trafficking system, which hinders the transport of MHC to the cell surface (Gordon et al., 2020; Keskinen et al., 1997; Kong et al., 2009). Recently, it has been observed that ORF8 protein of SARS-CoV-2 directly mediates autolysosomal degradation of MHC-I molecules (Y. Zhang et al., 2020). Thus, this characteristic of SARS-CoV-2 to infect macrophage and alter its activity is suggestive of its potential mechanism in its evasion of the innate and systemic immune response.

Both the resident and monocyte-derived dendritic cells take up the viral antigens, transport via lymph nodes, and present them to the adaptive immune cells to activate cell-mediated and humoral immune responses. A plethora of studies have reported SARS-CoV-2 infects the dendritic cells (DCs), which is responsible for insignificant antiviral cytokine production and upregulated expression of inflammatory chemokines like MIP-1α (R. Zhou et al., 2020; Yang et al., 2020; Sanchez-Cerrillo et al., 2020). Moreover, SARS-CoV-2 target DCs directly as the number and frequency of DCs were significantly reduced, ensuing impaired T cell activation in COVID-19 patients (Xu et al., 2020). Owing to a substantial loss in dendritic cell function due to SARS-CoV-2 infection significant delay in immune response in the host has been observed. Although DCs infection plays a vital role in SARS-CoV-2 mediated cytokine storm and T cell response, the exact mechanism by which it modulates the DCs are still under exploration.

Furthermore, in later stages, virus-infected cells encounter cell death releasing viral particles, which further triggers the innate immune cells to release a plethora of pro-inflammatory cytokines.

Thus, the adaptive immune cells come into action to process the pathogen. However, in a similar fashion as SARS-CoV, SARS-CoV-2 escapes the defense mediated by T cells also via induction of T cell apoptosis (Shah et al., 2020). The inactivation of the innate immune response in the infected epithelial cells and immune cells results in the propagation of SARS-CoV-2 in the host body without triggering any other immune response.

9.4.2 Cellular Immune Response/Adaptive Immunity in COVID-19

A transition from innate to adaptive immune response plays a critical role in the progression of any virus in the host. An active adaptive immune response is as responsible for clearance of the viral infection as the innate response. Further, the host's adaptive immune response helps prevent an encounter with the same virus by producing memory cells against the virus. Activation of adaptive immune response is crucial for the regulation of immune response to prevent overstimulation and host injury. Thus, a delicate balance and coordination between the antigen presentation by the innate immune cells and processing by the adaptive immune cells is the pre-requisite for preventing any viral infection. T lymphocyte, especially the CD4+ T cells, is protective as they mediate B cell-dependent antibody formation. In contrast, CD8+ cytotoxic T cells help in scavenging the infected cells from the host.

The adaptive immune response is conferred by various cell subset once they encounter the viral antigen presented by the antigen presenting cells (APCs). Like SARS-CoV and MERS-CoV, COVID-19 patients show variation in the white blood cell count ranging from a significant lymphopenia with a moderate decrease in B and NK cells and a significant decrease in total T cell count (O'Connell and Aldhamen, 2020). Among T lymphocyte subsets, the cytotoxic T cells (cytT), T helper cells (Th), and T regulatory cells (Treg) were lower in all cases. But a significant reduction in Th cells were observed in severe cases compared to non-severe. Abundant studies have reported a significant reduction in the CD4+ T cell and CD8+ T cells in severe cases. Variation in the functional ability of the APCs and inefficient dendritic cell relocation may reduce the antigen-specific T cells in SARS-CoV-2 patients (Yoshikawa et al., 2009; Zhao et al., 2009).

Additionally, while T cells are diminished, the residual T cells will adopt a hyperactivated or exhausted state as confirmed by the presence of exhaustion markers (i.e., Tim-3 and PD-1 co-expression) in severe cases (Zheng et al., 2020; Diao et al., 2020). The combination of depleted yet hyperactivated T cells was not uniformly seen in all patients, but it is a peculiar finding in severe cases.

A significant decrease in the IFN-γ, produced by the CD4+ T cells, further draws a parallel with the decrease in CD4+ T cells in SARS-CoV-2 infection. Along with reduced IFNγ, CD4+ T cell-mediated synthesis of antiviral cytokines such as granzyme B and TNFκ is also reduced during SARS-CoV-2 infection (Zheng et al., 2020). On the contrary, a significant rise in Th1 cytokines could be attributed to the NSP9 and NSP10-mediated upregulation of NFκB signaling (J. Li et al., 2020).

In contrast to the CD4+ T cells, the CD8+ T cells in SARS-CoV-2 infection appear in a hyperactive and exhausted state. A significant increase in granzyme B, perforin, and increased expression of PD-1, Tim-3, TIGIT, HLA-DR, CD38, CD25, and NKG2A on CD8+ T cells draws a parallel to the presence of hyperactivated as well as exhausted T cells (Westmeier et al., 2020).

Moreover, a concomitant decrease in the T lymphocytes could be attributed to the increased apoptosis of these cell in severe cases. Studies have reported enhanced p53 signaling is known to induce apoptosis, which may be responsible for the unexplained lymphopenia seen in SARS-CoV-2 cases (Xiong et al., 2020). Furthermore, a decrease in T cell significantly correlates with the increase in the cytokine levels. This increase in the cytokine could be explained by SARS-CoV-2-mediated predominance of inflammatory macrophages producing many inflammatory cytokines that further destroy the lymphocytes.

In addition to lymphocytes, among other immune cells, neutrophils also play a critical role in the pathogenesis of COVID-19. There is a significant increase in the neutrophil count as well as

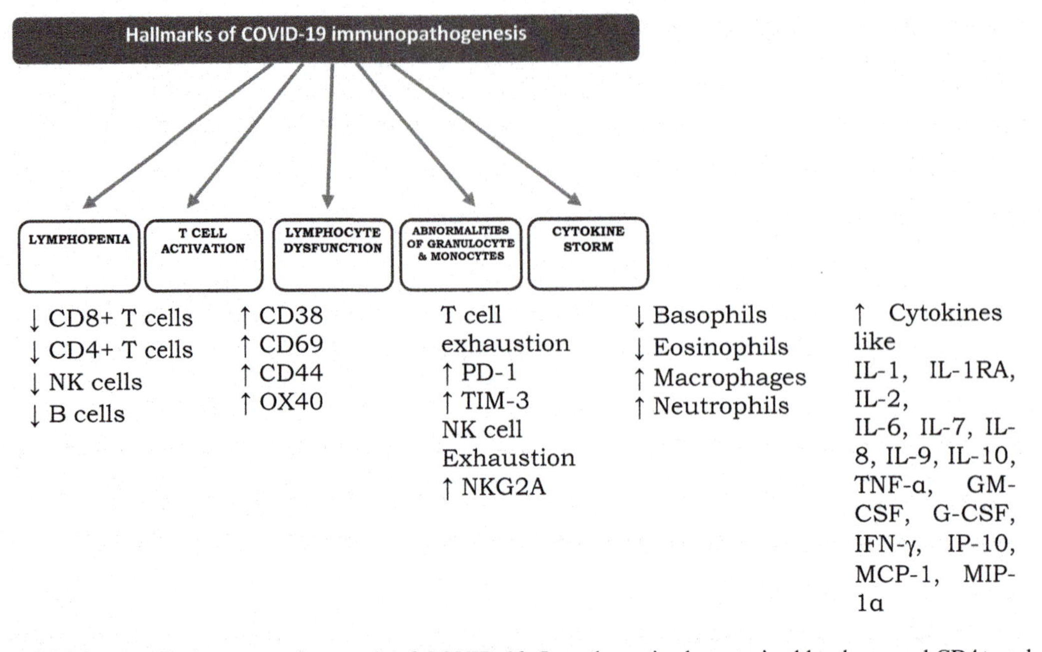

FIGURE 9.1 The immunopathogenesis of COVID-19: Lymphopenia characterized by decreased CD4+ and CD8+ T cells is a key feature of patients with COVID-19. Increased expression of activation markers CD69, CD38, and CD44 on the CD4+ and CD8+ T cells of patients. Lymphocytes also show an exhaustion phenotype with programmed cell death protein-1 (PD1), T cell immunoglobulin domain and mucin domain-3 (TIM3), and natural killer cell lectin-like receptor subfamily C member 1 (NKG2A) upregulation. Neutrophil and macrophage levels are significantly higher in severe patients, while the percentage of eosinophils, basophils, and monocytes are reduced. Cytokine storm with an increased expression of IL-1β, IL-6, IL-7, IL-9, and IL-10 is another key characteristic of severe COVID-19.

the cytokines such as CXCL2 and CXCL8 in COVID-19 patients. Neutrophil extracellular traps (NET) released by the neutrophils further attributes to severity and mortality in COVID-19 patients (Borges et al., 2020; Wang et al., 2020) (Figure 9.1).

9.4.3 Humoral Immune Response in COVID-19

The humoral immune response is the antibody-mediated response wherein CD4+T helper cells assist the B cells in producing antibodies specific to particular antigens. Just as the T cell-mediated immune response is essential for the viral clearance, B cell-mediated immune response also plays a critical role in viral clearance from the host. T cells assist the B cell-mediated synthesis of antibodies specific to the SARS-CoV-2 virus's surface glycoprotein, specifically the spike protein and the nucleocapsid protein.

A neutralizing antibody blocks the virus from entering the host cells and mediates a protective role in the late stages of infection and its recurrence. In a parallel fashion as T follicular helper (Tfh) cell response, B cell response starts around 1-week post symptom onset in SARS-CoV-2 patients. Tfh cells facilitate B cell differentiation into plasma cells producing significant levels of specific antibodies (Qin et al., 2018). In comparison to healthy controls, there is a higher frequency of Tfh cells in mild and recovering COVID-19 patients, which is suggestive of an effective humoral immune response. Studies have reported IgM-specific antibodies appear within a few days and IgG-specific antibodies appear within a week in patients infected with SARS-CoV-2 (Meyer, Drosten, and Müller, 2014; P. Zhou et al., 2020). Proteome microarray technology has revealed IgG antibodies in convalescent serum against the nucleocapsid protein, S1, ORF9b, NSP5, etc. (Jiang et al., 2020). Furthermore, the presence of IgG, IgM, and IgA antibodies specific to the nucleocapsid protein have also been reported in COVID-19 patients (Martinez-Fleta et al., 2020). Thus, the presence of virus-specific IgM, IgA, and IgG antibodies in COVID-19 patients and those recovering from the disease implicates antibody-mediated defensive immunity against SARS-CoV-2 (Ni et al., 2020; Padoan et al., 2020).

9.4.4 Cytokine Storm and Hyperinflammation

Cytokines and chemokines are among the most important immune mediators that coordinate and regulate the immune responses and help in mediating antiviral mechanism. In most infected individuals, the primary immune cells excise the pathogen resulting in recovery; however, in a particular proportion of infected subjects a hyperinflammatory response may be triggered that induces increased pro-inflammatory production cytokines at the infection site, thereby leading to severe lung injury and tissue damage. Extrapolation of knowledge from SARS-CoV and MERS-CoV has established that one of the root causes behind the disease's severity in SARS-CoV-2 is associated with hyperinflammation caused due to cytokine storm. Cytokine storm is defined as an unregulated immune response to various stimuli whereby auto-amplification of cytokines results in aggressive inflammatory condition (Shimizu, 2019). Cytokine storm is a state in which the individual experiences a hyperinflammatory reaction facilitated by a surge of large amounts of cytokines in response to infection. Cytokine storm is a key feature in severe cases of SARS-CoV-2 infection. Initially, the innate immune cells show a delay in the cytokine and chemokine secretion; however, lately activated macrophages and other immune cells result in a rapid increase in the various pro-inflammatory cytokines and chemokines (IL-6, TNF- α, IL-8, MCP-1, IL-1 β, CCL2, CCL5, and IFNs) at the infection site. Increased production of these immune mediators recruits and activate T cells, neutrophils, and NK cells. Activation of these cells subsequently results in severe insult and tissue damage in severe cases (Lingeswaran et al., 2020).

Dampened IFN-mediated antiviral response in SARS-CoV-2 induces cytokine storm, which further leads to macrophage activation syndrome in severe cases. ARDS and multiorgan failure in severe cases of COVID-19 are attributed to the cytokine storm. Clinical studies have reported significantly higher plasma levels of IL-2, IL-7, IL-4, IL-10, granulocyte colony-stimulating factor (G-CSF), IP-10, MCP1, macrophage inflammatory protein 1α (MIP1α) and tumor necrosis factor (TNF), IL-1β, and IFN-γ in patients requiring ICU care and mechanical ventilation. A significantly higher concentration of IL-6, IL-1, and TNF-α secreted by macrophages was found in severe cases compared to non-severe cases (Huang et al., 2020). Further, a bias towards Th1 cell function could be attributed to increased IL-1β, IFN-γ, CXCL10, and CCL2. Although SARS-CoV-2 patients showed an increase in both pro-as well as anti-inflammatory mediators, there is a predominance of G-CSF, TNFα, MCP1, IL-10, IL-2, and IL-7 in critically ill patients (Figure 9.2).

Thus, increased viral replication along with a hyperinflammatory host immune response severely damage the epithelial lining of lungs, which further deteriorates the alveolar cellular barriers. Moreover, the increased pro-inflammatory cytokines and activated immune cells may disrupt the endothelial cell that thereby results in alteration in the microvasculature. Such insult propagates tissue hypoxia and subsequently leads to ARDS (Herold et al., 2008; Högner et al., 2013).

9.4.5 Immunogenetics

A significant variation in susceptibility to viral infection, immune response development, clinical outcome in patients, and convalescence in any viral infection is attributed to heritable factors of

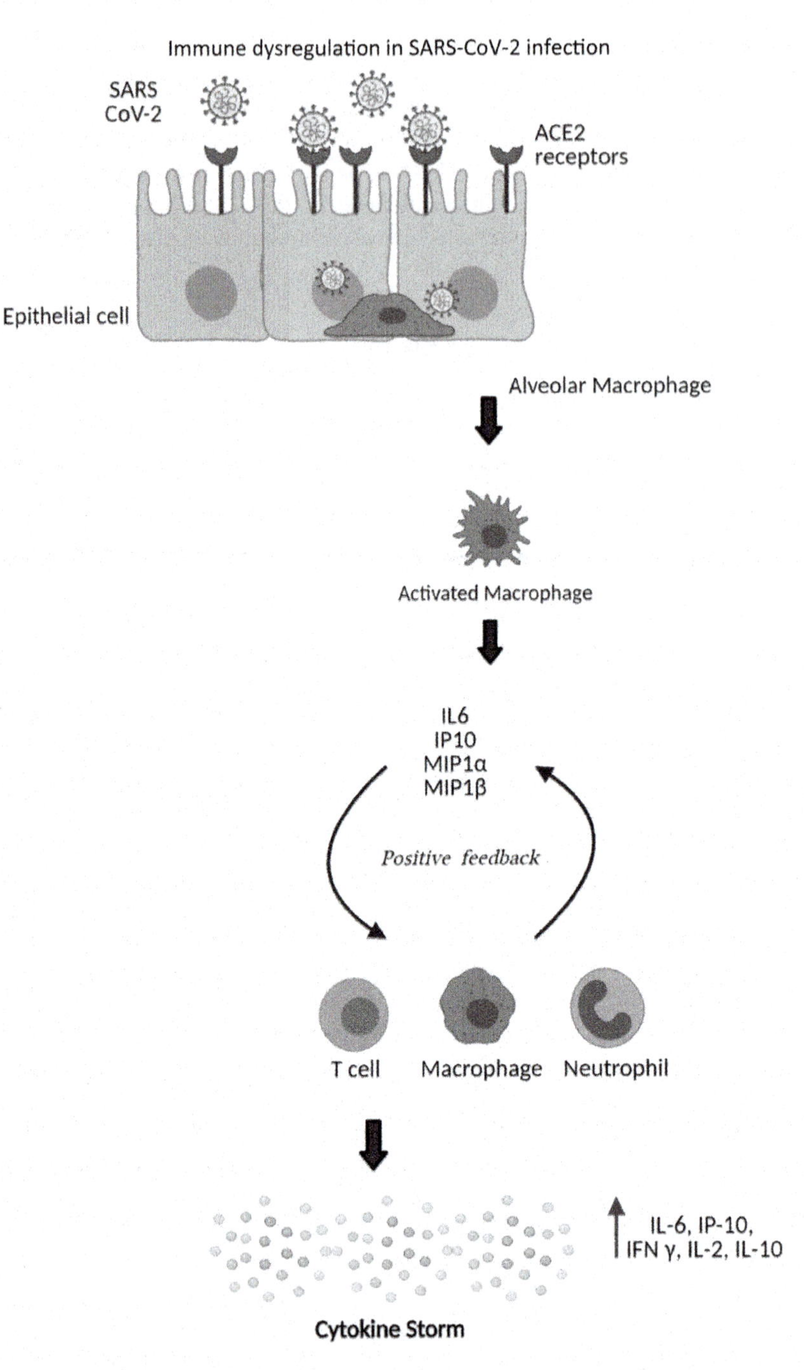

FIGURE 9.2 SARS-CoV-2 gain entry by binding to ACE2 Receptors on alveolar epithelial cells of the lung and trigger activation of localized alveolar macrophages that result in a cascade of immunomodulatory events. Pro-inflammatory mediators released from activated macrophages such as IL6, IP10, MIP1α, MIP1β, etc., recruit and activate other immune cells such as neutrophils, macrophages, monocytes, and certain T cells. On activation, these cells further release pro-inflammatory cytokines, thereby establishing a positive feedback to recruit more immune cells that ultimately cause extensive release of pro-inflammatory cytokines, resulting in the so called cytokine storm in severe COVID-19 infection.

the host. Individual response to the disease is directly linked to the mortality and morbidity of the disease. Genetic polymorphisms of the immune system and their respective regulatory systems influence disease susceptibility. Immune response to virus varies among individuals with different genetic make-up, resulting in varied disease severity among affected individuals (Mcweeney and Shannon, 2020).

The first step in immune response to virus includes recognition, processing, and prestation of virus by APCs to respective T cells. Presentation of virus by APCs requires a unique class of molecules known as Human Leukocyte Antigen (HLA) or Major Histocompatibility Complex (MHC). T-helper cell activation requires antigen presentation by MHC class I or HLA- A, B, and C, while MHC class II or HLA- DP, DQ, DR, DM, or DO mediate antigen presentation to cytotoxic T cells. A wide range of polymorphisms exist in HLA/MHC gene, enabling the recognition of a broad spectrum of pathogens. Genetic variation in HLA/MHC also results in differential T cell activation and response. Subsequently, HLA-b*46:01 genotype has been reported to have increased severity towards SARS-CoV (MacDonald et al., 2000). Further, in-silico analysis revealed poor binding affinity of HLA-A*25:01, B*46:01, C*01:02 genotype for SARS-CoV-2 and HLA-A*02:02, B*15:03, C*12:03 genotype to have most competent binding sequence for SARS-CoV-2 (Nguyen et al., 2020). Accordingly, the occurrence of different HLA haplotype may alter an individual's vulnerability to SARS-CoV-2 infection; thus, HLA genotyping could assist in determining an individual's risk towards the disease.

ACE2 gene polymorphism and TMPRSS2 gene polymorphism have also been linked to the difference in severity and the disease outcomes in COVID-19 patients (Choudhary et al., 2021). Two known variants of ACE2 receptors (i.e., K26R and I468V) have potentially lower binding affinities for the S protein that can alter the viral binding and internalization and thus is responsible for significant variability for COVID-19 infection (Q. Li, Cao, and Rahman, 2020).

A single nucleotide polymorphism (SNP) resulting in upregulation of TMPRSS2 and downregulation of the interferon-inducible gene MX1 in lung tissue is linked with increased susceptibility towards SARS-CoV-2 infections. This polymorphism mediates the increased expression of TMPRSS2 on the cell surface, thereby reducing antiviral response in individuals (Russo et al., 2020). Moreover, variations in these genes are also associated with increased susceptibility; however, the clinical outcomes in these individuals are still lacking.

9.5 CONCLUSION

Host immune evasion capacity of SARS-CoV-2 results in increased pathogenicity of the virus, which helps in production of large copy numbers in the infected cells. SARS-CoV-2 primarily infects the innate immune cells followed by recruitment of uninfected cells, which induces hyperinflammation and finally results in cytokine storm, consequently leading to multiple organ failure. Host immune response plays a pivotal role in progression of COVID-19 infection. Besides the host immune response, factors like genetic makeup of the host, type of the virus, titer of the virus, and load of virus are among the most significant factors that may result in increased progression and etiopathogenesis of SARS-CoV-2. Thus, future studies elucidating the immune pathology and mechanism of hyperinflammation to design therapeutic strategies for COVID-19 may result in better treatment modalities for SARS-CoV-2 infections.

REFERENCES

Alunno, Alessia, Ivan Padjen, Antonis Fanouriakis, and Dimitrios T. Boumpas. 2019. "Pathogenic and Therapeutic Relevance of JAK/STAT Signaling in systemic Lupus Erythematosus: Integration of Distinct Inflammatory Pathways and the Prospect of Their Inhibition with an Oral Agent." *Cells* 8 (8): 898. doi:10.3390/cells8080898.

Borges, Leandro, Tania Cristina Pithon-Curi, Rui Curi, and Elaine Hatanaka. 2020. "COVID-19 and Neutrophils: The Relationship between Hyperinflammation and Neutrophil Extracellular Traps." Edited by Juliana Vago. *Mediators of Inflammation* 2020 (December): 1–7. https://doi.org/10.1155/ 2020/ 8829674.

Chang, Feng-Yee, Hsiang-Cheng Chen, Pei-Jer Chen, Mei-Shang Ho, Shie-Liang Hsieh, Jung-Chung Lin, Fu-Tong Liu, and Huey-Kang Sytwu. 2020. "Immunologic Aspects of Characteristics, Diagnosis, and Treatment of Coronavirus Disease 2019 (COVID-19)." *Journal of Biomedical Science* 27 (1): 72. https://doi.org/10.1186/s12929-020-00663-w.

Channappanavar, Rudragouda, and Stanley Perlman. 2017. "Pathogenic Human Coronavirus Infections: Causes and Consequences of Cytokine Storm and Immunopathology." *Seminars in Immunopathology* 39 (5): 529–39. https://doi.org/10.1007/ s00281-017-0629-x.

Choudhary, Sarita, Karli Sreenivasulu, Prasenjit Mitra, Sanjeev Misra, and Praveen Sharma. 2021. "Role of Genetic Variants and Gene Expression in the Susceptibility and Severity of COVID-19." *Annals of Laboratory Medicine* 41 (2): 129–38. https://doi.org/10.3343/alm.2021.41.2.129.

Collange, Olivier, Charles Tacquard, Xavier Delabranche, Ian Leonard-Lorant, Mickaël Ohana, Mihaela Onea, Mathieu Anheim, et al. 2020. "Coronavirus Disease 2019: Associated Multiple Organ Damage." *Open Forum Infectious Diseases* 7 (7). https://doi.org/10.1093/ofid/ofaa249.

de Wit, Emmie, Neeltje van Doremalen, Darryl Falzarano, and Vincent J. Munster. 2016. "SARS and MERS: Recent Insights into Emerging Coronaviruses." *Nature Reviews Microbiology* 14 (8): 523–34. doi:10.1038/nrmicro.2016.81.

Diao, Bo, Chenhui Wang, Yingjun Tan, Xiewan Chen, Ying Liu, Lifen Ning, and Li Chen, et al. 2020. "Reduction and Functional Exhaustion of T Cells in Patients with Coronavirus Disease 2019 (COVID-19)." *Frontiers in Immunology* 11. doi:10.3389/fimmu.2020.00827.

Frieman, Matthew, Boyd Yount, Mark Heise, Sarah A. Kopecky-Bromberg, Peter Palese, and Ralph S. Baric. 2007. "Severe Acute Respiratory Syndrome Coronavirus ORF6 Antagonizes STAT1 Function by Sequestering Nuclear Import Factors on the Rough Endoplasmic Reticulum/Golgi Membrane." *Journal of Virology* 81 (18): 9812–24. https://doi.org/ 10.1128/JVI.01012-07.

Fung, To Sing, and Ding Xiang Liu. 2019. "Human Coronavirus: Host-Pathogen Interaction." *Annual Review of Microbiology* 73 (September): 529–57. https://doi.org/10.1146/annurev-micro-020518-115759.

García, Luis F. 2020. "Immune Response, Inflammation, and the Clinical Spectrum of COVID-19." *Frontiers in Immunology* 11. https://doi.org/10.3389/fimmu.2020.01441.

Gordon, David E., Gwendolyn M. Jang, Mehdi Bouhaddou, Jiewei Xu, Kirsten Obernier, Kris M. White, Matthew J. O'Meara, et al. 2020. "A SARS-CoV-2 Protein Interaction Map Reveals Targets for Drug Repurposing." *Nature* 583 (7816): 459–68. https://doi.org/10.1038/s41586-020-2286-9.

Hamming, I., W. Timens, M. L. C. Bulthuis, A. T. Lely, G. J. Navis, and H. van Goor. 2004. "Tissue Distribution of ACE2 Protein, the Functional Receptor for SARS Coronavirus. A First Step in Understanding SARS Pathogenesis." *The Journal of Pathology* 203 (2): 631–37. https://doi.org/10.1002/path.1570.

Herold, Susanne, Mirko Steinmueller, Werner von Wulffen, Lidija Cakarova, Ruth Pinto, Stephan Pleschka, Matthias Mack, et al. 2008. "Lung Epithelial Apoptosis in Influenza Virus Pneumonia: The Role of Macrophage-Expressed TNF-Related Apoptosis-Inducing Ligand." *The Journal of Experimental Medicine* 205 (13): 3065–77. https://doi.org/10.1084/jem.20080201.

Högner, Katrin, Thorsten Wolff, Stephan Pleschka, Stephanie Plog, Achim D. Gruber, Ulrich Kalinke, Hans-Dieter Walmrath, et al. 2013. "Macrophage-Expressed IFN-β Contributes to Apoptotic Alveolar Epithelial Cell Injury in Severe Influenza Virus Pneumonia." *PLoS Pathogens* 9 (2): e1003188. https://doi.org/10.1371/journal.ppat.1003188.

Huang, Chaolin, Yeming Wang, Xingwang Li, Lili Ren, Jianping Zhao, Yi Hu, Li Zhang, et al. 2020. "Clinical Features of Patients Infected with 2019 Novel Coronavirus in Wuhan, China." *The Lancet* 395 (10223): 497–506. https://doi.org/10.1016/S0140-6736(20)30183-5.

Jiang, He-wei, Yang Li, Hai-nan Zhang, Wei Wang, Dong Men, Xiao Yang, Huan Qi, Jie Zhou, and Sheng-ce Tao. 2020. "Global Profiling of SARS-CoV-2 Specific IgG/ IgM Responses of Convalescents Using a Proteome Microarray." *medRxiv*, March, 2020.03.20.20039495. https://doi.org/10.1101/2020.03.20.20039495.

Keskinen, P., T. Ronni, S. Matikainen, A. Lehtonen, and I. Julkunen. 1997. "Regulation of HLA Class I and II Expression by Interferons and Influenza A Virus in Human Peripheral Blood Mononuclear Cells." *Immunology* 91 (3): 421–29. https://doi.org/10.1046/j.1365-2567.1997.00258.x.

Kindler, E., V. Thiel, and F. Weber. 2016. "Interaction of SARS and MERS Coronaviruses with the Antiviral Interferon Response." *Advances in Virus Research* 96: 219–43. https://doi.org/10.1016/bs.aivir.2016.08.006.

Kong, Xiaocen, Mingming Fang, Ping Li, Fei Fang, and Yong Xu. 2009. "HDAC2 Deacetylates Class II Transactivator and Suppresses Its Activity in Macrophages and Smooth Muscle Cells." *Journal of Molecular and Cellular Cardiology* 46 (3): 292–99. https://doi.org/10.1016/j.yjmcc.2008.10.023.

Kuba, Keiji, Yumiko Imai, Shuan Rao, Hong Gao, Feng Guo, Bin Guan, and Yi Huan, et al. 2005. "A Crucial Role of Angiotensin Converting Enzyme 2 (ACE2) in SARS Coronavirus–Induced Lung Injury." *Nature Medicine* 11 (8): 875–79. doi:10.1038/nm1267.

Letko, Michael, Andrea Marzi, and Vincent Munster. 2020. "Functional Assessment of Cell Entry and Receptor Usage for SARS-Cov-2 and Other Lineage B Betacoronaviruses." *Nature Microbiology* 5 (4): 562–69. doi:10.1038/s41564-020-0688-y.

Li, Jingjiao, Mingquan Guo, Xiaoxu Tian, Xin Wang, Xing Yang, Ping Wu, Chengrong Liu, et al. 2020. "Virus-Host Interactome and Proteomic Survey Reveal Potential Virulence Factors Influencing SARS-CoV-2 Pathogenesis." *Med (New York, N.y.)*, July. https://doi.org/10.1016/j.medj.2020.07.002.

Li, Quan, Zanxia Cao, and Proton Rahman. 2020. "Genetic Variability of Human Angiotensin-Converting Enzyme 2 (HACE2) Among Various Ethnic Populations." Preprint. Genetics. https://doi.org/10.1101/2020.04.14.041434.

Liao, Mingfeng, Yang Liu, Jing Yuan, Yanling Wen, Gang Xu, Juanjuan Zhao, and Lin Cheng, et al. 2020. "Single-Cell Landscape of Bronchoalveolar Immune Cells in Patients with COVID-19." *Nature Medicine* 26 (6): 842–44. doi:10.1038/s41591-020-0901-9.

Lingeswaran, Malavika, Taru Goyal, Raghumoy Ghosh, Smriti Suri, Prasenjit Mitra, Sanjeev Misra, and Praveen Sharma. 2020. "Inflammation, Immunity and Immunogenetics in COVID-19: A Narrative Review." *Indian Journal of Clinical Biochemistry* 35 (3): 260–73. https://doi.org/10.1007/s12291-020-00897-3.

Lu, Xiaolu, Ji'an Pan, Jiali Tao, and Deyin Guo. 2010. "SARS-Cov Nucleocapsid Protein Antagonizes IFN-B Response by Targeting Initial Step of IFN-B Induction Pathway, and Its C-Terminal Region Is Critical for the Antagonism." *Virus Genes* 42 (1): 37–45. doi:10.1007/s11262-010-0544-x.

Lu, Roujian, Xiang Zhao, Juan Li, PeihuaNiu, Bo Yang, Honglong Wu, Wenling Wang, et al. 2020. "Genomic Characterisation and Epidemiology of 2019 Novel Coronavirus: Implications for Virus Origins and Receptor Binding." *The Lancet* 395 (10224): 565–74. https://doi.org/10.1016/S0140-6736(20) 30251-8.

MacDonald, K. S., K. R. Fowke, J. Kimani, V. A. Dunand, N. J. Nagelkerke, T. B. Ball, J. Oyugi, et al. 2000. "Influence of HLA Supertypes on Susceptibility and Resistance to Human Immunodeficiency Virus Type 1 Infection." *The Journal of Infectious Diseases* 181 (5): 1581–89. https://doi.org/10.1086/315472.

Martinez-Fleta, Pedro, Arantzazu Alfranca, Isidoro González-Álvaro, José M. Casasnovas, Daniel Fernández Soto, Gloria Esteso, Yaiza Cáceres-Martell, et al. 2020. "SARS-Cov-2 Cysteine-like Protease (Mpro) Is Immunogenic and Can Be Detected in Serum and Saliva of COVID-19-Seropositive Individuals." Preprint. Infectious Diseases (except HIV/AIDS). https://doi.org/10.1101/2020.07.16.20155853.

Mcweeney, Shannon. n.d. "Systems Immunogenetics and Bioinformatics." Accessed January 6, 2021. https://grantome.com/grant/NIH/U19-AI100625-06-7726.

Meyer, Benjamin, Christian Drosten, and Marcel A. Müller. 2014. "Serological Assays for Emerging Coronaviruses: Challenges and Pitfalls." *Virus Research* 194 (December): 175–83. https://doi.org/10.1016/j.virusres.2014.03.018.

Mitra, Prasenjit, Sanjeev Misra, and Praveen Sharma. 2020. "COVID-19 Pandemic in India: What Lies Ahead." *Indian Journal of Clinical Biochemistry* 35 (3): 257–59. https://doi.org/10.1007/s12291-020-00886-6.

Mitra, Prasenjit, Sanjeev Misra, and Praveen Sharma. 2021. "One Year of COVID-19: The 'New Normal'." *Indian Journal of Clinical Biochemistry* 36 (1): 1–2. https://doi.org/10.1007/s12291-020-00954-x.

Mitra, Prasenjit, Smriti Suri, Taru Goyal, Radhieka Misra, Kuldeep Singh, M. K. Garg, Sanjeev Misra, Praveen Sharma, and Abhilasha. 2020. "Association of Comorbidities with Coronavirus Disease 2019: A Review." *Annals of the National Academy of Medical Sciences (India)* 56 (02): 102–11. https://doi.org/10.1055/s-0040-1714159.

Nguyen, Austin, Julianne K. David, Sean K. Maden, Mary A. Wood, Benjamin R. Weeder, Abhinav Nellore, and Reid F. Thompson. 2020. "Human Leukocyte Antigen Susceptibility Map for SARS-CoV-2." *medRxiv*, April, 2020.03.22.2 0040600. https://doi.org/10.1101/2020.03.22.20040600.

Ni, Ling, Fang Ye, Meng-Li Cheng, Yu Feng, Yong-Qiang Deng, Hui Zhao, Peng Wei, et al. 2020. "Detection of SARS-CoV-2-Specific Humoral and Cellular Immunity in COVID-19 Convalescent Individuals." *Immunity* 52 (6): 971–77.e3. https://doi.org/10.1016/j.immuni.2020.04.023.

O'Connell, Patrick, and Yasser A. Aldhamen. 2020. "Systemic Innate and Adaptive Immune Responses to SARS-CoV-2 as It Relates to Other Coronaviruses." *Human Vaccines &Immunotherapeutics* 16 (12): 2980–91. https://doi.org/ 10.1080/21645515.2020.1802974.

Padoan, Andrea, Laura Sciacovelli, Daniela Basso, Davide Negrini, Silvia Zuin, Chiara Cosma, Diego Faggian, Paolo Matricardi, and Mario Plebani. 2020. "IgA-Ab Response to Spike Glycoprotein of SARS-CoV-2 in Patients with COVID-19: A Longitudinal Study." *ClinicaChimica Acta* 507 (August): 164–66. https://doi.org/10.1016/j.cca.2020.04.026.

Qin, Lei, Tayab C. Waseem, Anupama Sahoo, Shayahati Bieerkehazhi, Hong Zhou, Elena V. Galkina, and Roza Nurieva. 2018. "Insights into the Molecular Mechanisms of T Follicular Helper-Mediated Immunity and Pathology." *Frontiers in Immunology* 9 (August): 1884. https://doi.org/10.3389/fimmu.2018.01884.

Russo, Roberta, Immacolata Andolfo, Vito Alessandro Lasorsa, Achille Iolascon, and Mario Capasso. 2020. "Genetic Analysis of the Novel SARS-CoV-2 Host Receptor *TMPRSS2* in Different Populations." Preprint. Genetics. https://doi.org/10.1101/ 2020.04.23.057190.

Sanchez-Cerrillo, Ildefonso, Pedro Landete, Beatriz Aldave, Santiago Sanchez-Alonso, Ana Sanchez-Azofra, Ana Marcos-Jimenez, Elena Avalos, et al. 2020. "Differential Redistribution of Activated Monocyte and Dendritic Cell Subsets to the Lung Associates with Severity of COVID-19." *medRxiv*, May. https://doi.org/10.1101/2020.05.13.20100925.

Shah, Vibhuti Kumar, Priyanka Firmal, Aftab Alam, Dipyaman Ganguly, and Samit Chattopadhyay. 2020. "Overview of Immune Response During SARS-CoV-2 Infection: Lessons from the Past." *Frontiers in Immunology* 11 (August). https://doi.org/10.3389/fimmu.2020.01949.

Shang, Jian, Yushun Wan, Chang Liu, Boyd Yount, Kendra Gully, Yang Yang, Ashley Auerbach, Guiqing Peng, Ralph Baric, and Fang Li. 2020. "Structure of Mouse Coronavirus Spike Protein Complexed with Receptor Reveals Mechanism for Viral Entry." *PLOS Pathogens* 16 (3): e1008392. https://doi.org/10.1371/journal.ppat.1008392.

Shimizu, Masaki. 2019. "Clinical Features of Cytokine Storm Syndrome." In *Cytokine Storm Syndrome*, edited by Randy Q. Cron and Edward M. Behrens, 31–41. Cham: Springer International Publishing. https://doi.org/10.1007/978-3-030-22094-5_3.

Sims, Amy C., Ralph S. Baric, Boyd Yount, Susan E. Burkett, Peter L. Collins, and Raymond J. Pickles. 2005. "Severe Acute Respiratory Syndrome Coronavirus Infection of Human Ciliated Airway Epithelia: Role of Ciliated Cells in Viral Spread in the Conducting Airways of the Lungs." *Journal of Virology* 79 (24): 15511–24. https://doi.org/10.1128/ JVI.79.24.15511-15524.2005.

Spiegel, Martin, Andreas Pichlmair, Luis Martínez-Sobrido, Jerome Cros, Adolfo García-Sastre, Otto Haller, and Friedemann Weber. 2005. "Inhibition of Beta Interferon Induction by Severe Acute Respiratory Syndrome Coronavirus Suggests a Two-Step Model for Activation of Interferon Regulatory Factor 3." *Journal of Virology* 79 (4): 2079–2086. doi:10.1128/jvi.79.4.2079-2086.2005.

Sungnak, Waradon, Ni Huang, Christophe Bécavin, and Marijn Berg. 2020. "SARS-CoV-2 Entry Genes Are Most Highly Expressed in Nasal Goblet and Ciliated Cells within Human Airways." *arXiv*, March. https://www.ncbi.nlm.nih.gov/pmc/articles/ PMC7280877/.

Taefehshokr, Nima, Sina Taefehshokr, Nima Hemmat, and Bryan Heit. 2020. "COVID-19: Perspectives on Innate Immune Evasion." *Frontiers in Immunology* 11 (September). https://doi.org/10.3389/fimmu.2020.580641.

Tai, Wanbo, Lei He, Xiujuan Zhang, Jing Pu, Denis Voronin, Shibo Jiang, Yusen Zhou, and Lanying Du. 2020. "Characterization of the Receptor-Binding Domain (RBD) of 2019 Novel Coronavirus: Implication for Development of RBD Protein as a Viral Attachment Inhibitor and Vaccine." *Cellular & Molecular Immunology* 17 (6): 613–20. doi:10.1038/s41423-020-0400-4.

Walls, Alexandra C., Young-Jun Park, M. Alejandra Tortorici, Abigail Wall, Andrew T. McGuire, and David Veesler. 2020. "Structure, Function, and Antigenicity of the SARS-Cov-2 Spike Glycoprotein." *Cell* 183 (6): 1735. doi:10.1016/j.cell. 2020.11.032.

Wan, Yushun, Jian Shang, Rachel Graham, Ralph S. Baric, and Fang Li. 2020. "Receptor Recognition by the Novel Coronavirus from Wuhan: An Analysis Based on Decade- Long Structural Studies of SARS Coronavirus." Edited by Tom Gallagher. *Journal of Virology* 94 (7): e00127-20,/jvi/94/7/JVI.00127-20.atom. https://doi.org/10.1128/JVI.00127-20.

Wang, Jun, Qian Li, Yongmei Yin, Yingying Zhang, Yingying Cao, Xiaoming Lin, Lihua Huang, Daniel Hoffmann, Mengji Lu, and Yuanwang Qiu. 2020. "Excessive Neutrophils and Neutrophil Extracellular Traps in COVID-19." *Frontiers in Immunology* 11. https://doi.org/10.3389/fimmu.2020.02063.

Westmeier, Jaana, KrystalleniaPaniskaki, Zehra Karaköse, Tanja Werner, Kathrin Sutter, Sebastian Dolff, and Marvin Overbeck, et al. 2020. "Impaired Cytotoxic CD8+ T Cell Response in Elderly COVID-19 Patients." *Mbio* 11 (5). doi:10.1128/mbio.02243-20.

World Health Organization. Coronavirus disease 2019 (COVID-19) situation report—74. www.who.int/emergencies/diseases/novel-coronavirus-2019/situation-reports/. Accessed 4 Apr 2020.

Xiong, Yong, Yuan Liu, Liu Cao, Dehe Wang, Ming Guo, Ao Jiang, Dong Guo, et al. 2020. "Transcriptomic Characteristics of Bronchoalveolar Lavage Fluid and Peripheral Blood Mononuclear Cells in COVID-19 Patients." *Emerging Microbes & Infections* 9 (1): 761–70. https://doi.org/10.1080/22221751.2020.1747363.

Xu, Gang, Furong Qi, Hanjie Li, Qianting Yang, Haiyan Wang, Xin Wang, Xiaoju Liu, et al. 2020. "The Differential Immune Responses to COVID-19 in Peripheral and Lung Revealed by Single-Cell RNA Sequencing." *Cell Discovery* 6 (1): 1–14. https://doi.org/10.1038/s41421-020-00225-2.

Yang, Dong, Hin Chu, Yuxin Hou, Yue Chai, Huiping Shuai, Andrew Chak-Yiu Lee, Xi Zhang, et al. 2020. "Attenuated Interferon and Proinflammatory Response in SARS-CoV-2–Infected Human Dendritic Cells Is Associated with Viral Antagonism of STAT1 Phosphorylation." *The Journal of Infectious Diseases* 222 (5): 734–45. https://doi.org/10.1093/infdis/ jiaa356.

Yoshikawa, Tomoki, Terence Hill, Kui Li, Clarence J. Peters, and Chien-Te K. Tseng. 2009. "Severe Acute Respiratory Syndrome (SARS) Coronavirus-Induced Lung Epithelial Cytokines Exacerbate SARS Pathogenesis by Modulating Intrinsic Functions of Monocyte-Derived Macrophages and Dendritic Cells." *Journal of Virology* 83 (7): 3039–48. https://doi.org/10.1128/JVI.01792-08.

Zhang, Jinping, Peng Liu, Morong Wang, Jie Wang, Jie Chen, Wenling Yuan, Mei Li, et al. 2020. "The Clinical Data from 19 Critically Ill Patients with Coronavirus Disease 2019: A Single-Centered, Retrospective, Observational Study." *Zeitschrift Fur Gesundheitswissenschaften*, April, 1–4. https://doi.org/10.1007/s10389-020-01291-2.

Zhang, Wen, Yan Zhao, Fengchun Zhang, Qian Wang, Taisheng Li, Zhengyin Liu, Jinglan Wang, et al. 2020. "The Use of Anti-Inflammatory Drugs in the Treatment of People with Severe Coronavirus Disease 2019 (COVID-19): The Perspectives of Clinical Immunologists from China." *Clinical Immunology (Orlando, Fla.)* 214 (May): 108393. https://doi.org/10.1016/j.clim.2020.108393.

Zhang, Yiwen, Junsong Zhang, Yingshi Chen, Baohong Luo, Yaochang Yuan, Feng Huang, Tao Yang, et al. 2020. "The ORF8 Protein of SARS-CoV-2 Mediates Immune Evasion through Potently Downregulating MHC-I." Preprint. Microbiology. https://doi.org/10.1101/2020.05.24.111823.

Zhao, Jincun, Jingxian Zhao, Nico Van Rooijen, and Stanley Perlman. 2009. "Evasion by Stealth: Inefficient Immune Activation Underlies Poor T Cell Response and Severe Disease in SARS-CoV-Infected Mice." Edited by Michael Gale. *PLoS Pathogens* 5 (10): e1000636. https://doi.org/10.1371/journal.ppat.1000636.

Zheng, Hong-Yi, Mi Zhang, Cui-Xian Yang, Nian Zhang, Xi-Cheng Wang, Xin-Ping Yang, Xing-Qi Dong, and Yong-Tang Zheng. 2020. "Elevated Exhaustion Levels and Reduced Functional Diversity of T Cells in Peripheral Blood May Predict Severe Progression in COVID-19 Patients." *Cellular & Molecular Immunology* 17 (5): 541–43. doi:10.1038/s41423-020-0401-3.

Zhou, Peng, Xing-Lou Yang, Xian-Guang Wang, Ben Hu, Lei Zhang, Wei Zhang, Hao-Rui Si, et al. 2020. "A Pneumonia Outbreak Associated with a New Coronavirus of Probable Bat Origin." *Nature* 579 (7798): 270–73. https://doi.org/10.1038/s41586-020-2012-7.

Zhou, Runhong, Kelvin Kai-Wang To, Yik-Chun Wong, Li Liu, Biao Zhou, Xin Li, Haode Huang, et al. 2020. "Acute SARS-CoV-2 Infection Impairs Dendritic Cell and T Cell Responses." *Immunity* 53 (4): 864–77.e5. https://doi.org/10.1016/ j.immuni.2020.07.026.

10 Extrapulmonary Manifestations

After Effects of COVID-19 Post Recovery

*Shiv Bharadwaj,[1] Nikhil Kirtipal,[*2] and R. C. Sobti[*3]*

[1]Department of Biotechnology, Yeungnam University, College of life and applied sciences, Gyeongbuk-do, Republic of Korea
[2]School of Life Sciences, GIST, Gwangju, Republic of Korea
[3]Department of Biotechnology, Panjab University, Chandigarh, India
*Corresponding author: Nikhil Kirtipal, Email: Kirtipal.n@gmail.com and R. C. Sobti, Email: rcsobti@pu.ac.in

CONTENTS

10.1 INTRODUCTION

The ongoing pandemic of the coronavirus disease (Covid-19) resulted in a typical presentation of respiratory symptoms due to the affinity of the novel coronavirus (severe acute respiratory syndrome coronavirus-2 (SARS-CoV-2)) for the upper and lower respiratory tracts. Extrapulmonary areas are recognized as sites for the manifestation of disease and the spread of viruses (Sarkesh et al., 2020). All coronavirus species infections are thought to cause extrapulmonary symptoms in particular. Arabi and colleagues reported on clinical manifestations in patients with MERS and found significant extrapulmonary organ dysfunctions, which caused an increased mortality rate (Arabi

DOI: 10.1201/9781003358909-10

et al., 2014). They also found acute kidney injury (AKI), extrapulmonary manifestations in Arabic populations (Saudi Arabia) including circulatory instability requiring vasopressors, hepatic dysfunction with raised levels of liver enzymes, gastrointestinal disorders (including severe belly pain and diarrhea), and hematologic disorders (Arabi et al., 2014).

Other studies have also reported that MERS patients additionally have gastrointestinal discomfort and neurological sequelae (Arabi et al., 2017; Bradley and Bryan, 2019). Research on SARS pathogenesis also reported certain manifestations related to the hematological, neurological, renal, and gastrointestinal systems, leading to multiple organ dysfunctions (Gu et al., 2005). Comparative studies have suggested that SARS, MERS, and Covid-19 coronaviruses share a homologous sequence, because of this the major pathophysiology of Covid-19 can be predicted (Netland et al., 2008; Kui et al., 2020). Although Covid-19 shows similarities with SARS and MERS, the possibility of nonrespiratory manifestations, isolated extrapulmonary manifestations, and other complications must be included in the diagnostic and therapeutic management of patients with Covid-19 (Hosseiny et al., 2020). The gastrointestinal tract (GIT), nervous system, cardiovascular system (CVS), renal system, eyes, and symptoms resulting from hematological disorders are among the extrapulmonary manifestations in Covid-19 (Adukia et al., 2020) (Figure 10.1).

Although Covid-19 is mostly a respiratory disease, research reveal that SARS-CoV-2 can move to other body regions by targeting angiotensin-converting enzyme 2 receptors. As a result, the symptoms of the various affected organs differ (Sarkesh et al., 2020).

10.2 EXTRAPULMONARY MANIFESTATIONS OF CORONAVIRUS INFECTIONS

A new single-strand RNA coronavirus called SARS-CoV-2 is the causal agent of the coronavirus disease 2019 (Covid-19). The World Health Organization (WHO) has identified it as a global pandemic (Cui, Li and Shi, 2019; Sobh et al., 2020). SARS-CoV-2 primarily affects the lower respiratory system, resulting in viral pneumonia; however, it can also affect the heart, gastrointestinal tract, liver, kidneys, and central nervous system, finally leading to multiple organ failure (N. Zhu et al., 2020). A number of coexisting conditions have been recognized as risk factors for severe Covid-19 disease. The most common comorbidities found in several studies included obesity, respiratory disorders, diabetes mellitus, and hypertension (HTN) (T. Chen et al., 2020; Hu et al., 2020; Kang et al., 2020; Sun et al., 2020; Yang et al., 2020; Zimmerman et al., 2020). In this chapter, we will discuss extrapulmonary symptoms induced by Covid-19, as well as their impact on outcome and case management, indicating the requirement of a multidisciplinary team. The precise mechanism of extrapulmonary symptoms is still being studied; numerous mechanisms, either direct or indirect damage related to the inflammatory response to viral infection, have been hypothesized (Shi et al., 2020; W. Zhang et al., 2020). Following activation of the spike protein by transmembrane protease serine 2 (TMPRSS2), SARS-CoV-2 infection occurs when the viral surface spike protein (S) attaches to the human angiotensin-converting enzyme (ACE2) receptors (Hoffmann et al., 2020). All organs, especially the heart, intestinal epithelium, kidneys, vascular endothelium, and smooth muscle cells, express ACE2 receptors, which provides a pathway for Covid-19 to cause multi-organ failure (Hamming et al., 2004; Clerkin et al., 2020; Zou et al., 2020).

A second mechanism is the severe inflammatory reaction carried by the viral infection to the lungs and other organs (Kumar et al., 2009). SARS-CoV-2 enters cells and activates T lymphocytes, causing a severe immunological response, an inflammatory response, and an inflammatory cascade that results in the production of cytokines like interleukin (IL)-1, IL-6, tumor necrosis factor- α (TNF- α), interferon- γ (IFN- γ), and granulocyte-macrophage colony-stimulating factor (GM-CSF), which is known as a cytokine storm and results in tissue damage (Fan et al., 2020).

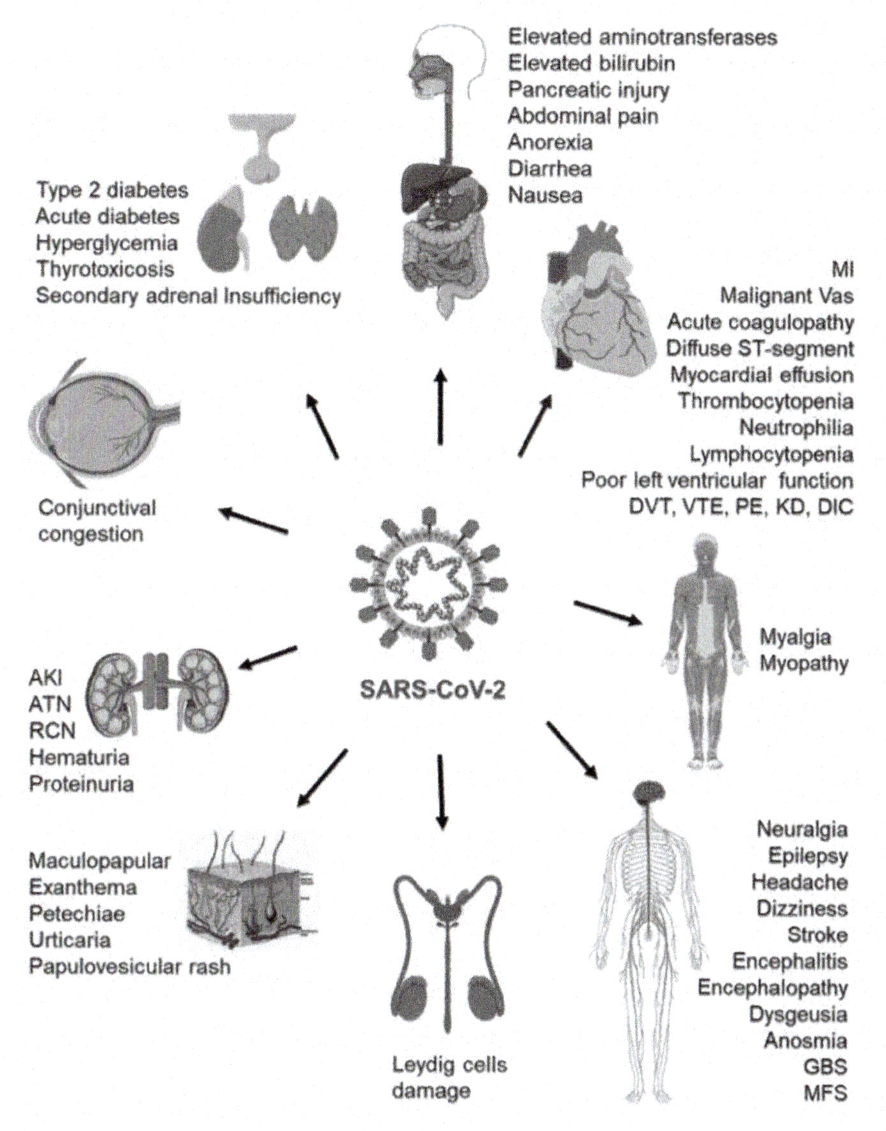

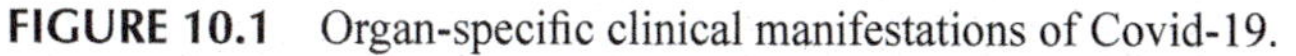

FIGURE 10.1 Organ-specific clinical manifestations of Covid-19.

10.2.1 Renal Dysfunction

Studies from China have reported acute kidney injury (AKI) in 25% to 29% of critically ill patients and the deceased of Covid-19 (T. Chen et al., 2020; W. Wang et al., 2020). Recent findings indicate an increased occurrence of renal problems. It was observed in a study of 59 Covid-19 patients that 34% experienced significant albuminuria on the first day of hospitalization, and 63% experienced proteinuria (Adukia et al., 2020). In another study on Covid-19 hospital patients proteinuria and hematuria were present in 44% of cases, while hematuria was present at admission in 26.7% of cases (Cheng et al., 2020). In 27% of Covid-19 patients as well as in 66% of those who passed away, blood urea nitrogen levels were high (W. Wang et al., 2020). On a CT scan, the density of the kidneys was reduced, suggesting edema and inflammation. Patients with Covid-19 who also have hypertension run the chance of developing a more severe illness (3.3 vs. 0.4%) (Guan et al., 2020; Henry and Lippi, 2020). The postulated causes of the kidney injury include cytokine storm syndrome (CSS) or direct cellular injury by the virus. SARS-CoV-2 was discovered in a patient's urine sample, and because renal tubular cells express the ACE2 receptor, the kidney was assumed to be the target organ (Li et al., 2003). Between 0.1% and 29% of patients with Covid-19 experience AKI, which is inconsistent throughout published research (Cummings et al., 2020; Farouk et al., 2020; Hirsch et al., 2020; Kunutsor and Laukkanen, 2020). In a sample of 32 confirmed Covid-19 cases without prior chronic kidney disease, Sun and colleagues reported the occurrence of subclinical AKI as indicated by increased urinary levels of β2-microglobulin, α1-microglobulin, N-acetyl- β-D-glucosaminidase, and retinol-binding protein (i.e., all biomarkers of kidney tubular damage) (Canatan, Vives Corrons and De Sanctis, 2020; Sun et al., 2020).

Although the mechanisms for the renal manifestations of Covid-19 are still unknown, a complex multifactorial pathway has been proposed and it includes the following characteristics: (a) Direct viral involvement and renal impairment caused by viral replication; (b) local disruption in renin-angiotensin-aldosterone system (RAAS) homeostasis; and (c) as a result of a systemic inflammatory response a "cytokine storm" (Diao et al., 2021; Hirsch et al., 2020; Kunutsor and Laukkanen, 2020). Monitoring kidney function markers may aid in identifying patients who are most at risk for negative outcomes (Diao et al., 2021). To identify efficient management approaches, more research is necessary to help us better understand the pathophysiology underlying the renal symptoms of Covid-19.

10.2.2 Gastrointestinal (GI) and Hepatic Manifestations

The most frequent gastrointestinal symptoms were found to be nausea, vomiting, abdominal pain, and diarrhea. In a few cases of Covid-19, GI symptoms preceded respiratory symptoms, and other patients only presented with digestive symptoms in the absence of respiratory symptoms (Huang et al., 2020; Tian et al., 2020; Tran et al., 2020). Therefore, the diagnosis of Covid-19 may be delayed as a result of these unusual GI clinical symptoms (Han et al., 2020).

SARS-CoV-2 infects the GI tract via the viral receptor angiotensin converting enzyme 2, which is expressed on ileum and colon enterocytes. A growing percentage of Covid-19 patients suffer from liver problems, ranging from mild to severe. Most liver injuries are minor and temporary, however significant liver damage can occur in patients with severe Covid-19 and is associated with poor outcomes (American Society for Gastrointestinal Endoscopy (ASGE), 2020). There is a higher risk of adverse outcomes in Covid-19 in patients with chronic liver diseases, especially in those who already have cirrhosis, which has been linked to their immunocompromised status (Zhang, Shi and Wang, 2020). The precise mechanism by which liver injury develops in Covid-19 patients is unknown.

Liver injury in Covid-19 patients could be caused by viral infection in liver cells or by other factors like as drug-induced liver injury, systemic inflammation caused by cytokine storms, or pneumonia-associated hypoxia. Furthermore, drug-induced hepatotoxicity, including drugs such as remdesivir (an RNA polymerase inhibitor) and hydroxychloroquine, may play a role in the rise of liver enzymes (Cascella et al., 2021).

10.2.3 Cardiac Manifestation

Cardiovascular arrhythmias or clinical heart failure with or without related hemodynamic instability, such as shock, can make Covid-19 more difficult to treat (Feng et al., 2020). These cardiac issues are increasingly becoming recognized as a kind of late complication that can occur after the patient's respiratory system has recovered. They can occur spontaneously at any moment while the patient is in the hospital (Fried et al., 2020; Lang et al., 2020). In certain patients of Covid-19, SARS-CoV-2 caused cardiac manifestations linked to the adrenergic drive, systemic inflammatory milieu, and cytokine-release syndrome with direct viral infection of myocardial and endothelial cells, hypoxia due to respiratory failure, electrolytic imbalances, fluid overload, and side effects of certain Covid-19 medications (Boukhris et al., 2020). Serial cardiac troponin and natriuretic peptide measurements, as well as fibrinogen, D-dimer, and inflammatory biomarkers, may all be used in monitoring (Canatan, Vives Corrons and De Sanctis, 2020). A multidisciplinary team with cardiologists, intensive care doctors, and experts in infectious diseases should be responsible for the management of acute Covid-19 cardiovascular syndrome.

10.2.4 Neurological Findings

A wide range of neurological complications were reported in patients with Covid-19 suggesting that SARS-CoV-2 may affect both the central and peripheral nervous system.

According to various studies, more over 35% of Covid-19 patients suffer neurological symptoms as the disease progresses. Covid-19's neurological symptoms and side effects can be divided into central (encephalopathy, acute hemorrhagic necrotizing encephalopathy, acute myelitis, cerebrovascular accident, encephalitis, headaches, and dizziness) and peripheral (other neurological manifestations and side effects) (anosmia and chemosensory dysfunction, Guillain-Barrè syndrome, and skeletal muscle damage) (Ahmad and Rathore, 2020; Filatov et al., 2020). It is unknown whether these consequences are caused directly by the viral infection, by postinfectious autoimmune reactions, or by hypoxia metabolic alterations (Ahmad and Rathore, 2020; Filatov et al., 2020).

10.2.4.1 Central Nervous System Manifestations

The most common central nervous system (CNS) symptoms are dizziness (16.8%) and headache (13.1%); other less frequent symptoms include impaired consciousness (7.5%), acute cerebrovascular disease (2.8%), ataxia (0.5%), seizure (0.5%), vision impairment (1.4%), and nerve pain (2.7%). The most common early neurological symptoms are headache, myalgia, and malaise. In severe Covid-19, altered sensorium ranges from disorientation, delirium, and stupor to coma (Ahmad and Rathore, 2020; Filatov et al., 2020). Patients with severe Covid-19 infection are more likely to experience neurological signs and symptoms, which could be a result of cerebral hypoxia caused by respiratory failure (Ahmad and Rathore, 2020; Filatov et al., 2020).

10.2.4.2 Peripheral Nervous System Manifestations

Peripheral nervous system (PNS) signs and symptoms of Covid-19 include hyposmia/anosmia, hypogeusia/ageusia, muscle pain, and Guillain-Barrè syndrome. Spinal cord involvement is uncommon (Mao et al., 2020). The most frequent PNS symptoms of SARS-CoV-2 include anosmia and/or ageusia. In a multicenter European study of 417 individuals with mild-to-moderate Covid-19, 85.6% and 88.0% showed olfactory and/or gustatory dysfunctions, respectively. Women were more likely to experience symptoms, and a 44% early olfactory recovery rate was seen, with olfactory symptoms continuing for up to 14 days after the Covid-19 symptoms had disappeared (Lechien et al., 2020). These symptoms appear abruptly and are frequently accompanied by less nasal symptoms such as nasal blockage or excessive nasal discharge. The majority of the time, anosmia and ageusia were present in asymptomatic people or as the disease's first presentation without any other symptoms.

10.2.5 Hematological Manifestations

Recent pathological studies on SARS-CoV-2 pathogenesis suggested that the course of the disease in Covid-19 was hypersensitivity pneumonitis rather than viral pneumonia (Liu et al., 2020).

It was concluded that in Covid-19 patients SARS-CoV-2 triggers a hyperactive immune response, also known as a cytokine storm. By releasing elevated amounts of cytokines into the circulatory system, the cytokine storm causes systemic problems in various organs.

One of the most significant side effects of SARS-CoV-2-infected pneumonia is multiorgan failure, which is caused by an excess of pro-inflammatory cytokines combined with a diminished oxygenation potential of the patient's blood. In extreme cases, additional consequences such as septic shock, difficult-to-cure metabolic acidosis, and coagulation malfunction are also observed (Kui et al., 2020; Lin et al., 2020; Liu et al., 2020; Rothan and Byrareddy, 2020). Patients with Covid-19 frequently also have early development of a normal or decreased white blood cell count or decreased lymphocyte count (lymphopenia) (Rothan and Byrareddy, 2020; Cao et al., 2020; Kui et al., 2020; Lin et al., 2020). A Covid-19 patient who experienced lower-extremity purpura, epistaxis, and neurological symptoms, including headaches, was also found to have isolated (presumably autoimmune) thrombocytopenia. Head CT results linked this condition to subarachnoid microhemorrhage (Zulfiqar et al., 2020).

10.2.6 Skin Manifestations

According to a recent study SARS-CoV-2 infection showed cutaneous involvement around skin manifestations of Covid-19 patients (Behzad et al., 2020). The study, based on information obtained from 88 Covid-19 patients, found that erythematous rash, widespread urticaria, and chickenpox-like vesicles were among the skin manifestations of the disease. It was also reported that the main area affected by the skin lesions was the trunk and that they were not particularly itchy (Recalcati, 2020).

10.2.7 Reproductive System Involvement

A recent study employing single-cell RNA sequencing revealed that the germ cells, Leydig cells, and Sertoli cells of testis contain the ACE2 receptor, a target for SARS-CoV-2 infection (Shen et al., 2020), showing that the SARS-CoV-2 virus may be a tropism site and reservoir in the testis. Leydig cells are involved in the generation of androgen, while Sertoli cells are critical to maintaining the balance of seminiferous tubules and spermatogenesis. According to the study, semen testing performed after contracting Covid-19 infection revealed reduced sperm concentration and slow motility for up to three months after infection (Q. Zhang et al., 2020). This indicated that the impact of Covid-19 infection on male fertility might be only transient. In addition, a study of 81 male Covid-19 patients revealed low testosterone, high luteinizing hormone (LH), and a low testosterone/LH ratio, suggesting potential viral testicular injury that directly effects the function of Leydig cells (Tsujimoto et al., 2020). Therefore, it has been recommended that after recovery, male Covid-19 survivors, particularly those with reproductive desires, should have their testicles and reproductive organs inspected (T. Chen et al., 2020; Tsuji et. al., 2020). Özveri and colleagues described an uncommon case of a male patient who had acute external genital pain, which was supposed to be the first clinical symptom of Covid-19 (Özveri et al., 2020).

10.2.8 Ophthalmic Manifestations

Conjunctivitis, anterior uveitis, retinitis, and optic neuritis are just a few of the many ophthalmic symptoms that can be caused by coronaviruses (L. Chen et al., 2020; Khavandi et al., 2020; Ulhaq and Soraya, 2020; Wu et al., 2020). Between 2% and 32% of people had ocular symptoms (Ulhaq and Soraya, 2020) and this was associated with the severity of the Covid-19 infection (L. Chen et al., 2020; Khavandi et al., 2020; Wu et al., 2020). Similar to other nonrespiratory systems, ophthalmic symptoms may show as the initial symptom without any other dysfunction. The nasolacrimal system, which acts as a conduit between the eye and the respiratory tract, and the lacrimal gland's function in hematogenous dissemination are two of the suggested mechanisms for the ocular transmission of the virus (Canatan, Vives Corrons and De Sanctis, 2020).

However, the virus probably does have an ocular tropism, just like other respiratory viruses. Additionally, it cannot be excluded that the ocular manifestation of a SARS-CoV-2 infection may in fact take the form of a localized, temporary vasculitis given the high vascularity of conjunctiva and the presence of ACE2 on the surface of endothelial cells (Gu and Korteweg, 2007; Ho et al., 2020).

10.2.9 Endocrinal Manifestations

It is generally known that there is a link between pulmonary diseases and blood sugar levels (Baker et al., 2006). According to studies, Covid-19 may make it worse for some diabetics to control their blood sugar levels (L. Zhu et al., 2020). According to preliminary data from a research by Ebekozien and colleagues, over half of the cases of Covid-19 that were either confirmed or suspected also had hyperglycemia, and approximately one-third also had diabetic ketoacidosis. In certain cases, new-onset diabetes also manifested (Ebekozien et al., 2020). Due to the high expression of ACE2 receptors in the pancreas and the intense inflammatory response, direct pancreatic involvement by SARS-CoV-2 was suspected in severe Covid-19 patients, which may account for pancreatic failure (Thaweerat, 2020). This may be related to elevated serum levels of lipase and/or amylase (Banks et al., 2013). To prevent the infection from flaring up and/or problems from uncontrolled diabetes, it is crucial to strictly regulate and monitor blood sugar levels in all cases of suspected or confirmed Covid-19. A panel of endocrine markers should be examined in severe situations.

10.2.10 Microbial Coinfections in COVID-19

Beyond the pathophysiology of SARS-CoV-2, microbial coinfection has a significant role in the initiation and progression of SARS-CoV-2 infection by making it difficult to diagnose, treat, and predict the prognosis of Covid-19, as well as by raising the disease's symptom and fatality rates (X. Chen et al., 2020). The reports on SARS-CoV-2 coinfections with bacteria, fungi, and other viruses are limited. Nevertheless, to treat Covid-19 with evidence-based care, the clinical data of SARS-CoV-2 coinfection are very helpful. Patients who have severe SARS-CoV-2 infections, which also included other viruses, bacteria, and fungus, have a considerably higher chance of coinfection than patients who are not in a critical condition (X. Chen et al., 2020). The study by X. Chen and colleagues revealed that Covid-19 patients have a significant chance of acquiring invasive pulmonary aspergillosis, which has been observed in France, where 9 out of 27 (33%) Covid-19 patients with invasive pulmonary aspergillosis were hospitalized in an intensive care unit (ICU) (Alanio et al., 2020) and 5 of 19 admitted (26%) in Germany as proven by histopathology of autopsy (Koehler et al., 2020). According to Zhou and colleagues, secondary bacterial infections caused the deaths of 50% of Covid-19 patients. Additionally, it was revealed that chronic obstructive pulmonary disease (COPD) patients had underlying, persistent bacterial infections prior to contracting the SARS-CoV-2 virus (Zhou et al., 2020).

Wang and colleagues described a case of a 37-year-old man from Wuhan who was simultaneously infected with SARS-CoV-2 and human immunodeficiency virus (HIV). They highlighted the possibility that coinfection could harm T lymphocytes, weaken the immune system, cause B-cell dysfunction and abnormal polyclonal activation, and lengthen the course of the disease (M. Wang et al., 2020; Thakur et al., 2021).

10.2.11 COVID-19 and Unusual Manifestations

As unexpected manifestations of Covid-19, subacute thyroiditis, oral lesions, large vascular stroke, rheumatologic skin disease, immune thrombocytopenia, endothelitis, pulmonary thromboembolism, and angiogenesis associated ARDS have also been documented. There is a chance that Covid-19 patients will develop viral arthritis and musculoskeletal pain because previous research shown that

SARS-CoV-2 antigen and antibodies react with those from people with rheumatoid arthritis, systemic sclerosis, and systemic lupus erythematosus (Arora et al., 2020).

Other unusual manifestations of Covid-19 have also been observed, including Covid toes (lesions), Covid tongue, long Covid (symptoms may persist for weeks or even months after virus recovery), Kawasaki disease, loss of taste or smell, severe appetite loss, headaches, dizziness and confusion, hallucinations, blood clots, hearing loss, high blood sugar, etc.

10.3 CONCLUSIONS

Every organ in the body is vulnerable to damage from the SARS-CoV-2 virus and many symptoms may be attributed to Covid-19 disease. To prevent deterioration and complications, every case must be evaluated immediately and with a high degree of suspicion. After recuperation, any afflicted organs should be monitored, and prompt management of problems is required.

Until the situation is resolved, doctors who are not involved in the administration of Covid-19 should be alert for any potential additional routes of infection transmission and practice infection control. To prevent complications and maintain proper control of any organ injury, multidisciplinary care is required.

REFERENCES

Adukia, S. A., R. S. Ruhatiya, H. M. Maheshwarappa, R. B. Manjunath and G. N. Jain (2020). "Extrapulmonary features of Covid-19: A concise review." *Indian Journal of Critical Care Medicine* 24(7): 575–580.

Ahmad, I. and F. A. Rathore (2020). "Neurological manifestations and complications of Covid-19: A literature review." *Journal of Clinical Neuroscience* 77: 8–12.

Alanio, A., S. Dellière, S. Fodil, S. Bretagne and B. Mégarbane (2020). "Prevalence of putative invasive pulmonary aspergillosis in critically ill patients with Covid-19." *The Lancet Respiratory Medicine* 8(6): e48–e49.

American Society for Gastrointestinal Endoscopy (ASGE) (2020). "Joint GI society message: Covid-19 clinical insights for our community of gastroenterologists and gastroenterology care providers." March 21.

Arabi, Y. M., A. A. Arifi, H. H. Balkhy, H. Najm, A. S. Aldawood, A. Ghabashi, H. Hawa, A. Alothman, A. Khaldi and B. Al Raiy (2014). "Clinical course and outcomes of critically ill patients with Middle East respiratory syndrome coronavirus infection." *Annals of Internal Medicine* 160(6): 389–397.

Arabi, Y. M., H. H. Balkhy, F. G. Hayden, A. Bouchama, T. Luke, J. K. Baillie, A. Al-Omari, A. H. Hajeer, M. Senga and M. R. Denison (2017). "Middle East respiratory syndrome." *New England Journal of Medicine* 376(6): 584–594.

Arora, G., M. Kassir, M. Jafferany, H. Galadari, T. Lotti, F. Satolli, R. Sadoughifar, Z. Sitkowska and M. Goldust (2020). "The Covid-19 outbreak and rheumatologic skin diseases." *Dermatologic Therapy* 33(4): e13357.

Baker, E. H., D. M. Wood, A. L. Brennan, N. Clark, D. L. Baines and B. J. Philips (2006). "Hyperglycaemia and pulmonary infection." *Proceedings of the Nutrition Society* 65(3): 227–235.

Banks, P. A., T. L. Bollen, C. Dervenis, H. G. Gooszen, C. D. Johnson, M. G. Sarr, G. G. Tsiotos and S. S. Vege (2013). "Classification of acute pancreatitis – 2012: Revision of the Atlanta classification and definitions by international consensus." *Gut* 62(1): 102–111.

Behzad, S., L. Aghaghazvini, A. R. Radmard and A. Gholamrezanezhad (2020). "Extrapulmonary manifestations of Covid-19: Radiologic and clinical overview." *Clinical Imaging* 66: 35–41.

Boukhris, M., A. Hillani, F. Moroni, M. S. Annabi, F. Addad, M. H. Ribeiro, S. Mansour, X. Zhao, L. F. Ybarra and A. Abbate (2020). "Cardiovascular implications of the Covid-19 pandemic: A global perspective." *Canadian Journal of Cardiology* 36(7): 1068–1080.

Bradley, B. T. and A. Bryan (2019). "Emerging respiratory infections: The infectious disease pathology of SARS, MERS, pandemic influenza, and Legionella." *Seminars in Diagnostic Pathology* 36(3): 152–159.

Canatan, D., J. L. Vives Corrons and V. De Sanctis (2020). "The multifacets of Covid-19 in adult patients: A concise clinical review on pulmonary and extrapulmonary manifestations for healthcare physicians: Covid-19 and pulmonary and extrapulmonary manifestations." *Acta Bio Medica Atenei Parmensis* 91(4): e2020173.

Cao, J., B. Wang, T. Tang, L. Lv, Z. Ding, Z. Li, R. Hu, Q. Wei, A. Shen, Y. Fu and B. Liu (2020). "Three-dimensional culture of MSCs produces exosomes with improved yield and enhanced therapeutic efficacy for cisplatin-induced acute kidney injury." *Stem Cell Research & Therapy* 11(1): 206.

Cascella, M., M. Rajnik, A. Cuomo, S. C. Dulebohn and R. Di Napoli (2021). "Features, evaluation, and treatment of coronavirus (Covid-19)." Statpearls.

Chen, L., C. Deng, X. Chen, X. Zhang, B. Chen, H. Yu, Y. Qin, K. Xiao, H. Zhang and X. Sun (2020). "Ocular manifestations and clinical characteristics of 535 cases of Covid-19 in Wuhan, China: A cross-sectional study." *Acta Ophthalmologica* 98(8): e951–e959.

Chen, T., D. Wu, H. Chen, W. Yan, D. Yang, G. Chen, K. Ma, D. Xu, H. Yu, H. Wang, T. Wang, W. Guo, J. Chen, C. Ding, X. Zhang, J. Huang, M. Han, S. Li, X. Luo, J. Zhao and Q. Ning (2020). "Clinical characteristics of 113 deceased patients with coronavirus disease 2019: Retrospective study." *British Medical Journal* 368: m1091.

Chen, X., B. Liao, L. Cheng, X. Peng, X. Xu, Y. Li, T. Hu, J. Li, X. Zhou and B. Ren (2020). "The microbial coinfection in Covid-19." *Applied Microbiology and Biotechnology* 104(18): 7777–7785.

Cheng, Y., R. Luo, K. Wang, M. Zhang, Z. Wang, L. Dong, J. Li, Y. Yao, S. Ge and G. Xu (2020). "Kidney disease is associated with in-hospital death of patients with Covid-19." *Kidney International* 97(5): 829–838.

Clerkin, K. J., J. A. Fried, J. Raikhelkar, G. Sayer, J. M. Griffin, A. Masoumi, S. S. Jain, D. Burkhoff, D. Kumaraiah, L. Rabbani, A. Schwartz and N. Uriel (2020). "Covid-19 and cardiovascular disease." *Circulation* 141(20): 1648–1655.

Cui, J., F. Li and Z.-L. Shi (2019). "Origin and evolution of pathogenic coronaviruses." *Nature Reviews Microbiology* 17(3): 181–192.

Cummings, M. J., M. R. Baldwin, D. Abrams, S. D. Jacobson, B. J. Meyer, E. M. Balough, J. G. Aaron, J. Claassen, L. E. Rabbani and J. Hastie (2020). "Epidemiology, clinical course, and outcomes of critically ill adults with Covid-19 in New York City: A prospective cohort study." *The Lancet* 395(10239): 1763–1770.

Diao, B., Z. Feng, C. Wang, H. Wang, L. Liu, C. Wang, R. Wang, Y. Liu, Y. Liu and G. Wang (2021). "Human kidney is a target for novel severe acute respiratory syndrome coronavirus 2 (SARS-CoV-2) infection." *Nature Communications* 12(1):2506.

Ebekozien, O. A., N. Noor, M. P. Gallagher and G. T. Alonso (2020). "Type 1 diabetes and covid-19: Preliminary findings from a multicenter surveillance study in the U.S." *Diabetes Care* 43(8): e83–e85.

Fan, R. F., Z. F. Li, D. Zhang and Z. Y. Wang (2020). "Involvement of Nrf2 and mitochondrial apoptotic signaling in trehalose protection against cadmium-induced kidney injury." *Metallomics* 12(12): 2098–2107.

Farouk, S. S., E. Fiaccadori, P. Cravedi and K. N. Campbell (2020). "Covid-19 and the kidney: What we think we know so far and what we don't." *Journal of Nephrology* 33(6): 1–6.

Feng, Y., X. Zhong, T. T. Tang, C. Wang, L. T. Wang, Z. L. Li, H. F. Ni, B. Wang, M. Wu, D. Liu, H. Liu, R. N. Tang, B. C. Liu and L. L. Lv (2020). "Rab27a dependent exosome releasing participated in albumin handling as a coordinated approach to lysosome in kidney disease." *Cell Death & Disease* 11(7): 513.

Filatov, A., P. Sharma, F. Hindi and P. S. Espinosa (2020). "Neurological complications of coronavirus disease (Covid-19): Encephalopathy." *Cureus* 12(3): e7352.

Fried, J. A., K. Ramasubbu, R. Bhatt, V. K. Topkara, K. J. Clerkin, E. Horn, L. Rabbani, D. Brodie, S. S. Jain and A. J. Kirtane (2020). "The variety of cardiovascular presentations of Covid-19." *Circulation* 141(23): 1930–1936.

Gu, J. and C. Korteweg (2007). "Pathology and pathogenesis of severe acute respiratory syndrome." *American Journal of Pathology* 170(4): 1136–1147.

Gu, J., E. Gong, B. Zhang, J. Zheng, Z. Gao, Y. Zhong, W. Zou, J. Zhan, S. Wang and Z. Xie (2005). "Multiple organ infection and the pathogenesis of SARS." *Journal of Experimental Medicine* 202(3): 415–424.

Guan, W. J., W. H. Liang, Y. Zhao, H. R. Liang, Z. S. Chen, Y. M. Li, X. Q. Liu, R. C. Chen, C. L. Tang, T. Wang, C. Q. Ou, L. Li, P. Y. Chen, L. Sang, W. Wang, J. F. Li, C. C. Li, L. M. Ou, B. Cheng, S. Xiong, Z. Y. Ni, J. Xiang, Y. Hu, L. Liu, H. Shan, C. L. Lei, Y. X. Peng, L. Wei, Y. Liu, Y. H. Hu, P. Peng, J. M. Wang, J. Y. Liu, Z. Chen, G. Li, Z. J. Zheng, S. Q. Qiu, J. Luo, C. J. Ye, S. Y. Zhu, L. L. Cheng, F. Ye, S. Y. Li, J. P. Zheng, N. F. Zhang, N. S. Zhong and J. X. He (2020). "Comorbidity and its impact on 1590 patients with Covid-19 in China: A nationwide analysis." *European Respiratory Journal* J55(5).

Hamming, I., W. Timens, M. Bulthuis, A. Lely, G. v. Navis and H. van Goor (2004). "Tissue distribution of ACE2 protein, the functional receptor for SARS coronavirus. A first step in understanding SARS pathogenesis." *Journal of Pathology: A Journal of the Pathological Society of Great Britain and Ireland* 203(2): 631–637.

Han, C., C. Duan, S. Zhang, B. Spiegel, H. Shi, W. Wang, L. Zhang, R. Lin, J. Liu and Z. Ding (2020). "Digestive symptoms in Covid-19 patients with mild disease severity: Clinical presentation, stool viral RNA testing, and outcomes." *American Journal of Gastroenterology* 115(6): 916–923.

Henry, B. M. and G. Lippi (2020). "Chronic kidney disease is associated with severe coronavirus disease 2019 (Covid-19) infection." *International Urology & Nephrology* 152(6): 1193–1194.

Hirsch, J. S., J. H. Ng, D. W. Ross, P. Sharma, H. H. Shah, R. L. Barnett, A. D. Hazzan, S. Fishbane, K. D. Jhaveri and M. Abate (2020). "Acute kidney injury in patients hospitalized with Covid-19." *Kidney International* 98(1): 209–218.

Ho, D., R. Low, L. Tong, V. Gupta, A. Veeraraghavan and R. Agrawal (2020). "Covid-19 and the ocular surface: A review of transmission and manifestations." *Ocular Immunology and Inflammation* 28(5): 726–734.

Hoffmann, M., H. Kleine-Weber, S. Schroeder, N. Krüger, T. Herrler, S. Erichsen, T. S. Schiergens, G. Herrler, N. H. Wu, A. Nitsche, M. A. Müller, C. Drosten and S. Pöhlmann (2020). "SARS-CoV-2 Cell Entry Depends on ACE2 and TMPRSS2 and Is Blocked by a Clinically Proven Protease Inhibitor." *Cell* 181(2): 271–280.e278.

Hosseiny, M., S. Kooraki, A. Gholamrezanezhad, S. Reddy and L. Myers (2020). "Radiology perspective of coronavirus disease 2019 (Covid-19): Lessons from severe acute respiratory syndrome and Middle East respiratory syndrome." *American Journal of Roentgenology* 214(5): 1078–1082.

Hu, Y., J. Sun, Z. Dai, H. Deng, X. Li, Q. Huang, Y. Wu, L. Sun and Y. Xu (2020). "Prevalence and severity of corona virus disease 2019 (Covid-19): A systematic review and meta-analysis." *Journal of Clinical Virology* 127: 104371.

Huang, C., Y. Wang, X. Li, L. Ren, J. Zhao, Y. Hu, L. Zhang, G. Fan, J. Xu and X. Gu (2020). "Clinical features of patients infected with 2019 novel coronavirus in Wuhan, China." *The Lancet* 395(10223): 497–506.

Kang, Y., T. Chen, D. Mui, V. Ferrari, D. Jagasia, M. Scherrer-Crosbie, Y. Chen and Y. Han (2020). "Cardiovascular manifestations and treatment considerations in Covid-19." *Heart* 106(15): 1132.

Khavandi, S., E. Tabibzadeh, M. Naderan and S. Shoar (2020). "Corona virus disease-19 (Covid-19) presenting as conjunctivitis: Atypically high-risk during a pandemic." *Contact Lens Anterior Eye* 43(3): 211.

Koehler, P., O. A. Cornely, B. W. Böttiger, F. Dusse, D. A. Eichenauer, F. Fuchs, M. Hallek, N. Jung, F. Klein and T. Persigehl (2020). "Covid-19 associated pulmonary aspergillosis." *Mycoses* 63(6): 528–534.

Kui, L., Y. Fang, Y. Deng, W. Liu, M. Wang and J. Ma (2020). "Clinical characteristics of novel coronavirus cases in tertiary hospitals in Hubei Province." *Chinese Medical Journal* (Engl) 133(9): 1025–1031.

Kumar, A., R. Zarychanski, R. Pinto, D. J. Cook, J. Marshall, J. Lacroix, T. Stelfox, S. Bagshaw, K. Choong and F. Lamontagne (2009). "Critically ill patients with 2009 influenza A (H1N1) infection in Canada." *JAMA* 302(17): 1872–1879.

Kunutsor, S. K. and J. A. Laukkanen (2020). "Renal complications in Covid-19: A systematic review and meta-analysis." *Annals of Medicine* 52(7): 345–353.

Lang, J. P., X. Wang, F. A. Moura, H. K. Siddiqi, D. A. Morrow and E. A. Bohula (2020). "A current review of Covid-19 for the cardiovascular specialist." *American Heart Journal* 226: 29–44.

Lechien, J. R., C. M. Chiesa-Estomba, D. R. De Siati, M. Horoi, S. D. Le Bon, A. Rodriguez, D. Dequanter, S. Blecic, F. El Afia and L. Distinguin (2020). "Olfactory and gustatory dysfunctions as a clinical presentation of mild-to-moderate forms of the coronavirus disease (Covid-19): A multicenter European study." *European Archives of Oto-Rhino-Laryngology* 277(8): 2251–2261.

Li, W., M. J. Moore, N. Vasilieva, J. Sui, S. K. Wong, M. A. Berne, M. Somasundaran, J. L. Sullivan, K. Luzuriaga and T. C. Greenough (2003). "Angiotensin-converting enzyme 2 is a functional receptor for the SARS coronavirus." *Nature* 426(6965): 450–454.

Lin, N., X. Zhou, X. Geng, C. Drewell, J. Hubner, Z. Li, Y. Zhang, M. Xue, U. Marx and B. Li (2020). "Repeated dose multi-drug testing using a microfluidic chip-based coculture of human liver and kidney proximal tubules equivalents." *Scientific Reports* 10(1): 8879.

Liu, J., X. Zheng, Q. Tong, W. Li, B. Wang, K. Sutter, M. Trilling, M. Lu, U. Dittmer and D. Yang (2020). "Overlapping and discrete aspects of the pathology and pathogenesis of the emerging human pathogenic coronaviruses SARS-CoV, MERS-CoV, and 2019-nCoV." *Journal of Medical Virology* 92(5): 491–494.

Mao, L., H. Jin, M. Wang, Y. Hu, S. Chen, Q. He, J. Chang, C. Hong, Y. Zhou, D. Wang, X. Miao, Y. Li and B. Hu (2020). "Neurologic Manifestations of Hospitalized Patients with Coronavirus Disease 2019 in Wuhan, China." *JAMA Neurology* 77(6): 683–690.

Netland, J., D. K. Meyerholz, S. Moore, M. Cassell and S. Perlman (2008). "Severe acute respiratory syndrome coronavirus infection causes neuronal death in the absence of encephalitis in mice transgenic for human ACE2." *Journal of Virology* 82(15): 7264–7275.

Özveri, H., M. T. Eren, C. E. Kırışoğlu and N. Sarıgüzel (2020). "Atypical presentation of SARS-CoV-2 infection in male genitalia." *Urology Case Reports* 33: 101349.

Recalcati, S. (2020). "Cutaneous manifestations in Covid-19: A first perspective." *Journal of the European Academy of Dermatology Venereology* 34(5): e212–e213.

Rothan, H. A. and S. N. Byrareddy (2020). "The epidemiology and pathogenesis of coronavirus disease (Covid-19) outbreak." *Journal of Autoimmunity* 109: 102433.

Sarkesh, A., A. Daei Sorkhabi, E. Sheykhsaran, F. Alinezhad, N. Mohammadzadeh, N. Hemmat and H. Bannazadeh Baghi (2020). "Extrapulmonary clinical manifestations in Covid-19 patients" *American Journal of Tropical Medicine and Hygiene* 103(5): 1783–1796.

Shen, Q., X. Xiao, A. Aierken, W. Yue, X. Wu, M. Liao and J. Hua (2020). "The ACE2 expression in Sertoli cells and germ cells may cause male reproductive disorder after SARS-CoV-2 infection." *Journal of Cellular Molecular Medicine* 24(16): 9472–9477.

Shi, Y., Y. Wang, C. Shao, J. Huang, J. Gan, X. Huang, E. Bucci, M. Piacentini, G. Ippolito and G. Melino (2020). "Covid-19 infection: The perspectives on immune responses." *Cell Death & Differentiation* 27(5): 1451–1454.

Sobh, E., E. Abuarrah, K. G. Abdelsalam, S. S. Awad, M. A. Badawy, M. A. Fathelbab, M. A. Aboulfotouh and M. F. Awadallah (2020). "Novel coronavirus disease 2019 (Covid-19) non-respiratory involvement." *Egyptian Journal of Bronchology* 14(1): 32.

Sun, D. Q., T. Y. Wang, K. I. Zheng, G. Targher, C. D. Byrne, Y. P. Chen and M. H. Zheng (2020). "Subclinical acute kidney injury in covid-19 patients: A retrospective cohort study." *Nephron* 144(7): 347–350.

Thakur, V., R. K. Ratho, P. Kumar, S. K. Bhatia, I. Bora, G. K. Mohi, S. K. Saxena, M. Devi, D. Yadav and S. Mehariya (2021). "Multi-organ involvement in covid-19: Beyond pulmonary manifestations." *Journal of Clinical Medicine* 10(3): 446.

Thaweerat, W. (2020). "Current evidence on pancreatic involvement in SARS-CoV-2 infection." *Pancreatology: Official Journal of the International Association of Pancreatology (IAP)* 20(5): 1013–1014.

Tian, Y., L. Rong, W. Nian and Y. He (2020). "Gastrointestinal features in Covid-19 and the possibility of faecal transmission." *Alimentary Pharmacology Therapeutics* 51(9): 843–851.

Tran, J., J. Glavis-Bloom, T. Bryan, K. Harding, C. Chahine and R. Houshyar (2020). "Covid-19 patient presenting with initial gastrointestinal symptoms." Eurorad.

Tsuji, A., Y. Ikeda, M. Murakami and S. Matsuda (2020). "COVID-19, an infertility risk?" *Clinical Obstetrics, Gynecology and Reproductive Medicine* 6: 1–1.

Tsujimoto, H., T. Kasahara, S. I. Sueta, T. Araoka, S. Sakamoto, C. Okada, S. I. Mae, T. Nakajima, N. Okamoto, D. Taura, M. Nasu, T. Shimizu, M. Ryosaka, Z. Li, M. Sone, M. Ikeya, A. Watanabe and K. Osafune (2020). "A modular differentiation system maps multiple human kidney lineages from pluripotent stem cells." *Cell Reports* 31(1): 107476.

Ulhaq, Z. S. and G. V. Soraya (2020). "The prevalence of ophthalmic manifestations in Covid-19 and the diagnostic value of ocular tissue/fluid." *Graefe's Archive for Clinical Experimental Ophthalmology* 258(6): 1351–1352.

Wang, M., L. Luo, H. Bu and H. Xia (2020). "One case of coronavirus disease 2019 (Covid-19) in a patient co-infected by HIV with a low CD4+ T-cell count." *International Journal of Infectious Diseases* 96: 148–150.

Wang, W., S. Zhang, F. Yang, J. Xie, J. Chen and Z. Li (2020). "Diosmetin alleviates acute kidney injury by promoting the TUG1/Nrf2/HO-1 pathway in sepsis rats." *International Immunopharmacology* l88: 106965.

Wu, P., F. Duan, C. Luo, Q. Liu, X. Qu, L. Liang and K. Wu (2020). "Characteristics of ocular findings of patients with coronavirus disease 2019 (Covid-19) in Hubei Province, China." *JAMA Ophthalmology* 138(5): 575–578.

Yang, J., Y. Zheng, X. Gou, K. Pu, Z. Chen, Q. Guo, R. Ji, H. Wang, Y. Wang and Y. Zhou (2020). "Prevalence of comorbidities and its effects in patients infected with SARS-CoV-2: A systematic review and meta-analysis." *International Journal of Infectious Diseases* 94: 91–95.

Zhang, C., L. Shi and F.-S. Wang (2020). "Liver injury in Covid-19: Management and challenges." *Lancet Gastroenterology Hepatology* 5(5): 428–430.

Zhang, Q., L. He, Y. Dong, Y. Fei, J. Wen, X. Li, J. Guan, F. Liu, T. Zhou, Z. Li, Y. Fan and N. Wang (2020). "Sitagliptin ameliorates renal tubular injury in diabetic kidney disease via STAT3-dependent mitochondrial homeostasis through SDF-1alpha/CXCR4 pathway." *FASEB* J34(6): 7500–7519.

Zhang, W., Y. Zhao, F. Zhang, Q. Wang, T. Li, Z. Liu, J. Wang, Y. Qin, X. Zhang, X. Yan, X. Zeng and S. Zhang (2020). "The use of anti-inflammatory drugs in the treatment of people with severe coronavirus disease 2019 (Covid-19): The perspectives of clinical immunologists from China." *Clinical Immunology* (Orlando, FL) 214: 108393–108393.

Zhou, F., T. Yu, R. Du, G. Fan, Y. Liu, Z. Liu, J. Xiang, Y. Wang, B. Song and X. Gu (2020). "Clinical course and risk factors for mortality of adult inpatients with Covid-19 in Wuhan, China: A retrospective cohort study." *The Lancet* 395(10229): 1054–1062.

Zhu, L., Z.-G. She, X. Cheng, J.-J. Qin, X.-J. Zhang, J. Cai, F. Lei, H. Wang, J. Xie, W. Wang, H. Li, P. Zhang, X. Song, X. Chen, M. Xiang, C. Zhang, L. Bai, D. Xiang, M.-M. Chen, Y. Liu, Y. Yan, M. Liu, W. Mao, J. Zou, L. Liu, G. Chen, P. Luo, B. Xiao, C. Zhang, Z. Zhang, Z. Lu, J. Wang, H. Lu, X. Xia, D. Wang, X. Liao, G. Peng, P. Ye, J. Yang, Y. Yuan, X. Huang, J. Guo, B.-H. Zhang and H. Li (2020). "Association of blood glucose control and outcomes in patients with covid-19 and pre-existing Type 2 diabetes." *Cell Metabolism* 31(6): 1068–1077.e1063.

Zhu, N., D. Zhang, W. Wang, X. Li, B. Yang, J. Song, X. Zhao, B. Huang, W. Shi, R. Lu, P. Niu, F. Zhan, X. Ma, D. Wang, W. Xu, G. Wu, G. F. Gao and W. Tan (2020). "A novel coronavirus from patients with pneumonia in China, 2019." *New England Journal of Medicine* 382(8): 727–733.

Zimmerman, K. A., J. Huang, L. He, D. Z. Revell, Z. Li, J. S. Hsu, W. R. Fitzgibbon, E. S. Hazard, G. Hardiman, M. Mrug, P. D. Bell, B. K. Yoder and T. Saigusa (2020). "Interferon regulatory factor-5 in resident macrophage promotes polycystic kidney disease." *Kidney* 3601(3): 179–190.

Zou, X., K. Chen, J. Zou, P. Han, J. Hao and Z. Han (2020). "Single-cell RNA-seq data analysis on the receptor ACE2 expression reveals the potential risk of different human organs vulnerable to 2019-nCoV infection." *Frontiers of Medicine* 14(2): 185–192.

Zulfiqar, A.-A., N. Lorenzo-Villalba, P. Hassler and E. Andrès (2020). "Immune thrombocytopenic purpura in a patient with Covid-19." *New England Journal of Medicine* 382(18): e43.

11 Update on Covid-19 Treatment

Vinant Bhargava,[1] Gaurav Bhandari,[1] Nikita Pawar,[1] and Priti Meena[2]
Consultant Nephrology, Institute of Renal Science, Sir Gangaram Hospital, New Delhi, India
[1]Email: vinant_bhargava@yahoo.com
[1]Institute of Renal Science, Sir Gangaram Hospital, New Delhi, India
[2]All India Institute of Medical Sciences, Bhubaneswar

CONTENTS

11.1 INTRODUCTION

Since the outbreak in late December 2019, the coronavirus disease (Covid-19) pandemic has affected around 633,601,048 individuals and killed approximately 6,596,542 patients as of 15 November 2022 (1). It has become a public health emergency affecting the global healthcare system and economy. Although understanding of the pathophysiology of Covid-19 has advanced greatly, safe and effective therapeutic measures to treat the disease are still lacking. Undoubtedly, early and appropriate treatment of Covid-19 can help in mitigating its progression to a more critical illness. However, to date, as per the WHO, there is no specific drug recommended to prevent the novel coronavirus (2). Nevertheless, continuous attempts to search for new potential antiviral agents are ongoing. Currently, there are many therapeutic options available including antiviral drugs (e.g., remdesivir), antiinflammatory drugs (e.g., dexamethasone), and immunomodulators agents (e.g., baricitinib, tocilizumab) available under FDA issued Emergency Use Authorization (EUA). Recently a cocktail of anti-SARS-CoV-2 monoclonal antibodies (e.g., bamlanivimab/etesevimab,

DOI: 10.1201/9781003358909-11

casirivimab/imdevimab) has been approved by the FDA (3). This chapter aims to describe the latest approaches to managing Covid-19. We summarize available antiviral agents focusing on their mechanism of action, efficacy, side effects, and pertinent clinical trials.

11.2 VIROLOGY AND POTENTIAL DRUG TARGETS

SARS-CoV-2 is a single-stranded RNA-enveloped virus. It attacks cells via the viral structural spike (S) protein that binds to the angiotensin-converting enzyme 2 (ACE2) receptor. Following receptor binding, virus enters the host cell through endosomes. It's entry facilited by another receptor, known as type 2 transmembrane serine protease, TMPRSS2. After entering the cell, synthesis of multiple viral polyproteins occurs, which encodes for the replicase-transcriptase complex. Viral RNA is then synthesized via RNA-dependent RNA polymerase. The next step is formation of structural proteins followed by assembly and release of viral particles (4). Understanding and insight into the viral lifecycle provide us potential targets for drug therapy. Figure 11.1 summarizes the mechanism of action of some antiviral agents used in Covid-19 treatment.

11.3 THERAPEUTIC APPROACHES

Treatment of Covid-19 approaches are based on the following (4,5):

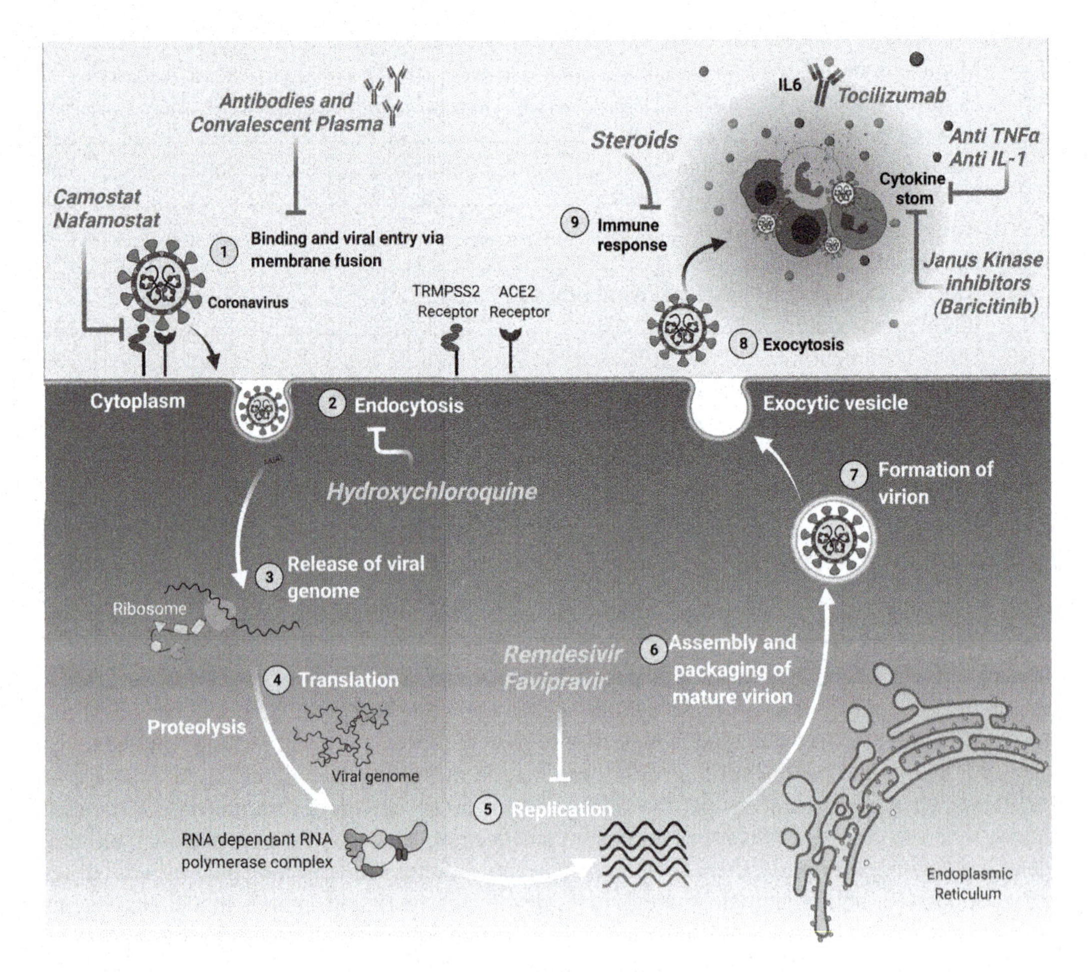

FIGURE 11.1 Mechanism of action of some therapeutic agents used in Covid-19 treatment.

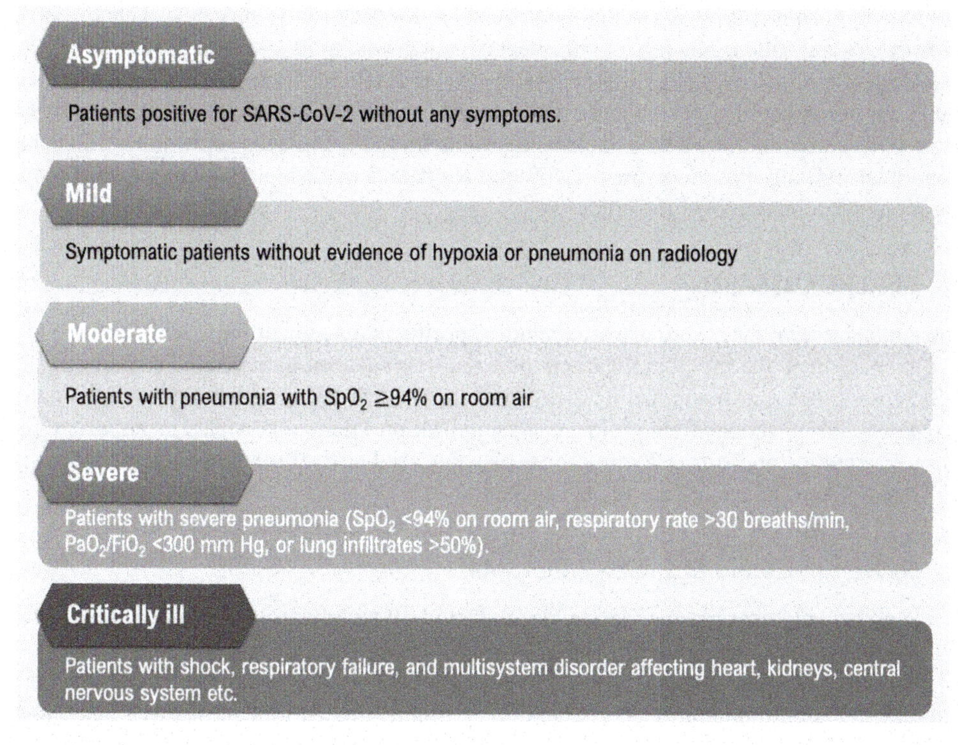

FIGURE 11.2 Categories of the severity of illness in Covid-19.

- Preventing virus entry into the host cells
- Inhibition of viral genome translation
- Suppression of viral genome replication
- Suppression of host inflammatory response against virus and cytokine storm

The disease course is mediated by two mechanisms. Disease manifestations are primarily due to viral replication in the early part of the illnesss, while dysregulated immune response causes tissue damage in the later clinical period (5). This helps to guide the appropriate treatment with use of antivirals during the initial period and immunosuppressants later.

11.4 MANAGEMENT OF COVID-19

The clinical spectrum of Covid-19 disease ranges from asymptomatic to critical illness requiring hospitalization (6) as shown in Figure 11.2. Clinical management of Covid-19 depends on the clinical status of the patient and severity of illness.

11.5 MILD-TO-MODERATE DISEASE

No specific treatment in clinically stable patients is required. Management consists of supportive care, symptomatic treatment, self-isolation, monitoring vitals, and counseling when to seek in-person evaluation in case of worsening symptoms (e.g., progressive dyspnea, SpO2 <95% at sea level, chest tightness, dizziness, confusion, etc.).

Symptomatic and supportive treatment consists of use of antipyretics, antitussives, analgesics, steam inhalation, breathing exercises, sleeping in prone position, etc. There are no recommendations for the use of steroids, chloroquine (or hydroxychloroquine), antibiotics (azithromycin or doxycycline), antivirals (remdesivir), or oral anticoagulation for outpatient treatment with mild-to-moderate disease (7).

Use of specific anti-Covid treatment consisting of combination of anti-SARS-CoV-2 monoclonal antibodies is recommended for patients with a high risk of clinical progression like the elderly or those with comorbidities. These include casirivimab 1,200 mg plus imdevimab 1,200 mg or bamlanivimab 700 mg plus etesevimab 1,400 mg. No data is available comparing the efficacy and safety of one combination over the other (8).

11.6 SEVERE DISEASE

Inpatient management for severe disease includes specific management for Covid disease and supportive management for the complications of Covid-19 such as hypoxemia respiratory failure/ARDS, sepsis/shock, cardiomyopathy, arrhythmia, and acute kidney injury (9). Additionally, in these patients, complications from prolonged hospitalization, such as secondary bacterial and fungal infections, thromboembolism, gastrointestinal bleeding, and critical illness polyneuropathy/myopathy also require proper attention and management (10).

11.7 ROLE OF CONVALESCENT PLASMA

Evidence on convalescent plasma therapy use in Covid-19 treatment is of low certainty. Recently, the FDA revised the convalescent plasma to limit the use of high-titre Covid-19 convalescent plasma in hospitalized patients during the early course of the disease or in hospitalized patients who have impaired humoral immunity (11). CONCOR-1 trial and REMAP-CAP showed that plasma is unlikely to benefit severe or critically ill patients (12). Preliminary review of double-blind, Phase 2 RCT in hospitalized adults with severe Covid-19 (n = 223) in the United States (n = 73) and Brazil (n = 150), which included 57% of the participants, required supplemental oxygen at baseline, 25% required high-flow oxygen or noninvasive ventilation, and 13% required Invasive Mechanical Ventilation or Extracorporeal Membrane Oxygenation and did not show any difference between convalescent plasma arm and control arm (13).

11.8 ROLE OF STEROIDS FOR COVID-19

Dexamethasone is a promising drug in Covid-19. Dexamethasone alone or in combination with remdesvir is recommended in severe disease (14). A multicentric randomized trial in the UK showed significantly decreased mortality and hospital stay in the dexamethasone group (14), where 23.3% of the participants in the dexamethasone group died within 28 days of enrollment compared with 26.2% in the standard of care arm.

Dexamethasone 6 mg intravenous or per-oral for 10 days (or until discharge) or equivalent glucocorticoid dose may be substituted if dexamethasone is unavailable. Equivalent total daily doses of alternative glucocorticoids to dexamethasone 6 mg daily are methylprednisolone 32 mg and prednisone 40 mg. A few studies have shown that high dose of dexamethasone (20 mg/day for 1–5 days followed by 10 mg/day for another 6–10 days) is associated with high ventilator-free survival (15).

Among hospitalized patients, 28-day mortality was 17% lower in the group that received dexamethasone than in the group that did not receive dexamethasone. Among critically ill hospitalized patients, the odds of mortality at 28 days was 34% less among patients treated with glucocorticoids than patients not treated with glucocorticoids.

11.9 ANTIVIRAL DRUGS

There are two different approaches to the use of antiviral drugs: repurposing already available conventional antiviral drugs, and developing novel therapeutic agents targeting the abovementioned steps. Already available conventional drugs such as hydroxychloroquine, nitazoxanide, azathioprine, lopinavir/ritonavir combination, and ivermectin have been tried; however, major RCTs have failed to demonstrate any efficacy of these drugs in the management of Covid-19 and there is insufficient evidence to recommend their use (7).

Remdesvir is an intravenous nucleotide analog prodrug that binds to the viral RNA-dependent RNA polymerase and inhibits viral replication through premature termination of RNA transcription. Currently, evidence to recommend either for or against the use of remdesivir is insufficient (16). In ACTT-1, a multinational RCT comprising 138 patients in this group, remdesivir showed no significant benefit in patients with mild-to-moderate disease (rate ratio for recovery 1.29; 95% CI, 0.91–1.83) (17). The results from the WHO SOLIDARITY trial conducted at 405 hospitals spanning across 40 countries involving 11,330 inpatients with Covid-19 who were randomized to receive remdesivir (2,750) or no trial drug (4,088) found that remdesivir had little or no effect on overall mortality, initiation of mechanical ventilation, and length of hospital stay (18). In most trials the dose of remdesivir used was 200 mg on day 1 followed by 100 mg daily on days 2–5. (17-19). According to an advisory for rational use of remdesivir for Covid-19 treatment from MOHFW, AIIMS, and ICMR, remdesivir is to be used only in selected moderate/severe hospitalized Covid-19 patients on supplemental oxygen as it is a reserve drug approved under Emergency Use Authorization only based on limited scientific evidence globally (19). Moreover, the role of remdesvir is uncertain in critical disease or in patients who are on ECMO and data on efficacy of remdesvir on new variants is also scarce.

Before starting remdesvir liver function tests and prothrombin time should be done and it should be discontinued if alanine transaminase is >10 times the upper limit of normal. Sulfobutylether beta-cyclodextrin sodium (SBECD), which acts as a vehicle, accumulates in patients with renal impairment and may result in liver and renal toxicities. Remdesivir is not recommended for patients with an eGFR <30 mL/min. But some studies have shown better tolerability of remdesvir in acute kidney injury and chronic kidney disease on hemodialysis patients (20).

Other antiviral drugs used are lopinavir/ritonavir during the early phase of the pandemic. But currently a combination is not recommended (21).

11.10 KINASE INHIBITORS

Janus kinase (JAK) inhibitors interfere with the phosphorylation of signal transducer and activator of transcription (STAT) proteins that are involved in vital cellular functions, including signaling, growth, and survival (22).

In a multinational, RCT-Adaptive Covid-19 Treatment Trial 2 (ACTT-2) on baricitinib use in hospitalized patients with Covid-19 pneumonia, 1,033 patients were randomized 1:1 to oral baricitinib 4 mg or placebo, for up to 14 days, in combination with remdesivir, for up to 10 days. It showed a shorter time to clinical recovery compared to placebo group (median recovery time of 7 vs. 8 days, respectively). Baricitinib plus remdesivir was superior to remdesivir therapy alone. It also accelerated clinical improvement in hospitalized patients with Covid-19, particularly those who were receiving high flow oxygen supplementation or noninvasive ventilation. But there was no significant effect on ventilated patients. Baricitinib use is not recommended in patients with impaired hepatic or renal function (estimated GFR <60 mL/min/1.73 m^2) (23). Patients should be screened for viral hepatitis and tuberculosis before treatment. It has been shown to be associated with thrombotic events.

Other kinase inhibitors, ruxolutinib to facitinib, have shown antiviral effects by blocking SARS-CoV-2 from entering and infecting lung cells. The data to support this is limited in the present literature (24). Newer drugs in the category of Bruton's tyrosine kinase inhibitor such as acalabrutinib, acalabrutinib, and zanubrutinib are not recommended for Covid-19, except in clinical trial(s) (25).

11.11 ANTI-SARS-COV-2 MONOCLONAL ANTIBODIES

The neutralizing antibodies to SARS-CoV-2 start developing about 10 days after disease onset. Antibody levels are higher in severe disease. The neutralizing activity of Covid-19 patients in plasma is correlated with the magnitude of antibody responses to SARS-CoV-2 S and N proteins. Monoclonal antibodies targeting the S protein are found to have the potential to prevent SARS-CoV-2 infection and alleviate symptoms and limit progression to severe disease (3).

Bamlanivimab (also known as LY-CoV555 and LY3819253) is a neutralizing monoclonal antibody that targets receptor-binding domain (RBD) of S protein of SARS-CoV-2. Etesevimab (also known as LY-CoV016 and LY3832479) is another monoclonal antibody that binds to a different epitope in RBD. Casirivimab (previously REGN10933) and imdevimab (previously REGN10987) are recombinant human monoclonal antibodies that bind to nonoverlapping epitopes of S protein RBD of SARS-CoV-2. These two combination products, bamlanivimab plus etesevimab and casirivimab plus imdevimab (REGN-COV2), are FDA-approved drugs used for mild-to-moderate disease who are progressing to severe category. Recommended doses are bamlanivimab 700 mg plus etesevimab 1,400 mg *or* casirivimab 1,200 mg plus imdevimab 1,200 mg. It should be started within 10 days of symptoms. It has not shown efficacy in severe Covid patients (ACTIV-3) (3).

11.12 IMMUNOMODULATORS

Interferon alfa 1b has been tried in moderate-to-severe disease, but it has not benefited acute respiratory distressed patients. Currently, there is no data available regarding efficacy of interferon β-1a on the three new SARS-CoV-2 variants (B.1.1.7; B.1.351; and P.1) (26).

Interleukin (IL)-1 antagonists such as anakinra have been used off-label in severe Covid and were assessed in a small case-control study trial. This trial revealed lower requirement of mechanical ventilation in severe Covid (27).

In contrast to other antiinflammatory antibodies, interleukin-6 receptor inhibitor is approved by the FDA. Tocilizumab is given as a single intravenous dose of 8 mg/kg actual body weight up to 800 mg in rapid decompensation of Covid-19 disease. The results of the RECOVERY trial and REMAP-CAP provide consistent evidence that tocilizumab, when administered with corticosteroids, offers a modest mortality benefit in certain patients with Covid-19 who are severely ill, rapidly deteriorating with increasing oxygen needs, and have a significant inflammatory response. It should be avoided in severely immunocompromised patients (28,29).

11.13 OTHER TREATMENTS

- High flow nasal cannula (HFNC) and noninvasive positive pressure ventilation (NIPPV) is useful in acute hypoxic episode and may avoid invasive mechanical ventilation. However, these modalities are associated with greater risk of aerolisation. Endotracheal intubation should be considered in patients with impending respiratory failure with preoxygenation with 100% oxygen for 5 mins by HNFC. Use of low tidal volume (4–8 ml/kg) with high PEEP to maintain driving pressure as low as possible targeting plateau pressure <30 cm H2O is recommended (30).
- In patients with refractory hypoxemia (PaO_2:FiO_2 of <150 mm Hg), prone ventilation for > 12 to 16 hours per day and the use of a conservative fluid management strategy for ARDS patients without tissue hypoperfusion are strongly emphasized. The National Institutes of Health (NIH) Covid-19 Treatment Guidelines Panel recommends against inhaled pulmonary vasodilators such as nitric oxide. Use of recruitment maneuvres are encouraged (12,13).
- Vasopressors should be started to maintain mean arterial pressure (MAP) between 60 mmHg and 65 mmHg. Norepinephrine is the preferred initial vasopressor (12).

- Empiric antibacterial therapy should be considered if there is a concern for a secondary bacterial infection (12,13).
- As Covid-19 itself is a procoagulant disease, all patients in the severe and critical category should be started on prophylactic anticoagulation unless contraindicated. Injectable low-molecular-weight heparin or unfractionated heparin is preferred over oral anticoagulation. Anticoagulation prophylaxis is not recommended on discharge, but for certain high-VTE risk patients, postdischarge prophylaxis has been shown to be beneficial. The FDA has approved the use of rivaroxaban 10 mg daily for 31 to 39 days in these patients (31).
- Three international trials (ACTIV-4, REMAP-CAP, and ATTACC) compared the effectiveness of therapeutic dose anticoagulation and prophylactic dose anticoagulation in reducing the need for organ support over 21 days in moderately ill or critically ill adults hospitalized for Covid-19 (31).
- In cases of acute kidney injury requiring renal replacement therapy (RRT), the Covid-19 Treatment Guidelines Panel recommends continuous renal replacement therapy (CRRT) (32).

11.14 TREATMENT OF COVID-19 IN SPECIAL POPULATIONS

11.14.1 Pregnancy (33)

- General measures as advised to nonpregnant patients like self-isolation, monitoring of symptoms, and counselling about the nature of the disease.
- In case of hospitalization, she should be admitted at a facility where maternal and fetal monitoring and multispeciality care including obstetric, pediatric, infectious, pulmonary, and critical care consultation is available.
- The timing of delivery is decided by obstetric indications and not on Covid-19 status.
- Pregnant females should be not be withheld from effective anti-Covid-19 treatment.
- Use of investigational agents in pregnant patients should be based on the maternal and fetal safety and the severity of disease.

11.14.2 Children (34)

Usually, children have mild and asymptomatic Covid-19 disease compared to adults (26,27). Thus, no specific treatment is required. In children with severe disease or with comorbidities, there is limited data regarding recommendations for treatment in pediatric populations.

Guidelines for treatment for Covid-19 in children is extrapolated from recommendations of treatment in adults. Remdesivir is approved for emergency use in pediatric population with risk factors for severe disease and increasing need of oxygen requirement with weight 3.5 kgs or more and age less than 12 years. Dexamethasone in the dose of 0.15 mg/kg/dose (maximum 6mg) once a day is recommended for children with Covid-19 who have very high oxygen requirement and those requiring noninvasive and invasive ventilation. Alternative steroids like prednisone, MPS, or hydrocortisone can be used in place of dexamethasone.

Use of anti-SARS-CoV-2 monoclonal antibodies can be extended in children at high-risk of severe disease with aged ≥16 years. There are no recommendations for the use of convalescent plasma, baricitinib in combination with remdesivir, or sarilumab.

REFERENCES

1. WHO Coronavirus (COVID-19) Dashboard. Available from https://covid19.who.int/. Last accessed on November 20, 2022.

2. Coronavirus Disease (COVID-19) Pandemic. World Health Organization (July 21, 2021). Available from www.who.int/emergencies/diseases/novel-coronavirus2019?gclid=Cj0KCQjwlvT8BRDeARIsAACRFiW_AoKe8cr7J7MyGq46ojb4LEDnznIAUBkEXISSPxIfKCLbfmxjYgUaAriBEALw_wcB.
3. Taylor, PC, Adams, AC, Hufford, MM, et al. Neutralizing monoclonal antibodies for treatment of COVID-19. *Nat Rev Immunol* 21, 382–393 (2021). https://doi.org/10.1038/s41577-021-00542-x
4. Trougakos IP, Stamatelopoulos K, Terpos E, Tsitsilonis OE, Aivalioti E, Paraskevis D, Kastritis E, Pavlakis GN, Dimopoulos MA. Insights to SARS-CoV-2 life cycle, pathophysiology, and rationalized treatments that target COVID-19 clinical complications. *J Biomed Sci.* January 12, 2021; 28(1):9. doi: 10.1186/s12929-020-00703-5. PMID: 33435929; PMCID: PMC7801873
5. Wiersinga WJ, Rhodes A, Cheng AC, Peacock SJ, Prescott HC. Pathophysiology, transmission, diagnosis, and treatment of coronavirus disease 2019 (COVID-19): A review. *J Am Med Assoc.* 2020;324(8):782–793. doi:10.1001/jama.2020.12839
6. Clinical Spectrum of SARS-CoV-2 Infection. Available from www.covid19treatmentguidelines.nih.gov/overview/clinical-spectrum/. Last accessed on July 28, 2021.
7. COVID-19 Treatment Guidelines Panel. Coronavirus Disease 2019 (COVID-19) Treatment Guidelines. Vol. 2019, National Institute of Health. 2020. 130 p. Available from https://covid19treatmentguidelines.nih.gov/
8. Hurt, AC, Wheatley, AK. Neutralizing antibody therapeutics for COVID-19. *Viruses* 2021;13:628. https://doi.org/10.3390/v13040628
9. Murthy S, Gomersall CD, Fowler RA. Care for critically ill patients with COVID-19. *J Am Med Assoc.* 2020;323(15):1499–1500. doi:10.1001/jama.2020.3633
10. Cheng VC, Edwards KM, Gandhi R, Gallagher J. Available from www.idsociety.org/COVID19guidelines. Last accessed on January 8, 2021.
11. Joyner MJ, Carter RE, Senefeld JW, Klassen SA, Mills JR, Johnson PW, et al. Convalescent plasma antibody levels and the risk of death from covid-19. *N Engl J Med* [Internet]. 2021;384(11):1015–1027.
12. Bégin P, Callum J, Heddle NM, Cook R, Zeller MP, Tinmouth A, et al. Convalescent plasma for adults with acute COVID-19 respiratory illness (CONCOR-1): Study protocol for an international, multicentre, randomized, open-label trial. *Trials* [Internet]. 2021;22(1):1–17.
13. Phua J, Weng L, Ling L, Egi M, Lim CM, Divatia JV, Shrestha BR, Arabi YM, Ng J, Gomersall CD, Nishimura M, Koh Y, Du B; Asian Critical Care Clinical Trials Group. Intensive care management of coronavirus disease 2019 (COVID-19): Challenges and recommendations. *Lancet Respir Med.* May 2020;8(5):506–517. doi: 10.1016/S2213-2600(20)30161-2. Epub April 6, 2020. Erratum in: *Lancet Respir Med.* May 2020;8(5):e42. PMID: 32272080; PMCID: PMC7198848.
14. Illar J, Ferrando C, Martínez D, Ambrós A, Muñoz T, Soler JA, et al. Dexamethasone treatment for the acute respiratory distress syndrome: A multicentre, randomised controlled trial. *Lancet Respir Med* [Internet]. 2020;8(3):267–276.
15. Tomazini BM, Maia IS, Cavalcanti AB, Berwanger O, Rosa RG, Veiga VC, et al. Effect of dexamethasone on days alive and ventilator-free in patients with moderate or severe acute respiratory distress syndrome and COVID-19: The CoDEX randomized clinical trial. *J Am Med Assoc* [Internet]. 2020;324(13):1307–1316.
16. Piscoya A, Ng-Sueng LF, Parra del Riego A, Cerna-Viacava R, Pasupuleti V, Roman YM, et al. Efficacy and harms of remdesivir for the treatment of COVID-19: A systematic review and meta-analysis. *PLoS ONE.* 2020; 15(12):e0243705. https://doi.org/10.1371/journal.pone.0243705
17. Beigel JH, Tomashek KM, Dodd LE, et al. Remdesivir for the treatment of covid-19 – Final report. *N Engl J Med.* 2020;383(19):1813–1826. doi:10.1056/NEJMoa2007764
18. WHO Solidarity Trial Consortium, Pan H, Peto R, Henao-Restrepo AM, Preziosi MP, Sathiyamoorthy V, Abdool Karim Q, et al. Repurposed antiviral drugs for COVID-19 – Interim WHO solidarity trial results. *N Engl J Med.* 2021;384(6):497–511. Available from www.ncbi.nlm.nih.gov/pubmed/33264556
19. Advisory for Rational use of Remdesivir for COVID-19 Treatment. Available from www.mohfw.gov.in/pdf/AdvisoryforRationaluseofRemdesivirforCOVID19Treatment.pdf. Last accessed on July 30, 2021.
20. Thakare S, Gandhi C, Modi T, Bose S, Deb S, Saxena N, et al. Safety of remdesivir in patients with acute kidney injury or CKD. *Kidney Int Reports.* 2021;6(1):206–210.

21. Cao B, Wang Y, Wen D, Liu W, Wang J, Fan G, et al. A trial of lopinavir–ritonavir in adults hospitalized with severe covid-19. *N Engl J Med.* 2020;382(19):1787–1799.
22. Spinelli FR, Conti F, Gadina M. HiJAKing SARS-CoV-2? The potential role of JAK inhibitors in the management of COVID-19. *Sci Immunol.* May 8, 2020;5(47):eabc5367. doi: 10.1126/sciimmunol.abc5367. PMID: 32385052.
23. Kalil AC, Patterson TF, Mehta AK, et al. Baricitinib plus remdesivir for hospitalized adults with COVID-19. *N Engl J Med.* 2021;384(9):795–807. Available from www.ncbi.nlm.nih.gov/pubmed/33306283
24. La Rosée, F, Bremer, HC, Gehrke, I et al. The Janus kinase 1/2 inhibitor ruxolitinib in COVID-19 with severe systemic hyperinflammation. *Leukemia* 2020, 34; 1805–1815. https://doi.org/10.1038/s41375-020-0891-0
25. Roschewski M, Lionakis MS, Sharman JP, Roswarski J, Goy A, Monticelli MA, Roshon M, Wrzesinski SH, Desai JV, Zarakas MA, Collen J, Rose K, Hamdy A, Izumi R, Wright GW, Chung KK, Baselga J, Staudt LM, Wilson WH. Inhibition of Bruton tyrosine kinase in patients with severe COVID-19. *Sci Immunol.* June 5, 2020;5(48):eabd0110. doi: 10.1126/sciimmunol.abd0110. Epub June 5, 2020. PMID: 32503877; PMCID: PMC7274761.
26. Haji Abdolvahab M, Moradi-Kalbolandi S, Zarei M, Bose D, Majidzadeh-A K, Farahmand L. Potential role of interferons in treating COVID-19 patients. *Int Immunopharmacol.* 2021;90:107171. doi:10.1016/j.intimp.2020.107171
27. Somagutta MKR, Lourdes Pormento MK, Hamid P, Hamdan A, Khan MA, Desir R, Vijayan R, Shirke S, Jeyakumar R, Dogar Z, Makkar SS, Guntipalli P, Ngardig NN, Nagineni MS, Paul T, Luvsannyam E, Riddick C, Sanchez-Gonzalez MA. The safety and efficacy of anakinra, an interleukin-1 antagonist in severe cases of COVID-19: A systematic review and meta-analysis. *Infect Chemother.* June 2021;53(2):221–237. doi: 10.3947/ic.2021.0016. PMID: 34216117; PMCID: PMC8258297.
28. RECOVERY Collaborative Group, Horby PW, Pessoa-Amorim G, et al. Tocilizumab in patients admitted to hospital with COVID-19 (RECOVERY): Preliminary results of a randomised, controlled, open-label, platform trial. *Lancet.* 2021;397(10285):1637–1645. Available from https://pubmed.ncbi.nlm.nih.gov/33933206/
29. REMAP-CAP Investigators, Gordon AC, Mouncey PR, et al. Interleukin-6 receptor antagonists in critically ill patients with COVID-19. *N Engl J Med.* 2021;384(16):1491–1502. Available from www.ncbi.nlm.nih.gov/pubmed/33631065
30. Grasselli G, Cattaneo E, Florio G, Ippolito M, Zanella A, Cortegiani A, Huang J, Pesenti A, Einav S. Mechanical ventilation parameters in critically ill COVID-19 patients: A scoping review. *Crit Care.* March 20, 2021;25(1):115. doi:10.1186/s13054-021-03536-2. PMID: 33743812; PMCID: PMC7980724.
31. Chandra A, Chakraborty U, Ghosh S, et al. Anticoagulation in COVID-19: Current concepts and controversies. *Postgraduate Medical Journal.* Published Online First: April 13, 2021. doi:10.1136/postgradmedj-2021-139923
32. Nadim, MK, Forni, LG, Mehta, RL, et al. COVID-19-associated acute kidney injury: Consensus report of the 25th Acute Disease Quality Initiative (ADQI) workgroup. *Nat Rev Nephrol* 2020; 16,747–764. https://doi.org/10.1038/s41581-020-00356-5
33. Guidance for Management of Pregnant Women in COVID-19 Pandemic. Available from www.icmr.gov.in/pdf/covid/techdoc/Guidance_for_Management_of_Pregnant_Women_in_COVID19_Pandemic_12042020.pdf
34. Comprehensive Guidelines for Management of COVID-19 in CHILDREN (below 18 years). Available from https://dghs.gov.in/WriteReadData/News/202106090337278932402DteGHSComprehensiveGuidelinesforManagementofCOVID-19inCHILDREN_9June2021.pdf

12 A Virus SARS-CoV-2

Vaccination Strategies

Vikas Kushwaha and Neena Capalash
Department of Biotechnology, Panjab University Chandigarh, India

CONTENTS

12.1 INTRODUCTION

In December 2019, the people of the city Wuhan in China reported pneumonia-like symptoms with unknown causative agents. Later on, it was reported that this disease had spread from a local seafood market in Wuhan, China (Hui et al., 2020; Zhu N et al., 2020). The infective agent was identified as a novel coronavirus belonging to family *Coronaviridae* and order *Nidovirales,* which had never been accounted for before in people but was related to those responsible for SARS and MERS. It was named as a Severe Acute Respiratory Syndrome Coronavirus 2 (SARS-CoV-2) (Spinney, 2020). This new acute respiratory infection changed the global perspective of health care. The World Health Organization (WHO) named the disease as 2019 Novel Coronavirus Disease (Covid-19) and declared the outbreak a global public health emergency on March 11, 2020. Covid-19 is mainly transmitted through respiratory microdroplets or from contaminated surfaces. It affects the respiratory tract with common symptoms such as fever, dry cough, and tiredness. Less common symptoms are aches and pains, sore throat, diarrhea, conjunctivitis, headache, loss of taste or smell, rash on the skin, or discoloration of fingers or toes. Severe symptoms include difficulty in breathing or shortness of breath, chest pain or pressure, loss of speech, or movement (WHO, 2020). Severe disease is characterized by a hyperinflammatory response and development of acute respiratory distress syndrome (ARDS), which leads to mechanical ventilation, kidney failure, and death (Huang et al., 2020).

SARS-CoV is a spherical enveloped virus, ranging from 50 to 200 nm in size, contain a ssRNA genome of approximately 30 kb. Two-thirds of the genome encodes the viral replicase genes, and one-third ORFs encode the structural and accessory proteins genes (Cascella et al., 2020). Coronaviruses consist of four structural proteins: Spike (S), membrane (M), envelop (E), and nucleocapsid (N). Spikes protruding from viral surface mediate the attachment to human cell surface receptor, the

DOI: 10.1201/9781003358909-12

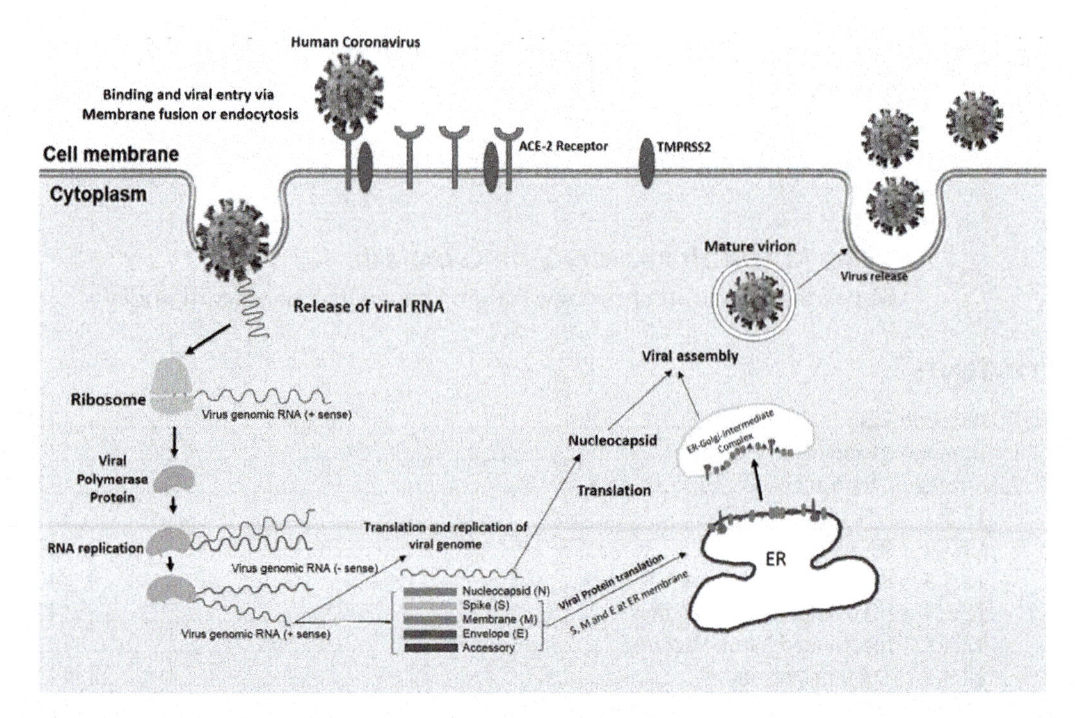

FIGURE 12.1 Life cycle of SARS-CoV-2.

angiotensin-converting enzyme 2 (ACE2), and allow the virus to enter the host cell. After binding with the host receptor, the spike protein undergoes two-step sequential protease cleavage for its activation (Yuki et al., 2020) (Figure 12.1). The University of Sheffield, England discovered a new mutated variant of the virus "D614G" (Asp614→Gly substitution in the spike glycoprotein of SARS-CoV-2), which is more infectious than the original strain but does not cause more severe disease (Korber et al., 2020). This strain exhibits increased infectivity in continuous cell lines and increased sensitivity to neutralization (Baric et al., 2020). Denmark has identified Cluster 5, a SARS-CoV-2 variant associated with farmed minks. This specific mink-associated variant has a moderately reduced susceptibility to neutralizing antibodies in both minks and humans.

Recently the UK announced the discovery of a new strain of SARS-CoV-2. This variant has a mutation in the receptor-binding domain (RBD) of the spike protein and has been named B.1.1.7. or SARS-CoV-2 VOC 202012/01. Experts have reported that this new strain is 70% more infectious than the other strains of the virus (Deccan Herald web desk, 2021). A new mutant known as B.1.617 has been discovered in India, and has been listed as a variant of conern because it includes two critical mutations: L452R and E484Q. Both these mutations are related to immune system evasion, and L452R mutation is also associated with spread of infection. Recently, a new B.1.617.2 subtype of B.1.617 variant was reported in UK (Vaidyanathan et al., 2021). Another strain B.1.618 with triple mutations L452R, P681R, and E484Q has been discovered in West Bengal, India (Vaidyanathan et al., 2021). Another variant, B.1.351, which has shown rapid spread in South Africa, carries a death risk of up to 60% and has three major mutations: E484K, N501Y, and K417N. A variant P.1 was first reported in Brazil, with N501Y, E484K, and K417T as the main mutations in the spike protein. This variant is 2.5 times more contagious than the original coronavirus and more resistant to antibodies (Thomsan Reuter Press release April 14, 2021). Another variant P.3 has been reported in the Philippines having two mutations E484K and N501Y. In California, a new variant of concern known as CAL.20C (B.1.427/B.1.429) with S13I, W152C, and L452R mutations has been discovered, which spreads more rapidly. These mutations play a role in immune evasion (McCallum et al., 2021).

12.2 IMMUNE RESPONSE TO COVID-19

B and T cells participate in immune-mediated protection against the viral infection. T cells identify cells that have been infected with a particular virus and quickly multiply to combat the infection. CD8+ cytotoxic T cells destroy the virus-infected cells and help to delay or avoid the infection, while CD4+ helper T cells bring in other immune system cells and activate B-cells to develop virus-specific antibodies. Antibodies unique to that virus are generated by B cells (WHO, 2020). Humoral immunity against any foreign antigen is recognized by the rapid production of IgM antibodies and slower production of IgG antibodies that persist in the body for a long time (Liu et al., 2006). The humoral immune response against the Covid-19 has been identified by the production of specific neutralizing antibodies. High titer of IgM and IgG antibodies specific to SARS-CoV-2 nucleocapsid protein and S protein's receptor-binding domain (S-RBD) has been reported in newly discharged patients and these antibodies remain present in the discharged patient for two weeks (Ni et al., 2020). Covid-19 patients exhibited nucleocapsid protein (NP)-specific antibody response, with higher IgM at day 9 after disease and then switched to IgG by second week (Zhou et al., 2020).

It has been observed that in severe Covid-19 infection, the number of CD4+ and CD8+ T lymphocytes are low in the peripheral blood but with increased levels of proinflammatory cytokines [IL-6, IL-10, TNF-α, Granulocyte-Colony Stimulating Factor (G-CSF), Monocyte Chemoattractant Protein 1 (MCP1), Macrophage Inflammatory Protein (MIP)1α, and chemokine (IL-8)] (Zhou et al., 2020). Discharged patients with a high titer of neutralizing antibodies also showed a higher concentration of T cells that secreted IFN-γ in response to the SARS-CoV-2 nucleocapsid protein (Ni et al., 2020). The WHO has summarized the immune responses to Covid-19, stating that the majority of Covid-19 recovered patients have SARS-CoV-2 virus antibodies detectable in their blood. About 1–3 weeks after beginning of symptoms, several Covid-19 patients develop antibodies. Patients with more serious diseases have large neutralizing antibodies that tend to be higher. Low levels of neutralizing antibodies are found in patients with moderate or asymptomatic Covid-19 (or even undetectable levels). In these persons it is possible that the innate immune response and the T cell response cleared the virus. Recent studies have shown that neutralizing antibodies may disappear after 3 months (WHO, 2020).

12.3 STRATEGIES FOR VACCINE AGAINST SARS-COV-2

Earlier it was speculated that Covid-19 is a weak pathogen that has little effect on humans, but the severity of disease seen in outbreaks like SARS, MERS, and now Covid-19 has proven it otherwise and has compelled researchers to explore multiple approaches in vaccine technology like DNA and RNA vaccines, recombinant protein vaccines, attenuated vaccines, synthetic vaccines, and subunit vaccines to produce safe, effective, and tolerable vaccines against SARS-CoV-2 (Figure 12.2). SARS-CoV-2 infected patients show a high level of the humoral immune response against the surface spike proteins, which suggests that there is a vital role of T cell immunity and neutralizing antibodies in the prevention of early infection (Corey et al., 2020). The primary goal is to develop an effective vaccine that could upregulate the immune responses with a high degree of safety and a low level of risk. S protein has emerged as an ideal target for a vaccine against SARS-CoV-2 due to structural similarities between SARS-CoV and MERS-CoV, as well as variation in the S protein sequence and its role in attachment to the host cell. The development of an effective SARS-CoV-2 vaccine requires large-scale production followed by a vaccination program targeting the populations of different geographical regions (Figure 12.3).

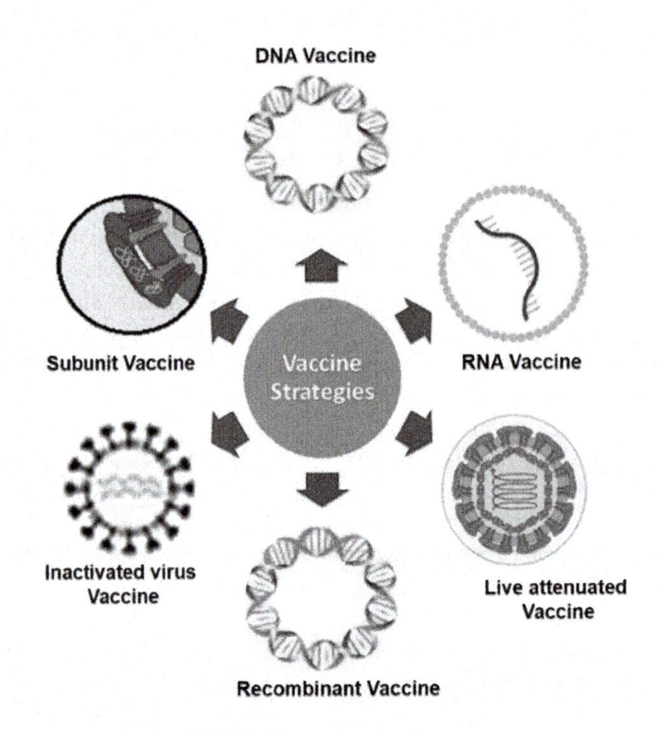

FIGURE 12.2 Various strategies for COVID-19 vaccine development.

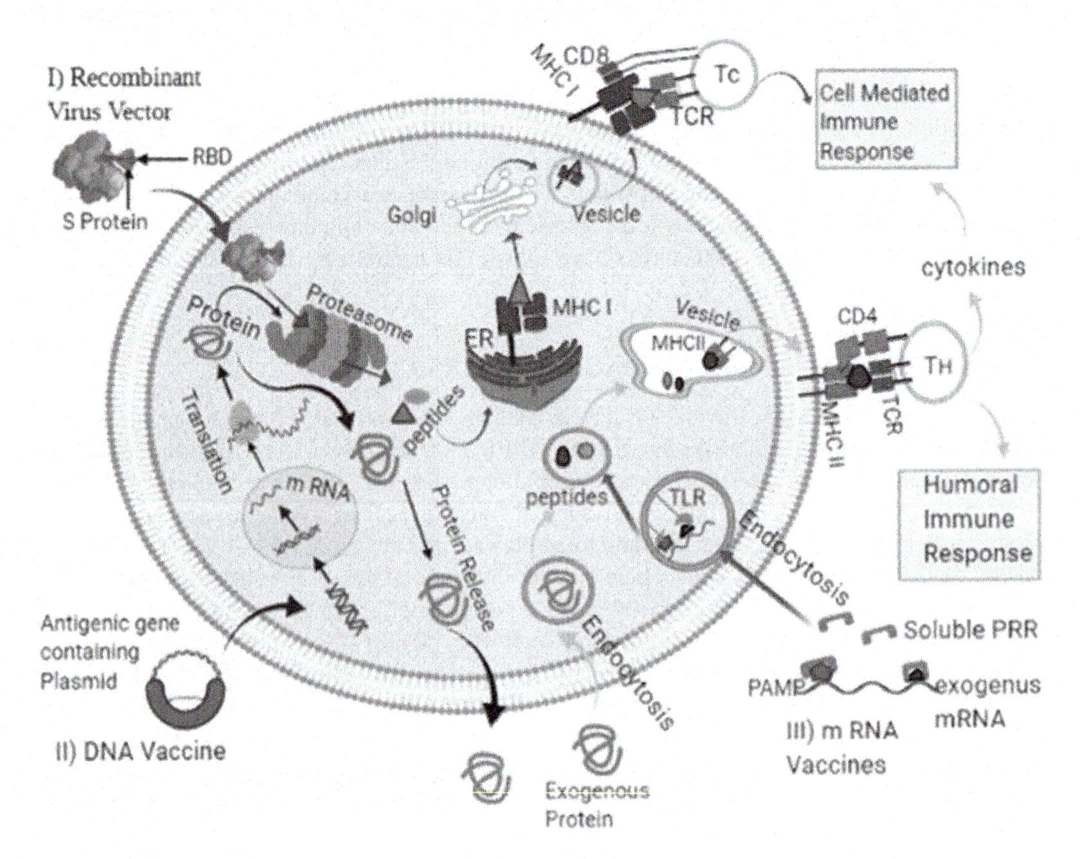

FIGURE 12.3 Mechanism of action of various vaccine candidates.

12.3.1 DNA Vaccines

DNA-based vaccines (plasmid DNA-encoding proteins from the pathogens) is an innovative approach to inducing a humoral and cell-mediated antigen-specific immune response. In this approach, the DNA sequence of a specific gene is transferred into the microorganism with the help of a plasmid or viral vector for the synthesis of the immunogenic protein. Many SARS-CoV proteins like spike protein (S), membrane protein (M), and nucleocapsid protein (N) have been utilized as DNA vaccines because they increase cellular and humoral responses (Wang et al., 2005). It has been shown that INO-4800 as a promising Covid-19 vaccine candidate targets the major surface antigen spike protein of SARS-CoV-2 and increases the production of neutralizing antibodies as well as T cell response against SARS-CoV-2 in mice and guinea pigs (Smith et al., 2020). Inovio pharmaceuticals of South Korea have developed a DNA vaccine candidate (INO-4800 containing plasmid pGX9501) encoding for the entire length of the SARS-CoV-2 spike glycoprotein. This nucleic acid-based vaccine is stable for over a year and requires no cold chain for transport. This is going to be of advantage when mass vaccination is implemented against SARS-CoV-2. Inovio in collaboration with the International Vaccine Institute reported that macaques received two doses of INO-4800 (1 mg) 4 weeks apart and were subsequently challenged with the live virus 13 weeks after the second dose and showed reduced viral load both in the lower lungs and nasal passages (Patel et al., 2020). Phase I clinical trial of INO-4800 to assess its safety, tolerability, and immunogenicity has been completed (Clinicaltrials.gov/NCT03721718). Inovio announced that 94% participants of the phase I trial elicited both humoral and cellular immune response at week 6 after the second dose of INO-4800 (Tebas et al., 2020). INO-4800 was found to be stable, well-tolerated, and immunogenic in all age groups studied in phase II trials, and phase III trials are currently underway. In preclinical models, INO-4802, a next-generation Pan-Covid-19 vaccine candidate, induced potent neutralizing antibodies and T cell responses against B.1.1.7, B.1.351, and P.1 variants (Inovio Pharmaceuticals, 2021).

12.3.2 RNA Vaccines

Traditionally the development of an active vaccine candidate against any disease takes several years. To shorten the time period, the mRNA vaccine strategy was adopted for the production of an active vaccine candidate against SARS-CoV-2. A short synthetic viral mRNA is introduced into the host cell, which produces surface antigen. This strategy shortens the lengthy process of purification of viral proteins and also time duration required for the mass production of vaccines. American biotech firm Moderna in collaboration with National Institute of Health (NIH) and Coalition for Epidemic Preparedness Innovations (CEPI) have developed a lipid nanoparticle-encapsulated mRNA-based vaccine mRNA-1273 that expresses stabilized full-length spike protein (S) of the SARS-CoV-2. mRNA-1273 induced robust neutralizing antibodies and increases T cell response in non-human primates during preclinical study (Corbett et al., 2020). Phase I clinical study demonstrated high neutralizing antibodies with Th1-biased CD4 T cell response in the participants (Jackson et al., 2020). Moderna completed the phase II and phase III trials in collaboration with the National Institutes of Health (NIH) and the Biomedical Advanced Research and Development Authority (BARDA). Moderna announced that the vaccine had 94.1% protective efficacy against severe Covid-19 cases and 100% protective efficacy against moderate Covid-19 cases (Moderna. Press release November 30, 2021) (Clinicaltrials.gov/NCT04283461). The US Food and Drug Administration (FDA) and WHO Strategic Advisory Group of Experts (SAGE) approved emergency use of Moderna's vaccine mRNA-1273 against Covid-19 in individuals 18 years of age or older. The Moderna mRNA-1273 vaccine is given in two doses (100 μg, 0.5 ml each) 28 days apart, according to SAGE. The time between doses can be increased to 42 days if desired. Moderna mRNA-1273 vaccine is effective against new variants of SARS-CoV-2, B.1.1.7 and the B.1.351. mRNA-1273 remains stable at –20°C for up to

6 months, at refrigerated conditions for up to 30 days and at room temperature for up to 12 hours. Moderna has developed multivalent booster candidates mRNA-1273.351 and mRNA-1273.211 that have shown increased neutralizing titers against SARS-CoV-2 variants of concern in Balb/c mice (Kai et al., 2021). Another mRNA-based vaccine candidate BNT162 (BNT162b1 encodes an optimized SARS-CoV-2 receptor-binding domain (RBD) antigen and BNT162b2: encodes an optimized SARS-CoV-2 full-length spike protein antigen) has been developed by BioNTech with Pfizer. The US FDA demonstrated that this vaccine is stable when stored at temperatures ranging from –25°C to –15°C. The phase I/II clinical trials of vaccine candidate BNT162 showed 2–4 fold increase in functional antibody titers (Clinicaltrials.gov/NCT04380701). Robust immunogenicity (8–50 fold higher RBD binding IgG antibodies) was also observed after vaccination with BNT162b1 during the phase I/II clinical trial (Mulligan et al., 2020). Data analysis of the phase III trial indicates that the vaccine efficacy rate is 95% against Covid-19. The US FDA gave Emergency Use Authorization to BioNTech BNT162. A single dose of BNT162 decreases the risk of infection by more than 70%, increasing to 85% after the second dose. Vaccine will also help to stop virus transmission. The BNT162b2 vaccine was found effective (>90%) against the B.1.1.7 and B.1.351 variants (Abu-Raddad et al., 2021). Arcturus Therapeutics Vaccines Company is working on self-transcribing and replicating mRNA (STARR™) technology to develop the LUNAR-COV19 (ARCT-021) vaccine with lipid-mediated delivery system to increase the protective efficacy of mRNA vaccines. Arcturus announced preclinical data of LUNAR-COV19 vaccine that showed strong IgG antibody response, with balanced Th1/Th2 CD4+ T-cell response and 100% virus neutralization at a very low dose of vaccine (Arcturus Therapeutics. Press release May 8, 2020). The phase I/II clinical trial has been completed in Singapore to evaluate the safety, tolerability, and the extent and duration of the humoral and cellular immune response in 100 healthy volunteers aged between 21 and 55. Arcturus Therapeutics received FDA allowance to proceed with Phase II Study of ARCT-021 (LUNAR-COV19) vaccine candidate in the United States. Phase III of the clinical trial was completed in March 2022. The Sanofi and Translate Bio is producing multiple mRNA constructs as vaccine candidates for Covid-19 and initiate phase I/II clinical trial (Sanofi and Translate Bio. Press release March 12, 2021). Imperial College London and Enesi Pharma are working on a thermostable and immunogenic vaccine based on self-amplifying RNA technology (saRNA) and DNA/RNA stabilization technologies (stable at ambient temperatures and up to 40°C), which is currently being tested in over 400 healthy volunteers in a phase I/II clinical trial (Imperial College London. Press release June 15, 2020). German Health Authority Paul-Ehrlich-Institute (PEI) and the Belgian Federal Agency for Medicines and Health Products (FAMHP) approved the CureVac Company for clinical trials of mRNA vaccine (CVnCoV), encoding spike protein of SARS-CoV-2 (CureVac. Press release June 18, 2020). Based on promising results of phase I/II clinical trials, CureVac Company has started phase III clinical trials (a two-dose schedule of 12 μg) to evaluate the safety and immunogenicity of CVnCoV. CVnCoV remained stable for at least 3 months when stored at a standard temperature of +5°C and up to 24 hours as a ready-to-use vaccine when stored at room temperature (Cure Vac. Press release March 11, 2021). CVnCoV showed protection against the SARS-CoV-2 variant B.1.351 and strain of the original SARS-CoV-2 B1 lineage (BavPat1) in a transgenic mouse model (Hoffmann et al., 2021).

12.3.3 Recombinant Vector Vaccines

Recombinant vector vaccines are genetically engineered vaccines having a viral vector backbone expressing an antigen from a foreign transgene, which enhanced the humoral and cell-mediated immune responses for long-term immunity (Bull et al., 2019). Traditional recombinant DNA technology can be used to express the viral gene of interest like spike protein (S), membrane protein (M), and an envelope protein (E) into well-characterized vectors.

A recombinant protein-based vaccine (CoV-RBD219N1) from the receptor-binding domain (RBD219-N1) of the SARS-CoV spike (S) protein induced a high-level of neutralizing antibodies with 100% survival of mice and rhesus macaques after SARS-CoV challenge (Chen et al., 2020). Viral vector-based recombinants are easy to construct and do not require adjuvants. China-based CanSino Biologics is working on a recombinant adenovirus type-5 (Ad5) vectored Covid-19 vaccine Convidecia (Ad5-nCoV) expressing the spike glycoprotein of SARS-CoV-2 and reported increased humoral response at day 28 postvaccination and rapid specific T-cell response from day 14 postvaccination during phase I/II clinical trial (Zhu FC et al., 2020) (Clinicaltrials.gov/NCT04398147), but the production of neutralizing antibody titer was below the range reported in human convalescent plasma (Robbiani et al., 2020). Convidecia interim analysis data for phase III clinical trial showed that the vaccine has a total efficacy of 65.28% after 28 days of single-dose vaccination to prevent Covid-19 symptomatic disease, and 68.83% after 14 days of single-dose vaccination to prevent Covid-19 symptomatic disease (CanSino Biologics. Press release March 01, 2021). Convidecia got approval for single-dose Covid-19 vaccine as emergency use in Chile (CanSino Biologics. Press release April 8, 2021).

Another adenovirus-vectored vaccine (ChAdOx1 nCoV-19 or AZD1222) encoding the spike protein of SARS-CoV was found to be immunogenic in mice and rhesus macaques. It elicited neutralizing antibody titers and enhanced humoral and cell-mediated response with partial protection (Van et al., 2020). The University of Oxford and AstraZeneca produced a recombinant vaccine AZD1222 (ChAdOx1 nCoV-19 adenovirus) expressing the SARS-CoV-2 Spike protein and can remain stable for 6 months when refrigerated at 2–8°C. The immunogenic property of adenovirus could generate robust memory B cell and T cell response that might result in better prophylaxis (Mullard, 2020). The phase I/II trial of ChAdOx1 nCoV-19 was performed on healthy adults aged 18–55 years with no history of SARS-CoV-2 infection or of Covid-19-like symptoms and it induced both humoral and cellular immune responses. It has been shown that ChAdOx1 nCoV-19 is safe, well-tolerated, and immunogenic (Folegatti et al., 2020). The ChAdOx1 nCoV-19 vaccine was found to have 82.5% efficacy in all cohorts in Brazil, South Africa, and the United Kingdom during phase III clinical trial (Voysey et al., 2020) (Clinicaltrials.gov/NCT04400838). The UK Medicines and Healthcare products Regulatory Agency (MHRA) has authorized AstraZeneca for an emergency supply of Covid-19 vaccine for the active immunization of individuals 18 years or older. The Central Drugs and Standards Committee (CDSCO) of India has approved the Oxford-AstraZeneca vaccine known as Covishield in India, which is being produced locally by the Serum Institute of India, the world's largest vaccine manufacturer. Gamaleya Research Institute of Epidemiology and Microbiology and Health Ministry of the Russian Federation are working on a heterologous Covid-19 vaccine Gam-Covid-Vac Lyo vaccine (consist of attenuated recombinant adenovirus type 5 and type 26) expressing the SARS-CoV-2 sSpike protein in two formulations (frozen and lyophilized). The Gam-Covid-Vac Lyo vaccine showed good safety profile and induced strong humoral and cellular immune response during phase I/II studies with common adverse effects like pain at the injection site, hyperthermia, headache, asthenia, and muscle and joint pain (Logunov et al., 2020). Russia became the first country to license a Covid-19 vaccine, "Sputnik V." The phase III Sputnik V clinical trial was highly efficient (91.6%) and showed a robust humoral and a long-term cell-mediated immune response (Logunov et al., 2021). The vaccine showed excellent safety profile with no serious adverse events, no strong allergies, and no anaphylactic shock. Indian pharma giant Dr Reddy's Laboratories Ltd. and the Russian Direct Investment Fund (RDIF) got approval from CDSCO, India for distribution of 100 million doses of Sputnik V vaccine in India.

Novavax, USA reported that vaccination using prefusion protein coronavirus vaccine candidate (NVX-CoV2373) with saponin-based Matrix-M™ adjuvant enhanced the immunogenicity and neutralizing antibodies during the preclinical trial. Novavax phase I/II randomized study showed that NVX-CoV2373 induced neutralizing anitbody titers in 100% of participants with robust polyfunctional CD4+ T cell response (release of IFN-γ, IL-2, and TNF-α) (Clinicaltrials.gov/NCT04368988) (Keech et al., 2020). In the phase III trial of NVX-CoV2373 in the UK, the company enrolled over 15,000 people aged 18 to 84, with 27% of those over 65. Novavax reported that

NVX-CoV2373 reached the primary endpoint of its phase III clinical trial in the United Kingdom, with a vaccine efficacy of 89.3%. The company has developed NanoFlu/NVX-CoV2373, which is a combination of quadrivalent seasonal flu vaccine (NanoFlu™) and Covid-19 vaccine candidate (NVX-CoV2373) that enhanced positive immune response against both influenza and SARS-CoV-2 (Novavax. Press release May10, 2021). Clover Biopharmaceuticals has developed a recombinant protein vaccine SCB-2019 containing a stabilized trimeric form of the spike (S)-protein (S-Trimer) combined with two different adjuvants (Clinicaltrials.gov/NCT04405908). The SCB-2019 vaccine was well tolerated and elicited robust humoral and cellular immune response with high viral neutralizing activity against SARS-CoV-2 when formulated with either AS03 or CpG/Alum adjuvants (Lara et al., 2021). The bacTRL-Spike vaccine is an oral vaccine produced by Symvivo Corporation using genetically modified *Bifidobacterium longum* with the DNA sequence of SARS-CoV-2 spike pProtein. To evaluate the efficacy of the orally delivered bacTRL-Spike vaccine in healthy adults, phase I trial has been completed (Clinicaltrials.gov/NCT04334980). Janssen Covid-19 (JNJ-78436735/Ad26.COV2.S), a single-shot AdVac® vectored vaccine developed by Johnson and Johnson, induced a potent and long-lasting humoral and cellular immune response. The US Food and Drug Administration granted Emergency Use Authorization for the Janssen Covid-19 vaccine, allowing it to be administered in the US to people aged 18 and up (Johnson & Johnson. Press release February 27, 2021). Spain's Health Ministry approved the use of the Janssen single-dose Covid-19 vaccine among people under 60 (Thomsan Reuter. Press release April 13, 2021).

12.3.4 Live Attenuated Vaccines

Live attenuated vaccines have shown promising results against various pathogens because they elicit long-lasting protective immune response. These vaccines are developed by the cultivation of microorganisms under suboptimal conditions or techniques that attenuate their virulence. The reversion of the pathogen to its virulent form is one of the problems with this type of approach. Mice immunized with attenuated mutants (coronavirus lacking the E gene or rSARS-CoV-ΔE) showed higher titer of neutralizing antibodies and reduced replication of SARS-CoV genes in the respiratory tract (Lamirande et al., 2008). US-based Codagenix, in collaboration with the Serum Institute of India, announced the development of an intranasal, live-attenuated Covid-19 vaccine COVI-VAC through synthetic biology approach, and phase I trial has been initiated in the UK. COVI-VAC would not need a needle and syringe, nor ultra-low temperature freezers for storage. In comparison to other vaccines for SARS-CoV-2, live attenuated SARS-CoV-2 vaccine can induce a more robust immune response and long-lasting cellular immunity. COVI-VAC has been found to be effective against new SARS-CoV-2 variants (Codagenix. Press release January 11, 2021).

12.3.5 Inactivated Virus Vaccines

The inactivated virus or whole killed virus (WKV) vaccine approach is based on neutralization of infectivity of pathogens by chemicals, heat, or radiation while retaining their stability, and immunogenicity. In preclinical studies, WKV vaccine immunization reduces the multiplication of pulmonary SARS-CoV-2 in BALB/c mice (Pandey et al., 2020). The Chinese Academy of Medical Sciences in collaboration with West China Second University Hospital and Yunnan Center for Disease Control and Prevention developed an inactivated SARS-CoV-2 vaccine and successfully completed a phase I/II clinical trial, demonstrating that the vaccine is safe and immunogenic in healthy people. A phase III trial is also underway (Clinicaltrials.gov/NCT04412538). The Inactivated Novel Coronavirus Pneumonia vaccine (ChiCTR2000031809) developed by the Wuhan Institute of Biological Products Co., Ltd. has also entered into clinical trial phase I/II (ChiCTR. Press release September, 03, 2020). The Beijing-based Sinovac Biotech Ltd. has developed formalin-inactivated SARS-CoV-2 and alum adjuvanted virus vaccine (PiCoVacc), which induced SARS-CoV-2-specific neutralizing antibodies in mice, rats, and nonhuman primates (macaques). The vaccine elicited neutralizing antibodies 14 days after vaccination in a phase I/II study, with favorable immunogenicity and protection profiles (Gao et al., 2020). According to data from a phase III trial in Brazil and Turkey, the vaccine has a protective efficacy of 50.65%, and it has been approved for emergency use. However, a study based on the use of aluminium adjuvants with PiCoVacc showed high levels of neutralizing antibody, which is recognized as a cornerstone of the protection against Covid-19 (Hotez et al., 2020). Bharat Biotech, in association with the Indian Council for Medical Research-National Institute of Virology (ICMR-NIV), has developed whole virion inactivated vaccine candidate COVAXIN/BBV152 that has shown promising results in phase I/II clinical studies and is found effective and safe. The Central Drugs and Standards Committee (CDSCO) of India approved COVAXIN for emergency use in India. Covaxin announced that COVAXIN showed an interim efficacy of 81% with higher titer of neutralizing antibodies in phase III clinical trials (Deshpande et al., 2020). COVAXIN is effective against the UK-variant B.1.1.7 and Indian-variant B.1.617 (Sapkal et al., 2021). Adjuvant Alhydroxiquim-II is added to the vaccine to increase immune response and provide longer-lasting immunity. Bharat Biotech is also working with Washington University School of Medicine in St. Louis, Missouri to produce a nasal Covid-19 adenovirus vaccine, which is currently in phase I trial.

12.3.6 Subunit Vaccines

Subunit vaccines contain surface antigens of the pathogens as antigenic determinants that activate the immune response and reduce the risk of side effects. These vaccines are administrated with adjuvants or aluminium salts to avoid the loss of immunogenicity of antigens (Wang et al., 2016). Previously it was reported that vaccination with Receptor-Binding Domain (RBD; a segment of S protein) of SARS-CoV in mice and rabbits induced higher neutralizing antibodies production as compared to the full length of S protein (He et al., 2004). To develop an effective vaccine against SARS-CoV-2, the structural spike proteins have been targeted and a multiepitope vaccine has been desigend that triggered both CD4+ and CD8+ T-cell immune response. The construct consisted of B and T cell epitopes that could act as immunogens and elicited robust immune response in the host system. A vaccine construct of 425 amino acids (including the 50S ribosomal protein adjuvant) showed better protein-protein interactions with TLR-3 immune receptors and elicited immune response against Covid-19 infection (Abraham et al., 2020). A vaccine candidate expressing the SARS-CoV-2 RBD protein subunit (fused with Fc of IgG) developed by Lindsley F. Kimball Research Institute, New York and Beijing Institute of Microbiology and Epidemiology, Beijing, China showed higher neutralizing antibody responses during preclinical studies (Tai et al., 2020). The National Institute of Infectious Disease, Japan worked on recombinant S protein with gold nanoparticles (AuNPs) because their physicochemical properties prevented antibody production against the platform material. They were able to elicit a strong antigen-specific IgG response against SARS-CoV-2 (Sekimukai et al., 2020). The University of Pittsburgh developed a recombinant protein subunit vaccine (PittCoVacc) expressing S1 subunit of spike protein, which generated a surge of antibodies against SARS-CoV-2 within two weeks of Microneedle (MNA array) delivery in mice (Kim et al., 2020). Sanofi and GSK announced that they had developed an adjuvanted vaccine for Covid-19 expressing S-protein of SARS-CoV-2. Preclinical studies showed the vaccine candidate had promising immunogenicity. Phase I/II clinical trials on 440 healthy adults across 11 investigational sites in the United States showed a higher immune response comparable to patients who recovered from Covid-19 at ages between 18 and 49 years, but a low immune response in older adults (Sanofi and GSK. Press release September 03, 2020). The University of Queensland with GSK and Dynavax performed a preclinical trial of a subunit vaccine consisting of a SARSCoV-2 spike protein stabilized with a S protein "molecular clamp." The phase I trial of the UQ-CSL v451 Covid-19 vaccine was able to elicit a robust response towards the virus and a strong safety profile (University of Queensland. Press release December 11, 2020). GSK's collaboration with

TABLE 12.1
Companies Currently Developing Vaccines Against Covid-19

Company	Stage	Strategy	Outcomes
DNA Vaccine			
Inovio Pharmaceuticals	Phase III	DNA Vaccine (GLS-5300) (NCT03721718)	Higher immune responses during phase I trial; 94% participants of the phase I trial elicited both humoral and cellular immune response at week 6 after second dose. https://clinicaltrials.gov/ct2/show/NCT03721718. ClinicalTrials.gov NCT03721718 (2020).
RNA Vaccine			
Moderna Therapeutics	Approved for use	mRNA -1273 (NCT04283461)	Robust neutralizing antibodies and higer T cell responses in non-primates and during phase I clinical trial. 94.1% protective efficacy against severe Covid-19 cases and 100% protective efficacy against moderate Covid-19 cases https://clinicaltrials.gov/ct2/show/NCT0 4283461. ClinicalTrials.gov NCT04283461 (2020). www.nih.gov/news-events/news-releases/nih-clinical-trialinvestigational-vaccine-covid-19-begins. Moderna (2020).
BioNTech/Fosun Pharma/Pfizer	Approved for use	mRNA-BNT162 (NCT04380701)	Preclinical data showed 2–4 fold increase in functional antibody titers, 8–50 fold hogher RBD binding IgG antibodies during phase I/II clinical studies. 95% protective efficacy against COVID-19. ClinicalTrials.gov NCT04380701 (2020). https://clinicaltrials.gov/ct2/show/NCT04380701
Arcturus Therapeutics	Phase III	Engineering RNA with nanoparticle (LUNAR-COV19)	Preclinical data showed strong anti-spike protein IgG antibody response, a dose-dependent CD8+ T cell response, with balanced Th1/Th2 CD4+ T-cell response and 100% virus neutralization. Phase II data showed that its safe, tolerable, and induced humoral and cellular response. www.catalent.com/catalent-news/arcturus-therapeutics-and-catalent-announce-partnership-to-manufacture-mrna-based-covid-19-vaccine. Arcturus Therapeutics (2020).
Sanofi and Translate Bio	Phase I/II	Multiple mRNA construct	Under clinical trial https://investors.translate.bio/news-releases/news-release-details/sanofi-and-translate-bio-initiate-phase-12-clinical-trial-mrna. Sanofi and Translate Bio (2021).
Imperial college, London	Phase I/II	self-amplifying RNA technology (saRNA)	Under clinical trial www.imperial.ac.uk/news/198314/imperial-begin-first-human-trials-covid-19/. Imperial College London (2020)
CureVac	Phase III	CVnCoV	Promising results of phase I/II trial with robust immune response. CureVac (2020) www.clinicaltrialsarena.com/news/curevac-covid-19-vaccine-triawww.precisionvaccinations.com/vaccines/cvncov-vaccine
Recombinant Vector Vaccines			
CanSino Biologics	Phase III	Genetically engineered adenovirus, recombinant adenovirus type-5 (Ad5) Covidecia (Ad50nCoV)	Preclinical studies showed elevated levels of IFN-γ, TNF-α, and IL-2 expressed by CD4+ and CD8+ T lymphocytes, production of IgG and neutralizing antibodies. Phase III clinical trial showed that vaccine has a total efficacy of 65.28% after 28 days of single-dose and 68.83% 14 days after a single dose. https://clinicaltrials.gov/ct2/show/NCT04398147. ClinicalTrials.gov NCT04398147 (2020). www.precisionvaccinations.com/vaccines/convidicea-vaccine

TABLE 12.1 (Continued)

Company	Stage	Strategy	Outcomes
University of Oxford and AstraZeneca	Approved for use	Recombinant vaccine (AZD1222/ ChAdOx1 nCoV-19) (NCT04400838)	Induce both humoral and cellular immune responses during phase I/II clinical trials. 82.5% efficacy in all cohorts in Brazil, South Africa, and the UK during phase III clinical trial. www.clinicaltrials.gov/ct2/show/NCT04400838. ClinicalTrials.gov NCT04400838 (2020). (Voysey et al., 2020).
Gamaleya Research Institute of Epidemiology and Microbiology	Approved for use	Gam-COVID-Vac and Gam-COVID-Vac Lyo (NCT04437875)	Induce strong humoral and cellular immune responses during phase I/II studies, first registered vaccine for COVID-19. Highly efficient about 91.6% and showed a robust humoral and a long-term cell-mediated immune response. (Logunov et al., 2021)
Novavax's Matrix-M technology	Phase III	Prefusion vaccine NVX-COV2373 vaccine (NCT04368988)	Preclinical studies showed high immunogenicity and high levels of neutralizing antibodies, induced neutralization titers with robust CD4+ T cell response (IFN-γ, IL-2, and TNF-α). Induced neutralizing anitbody titers in 100% of participants with robust polyfunctional CD4+ T cell response during phase II trial. www.clinicaltrials.gov/ct2/show/NCT0 4368988. ClinicalTrials.gov NCT04368988 (2020). https://ir.novavax.com/news-releases/news-release-deta ils/novavax-announces-positive-preclinical-data-comb ination. Novavax. (2021). Press release on May 10, 2021. (Keech et al., 2020)
Clover Biopharmaceuticals AUS Pty Ltd	Phase I	Recombinant protein vaccine (SCB-2019) (NCT04405908)	Under clinical trial. www.clinicaltrials.gov/ct2/show/NCT04405908. ClinicalTrials.gov NCT04405908 (2020).
Symvivo Corporation	Phase II	bacTRL-Spike vaccine (NCT04334980)	Completed. https://clinicaltrials.gov/ct2/show/NCT04334980. ClinicalTrials.gov NCT04334980 (2020)
Johnson & Johnson	Approved for use	JNJ-78436735 or Ad26.COV2.S	www.fda.gov/emergency-preparedness-and-response/ coronavirus-disease-2019-covid-19/janssen-covid-19-vaccine (Johnson & Johnson. Press release February 27, 2021).
Live Attenuated Vaccine			
Serum Institute of India+ codagenix	Phase I	Live attenuated vaccine (COVI-VAC)	Phase I trial has been initiated in the UK www.businesstoday.in/sectors/pharma/coronavirus-serum-institute-claims-covid-19-vaccine-to-be-market-ready-by-2022/story/397920 (Codagenix. Press release January 11, 2021). www.prnewswire.com/news-releases/ codagenix-and-serum-institute-of-india-initiate-dosing-in-phase-1-trial-of-covi-vac-a-single-dose-intranasal-live-attenuated-vaccine-for-covid-19-301203130.html
Inactivated Virus Vaccines			
Chinese Academy of Medical Sciences	Phase III	an Inactivated SARS-CoV-2 Vaccine (NCT04412538)	Phase I/II clinical trial demonstrated that the vaccine is safe and immunogenic. https://clinicaltrials.gov/ct2/ show/NCT04412538

(continued)

TABLE 12.1 (Continued)

Company	Stage	Strategy	Outcomes
Wuhan Institute of Biological Products Co., Ltd., China	Phase I/II	Inactivated Novel Coronavirus Pneumonia (COVID-19) vaccine (Vero cells) (ChiCTR2000031809)	Under clinical trials. www.chictr.org.cn/showprojen.aspx?proj=52227 (ChiCTR. Press release September 03, 2020).
Sinovac Biotech Ltd.	Phase III	Chemically inactivated virus vaccine (PiCoVacc)	Phase I/II studies showed that it is immunogenic. In phase III trial in Brazil and Turkey, the vaccine showed a protective efficacy of 50.65% (Gao et al., 2020; Hotez et al., 2020).
Bharat Biotech	Approved for use	Inactivated vaccine candidate (COVAXIN)	Phase I/II clinical trial showed vaccine is immunogenic and has no adverse events. 81% protective efficacy in phase III clinical trials Deshpande et al., 2020). Effective against UK strain (Sapkal et al., 2021).
Subunit Vaccines			
GlaxoSmithkline+ Clover Biopharmaceuticals	Phase II	Engineering adjuvants with proteins	Phase I studies demonstrated that vaccine-elicited the neutralizing immune responses. www.worldpharmanews.com/gsk/5305-gsk-covid-19-vaccine-development-collaboration-with-clover-biopharmaceuticals-begins-clinical-trials GlaxoSmithKline (2020).
Chulalongkorn University, Thailand	Preclinical	RBD protein fused with Fc of IgG and adjuvant	(Tai et al., 2020)
National Institute of Infectious Disease, Japan	Phase I/II	recombinant S protein with gold nanoparticles (AuNPs)	Elicits a strong antigen-specific IgG response against SARS-CoV-2 (Sekimukai et al., 2020).
University of Pittsburgh microneedle (MNA) arrays	Preclinical	Microneedle array (MNA) – S1 subunit of spike protein – CoV-2	Generates a surge of antibodies against SARS-CoV-2 within two weeks (Kim et al., 2020).
Sanofi Pasteur + GSK	Phase III	S protein	Preclinical studies showed the vaccine candidate has promising immunogenicity Phase I/II trial showed a higher immune response. www.sanofi.com/en/media-room/press-releases/. Sanofi and GSK (2020).
University of Queensland/with GSK & Dynavax	Phase I	Molecular clamp stabilized Spike protein	www.uq.edu.au/news/article/2020/04/international-partnership-progresses-uq-covid-19-vaccine-project. University of Queensland, Australia (2020). www.uq.edu.au/news/article/2020/12/update-uq-covid-19-vaccine. University of Queensland (2020).
Vaxart company	Phase I	COVID-19 Oral Vaccine Candidate encoder various configurations and combinations of the spike protein	https://investors.vaxart.com/news-releases/news-release-details/vaxarts-oral-covid-19-vaccine-candidate-induces-potent-systemic. Vaxart company (2020).
Convalescent Plasma Therapy			
Immunitor LLC	Emergency use	Heat-inactivated plasma from donors with COVID-19 V-(SARS) (NCT04380532)	Follow-up is currently ongoing. clinicaltrials.gov/ct2/show/NCT04380532. ClinicalTrials.gov NCT04380532 (2020).

Clover Biopharmaceuticals started phase I clinical trial of Covid-19 S-Trimer vaccine (SCB-2019) candidate in combination with their pandemic adjuvant system. Preliminary results from the phase I studies demonstrated that the vaccine elicited the neutralizing immune responses and is stable at 2–8°C for at least 6 months (longer-term stability studies are ongoing) and stable at room temperature and 40°C for at least one month (GlaxoSmithkline. Press release March 11, 2021). Vaxart developed an oral Covid-19 vaccine candidate on the company's Vector-Adjuvant-Antigen Standardized Technology (VAAST) Platform encoding various configurations and combinations of the spike protein. The vaccine generates strong immune response after a single dose of vaccine as well as induces potent systemic and mucosal immune response (Vaxart company. Press release September 08, 2020).

12.4 PLASMA THERAPY

To date, there is no specific treatment for SARS-CoV-2 infection. The plasma therapy or convalescent plasma treatment uses antibody-rich blood plasma isolated from cured patients to treat infected patients. This is a promising approach to help patients whose immune system is not producing enough antibodies to control the disease. Convalescent plasma therapy has shown a promising role in treating measles, chickenpox, and rabies (Garraud et al., 2016). Guy's and St Thomas' NHS Foundation Trust, in a joint effort with UK National Institute for Health Research (NIHR) started a clinical trial to assess blood plasma therapy for Covid-19 treatment. There is reduction in the mortality rate of Covid-19 infected patients treated with convalescent plasma therapy (Chen et al., 2020). Immunitor Company developed a pill of therapeutic vaccine containing heat-inactivated plasma from donors with Covid-19 (Clinicaltrials.gov/ NCT04380532).

The Indian Council of Medical Research, India, has approved 21 institutions to conduct phase II trials using convalescent plasma in Covid-19 patients with moderate illness (ICMR, 2020). Central Drugs Standard Control organization (CDSO), India issued the guidelines on the use of Convalescent Plasma therapy in Covid-19 patients with moderate symptoms (https://cdsco.gov.in/opencms/opencms/en/Notifications/Public-Notices/). The FDA has also given emergency authorization to use plasma to treat Covid patients (U.S. Food and Drug Administration. Press release September 02, 2020).

12.5 ADVERESE EFFECTS

Serious or long-lasting side effects to vaccines are extremely rare. The majority of side effects seen in clinical trials of SARS-CoV-2 vaccines were mild including pain at the injection site, redness, swollen lymph nodes in the arm at the site of vaccination, tiredness, headache, muscle or joint aches, nausea, diarrhea and vomiting, fever, or chills. Vaccines are continually monitored for any adverse event (WHO, 2020).

A small percentage of people have experienced an allergic reaction after their first dose. Anaphylaxis (acute allergic reaction) to the Covid-19 vaccines is estimated to occur in 2.5 to 11.1 cases per million doses, with the majority of cases occurring in people who have a history of allergy (Centers for Disease Control and Prevention Press release January 06, 2021).

12.6 CONCLUSION

More than 300 million doses of coronavirus vaccine have been administered in over 100 countries around the world. Some countries have already secured and administrated doses to a significant portion of their population. India is the largest manufacturer of Covid-19 vaccines, providing vaccines to 69 countries of the world. On January 16, 2021, India launched a national vaccination campaign involving 3,006 vaccination centers, each of which will deliver either Covishield or Covaxin. Ten million healthcare staff, who are at high risk of being infected with the Covid-19, are among the first to be vaccinated. The vaccine campaign will protect 300 million target groups that are at higher risk of infection over the next six months. There are approximately 270 million people over the age of 50 and/or with co-morbidities among the 10 million health professionals, 20 million frontline workers, and 10 million health workers. Covid-19 vaccination will help save lives, stabilize health systems, and stimulate economic growth.

REFERENCES

Abraham PK, Srihansa T, Krupanidhi S, Vijaya SA (2020). Venkateswarulu TC. Design of multi-epitope vaccine candidate against SARS-CoV-2: A in-silico study. *J Biomol Struct Dyn*1–9.

Abu-Raddad LJ, Chemaitelly H, Butt AA (2021). National Study Group for COVID-19 Vaccination. Effectiveness of the BNT162b2 Covid-19 Vaccine against the B.1.1.7 and B.1.351 Variants. *N Engl J Med*. doi: 10.1056/NEJMc2104974.

Arcturus Therapeutics (2020). Press release on 8 May 2020 about preclinical data for Covid-19. www.catalent.com/catalent-news/arcturus-therapeutics-and-catalent-announce-partnership-to-manufacture-mrna-based-covid-19-vaccine

Baric RS (2020). Emergence of a highly fit SARS-CoV-2 variant. *New England Journal of Medicine* 383(27):2684–2686.

Bull JJ, Nuismer SL, Antia R (2019). Recombinant vector vaccine evolution. *PLoS Comput Biol* 5(7):e1006857.

CanSino Biologics (2021). Press release on March 01, 2021. "Convidicea Vaccine." www.precision vaccinations. com/vaccines/convidicea-vaccine

CanSino Biologics (2021). Press release on April 08, 2021. CanSinoBIO Announces Approval for its Single-Dose COVID-19 Vaccine Convidecia™ in Chile. www.prnewswire.com/news-releases/cansinobio-announces-approval-for-its-single-dose-covid-19-vaccine-convidecia-in-chile-301265147.html.

Cascella M, Rajnik M, Cuomo A, Dulebohn SC, Di Napoli R (2020). Features, Evaluation and Treatment Coronavirus (COVID-19). In: *StatPearls*. Treasure Island (FL): StatPearls Publishing.

Centers for Disease Control and Prevention (CDC), USA (2021). Press release on January 6, 2021. Allergic reactions including anaphylaxis after receipt of the first dose of Pfizer-BioNTech COVID-19 Vaccine, United States. www.cdc.gov/ mmwr/volumes /70/wr/mm7002e1.htm

Chen L, Xiong J, Bao L, Shi Y (2020). Convalescent plasma as a potential therapy for COVID-19. *Lancet Infect Dis* 20(4):398–400.

Chen WH, Tao X, Agrawal A, et al. (2020). Yeast-expressed SARS-CoV recombinant receptor-binding domain (RBD219-N1) formulated with alum induces protective immunity and reduces immune enhancement. *bioRxiv* 2020.05.15.098079.

ChiCTR (2021). Press release on November 10, 2021. A randomized, double-blind, placebo parallel-controlled phase I/II clinical trial for inactivated Novel Coronavirus Pneumonia vaccine (Vero cells). www.chictr.org.cn/showprojen.aspx?proj=52227

ClinicalTrials.gov NCT03721718 (2020). Evaluate the Safety, Tolerability and Immunogenicity Study of GLS-5300 in Healthy Volunteers. www.clinicaltrials.gov/ct2/show/NCT03721718

ClinicalTrials.gov NCT04283461 (2020). Safety and Immunogenicity Study of 2019-nCoV Vaccine (mRNA-1273) for Prophylaxis of SARS-CoV-2 Infection (COVID-19). www.clinicaltrials.gov/ct2/show/NCT04283461

ClinicalTrials.gov NCT04334980 (2020). Evaluating the Safety, Tolerability and Immunogenicity of bacTRL-Spike Vaccine for Prevention of COVID-19. www.clinicaltrials.gov/ct2/show/NCT04334980

ClinicalTrials.gov NCT04368988 (2020). Evaluation of the Safety and Immunogenicity of a SARS-CoV-2 rS (COVID-19) Nanoparticle Vaccine With/Without Matrix-M Adjuvant. www.clinicaltrials.gov/ct2/show/NCT04368988

ClinicalTrials.gov NCT04380532 (2020). Tableted COVID-19 Therapeutic Vaccine (COVID-19). Accessed June 25, 2020. NCT04380532. www.clinicaltrials.gov/ct2/show/NCT04380532

ClinicalTrials.gov NCT04380701 (2020). A Trial Investigating the Safety and Effects of Four BNT162 Vaccines against COVID-2019 in Healthy Adults. www.clinicaltrials.gov/ct2/show/NCT04380701

ClinicalTrials.gov NCT04398147 (2020). Phase I/II Clinical Trial of Recombinant Novel Coronavirus Vaccine (Adenovirus Type 5 Vector) in Canada. www.clinicaltrials.gov/ct2/show/NCT04398147

ClinicalTrials.gov NCT04400838 (2020). Investigating a Vaccine against COVID-19. www.clinicaltrials.gov/ct2/show/NCT04400838

ClinicalTrials.gov NCT04405908 (2020). SCB-2019 as COVID-19 Vaccine. www.clinicaltrials.gov/ct2/show/NCT0440 5908

ClinicalTrials.gov NCT04412538 (2020). Safety and Immunogenicity Study of an Inactivated SARS-CoV-2 Vaccine for Preventing Against COVID-19. www.clinicaltrials.gov/ct2/show/NCT04412538

Codagenix (2021). Press release on January 11, 2021. Codagenix and Serum Institute of India Initiate Dosing in Phase 1 Trial of COVI-VAC, a Single Dose, Intranasal, Live Attenuated Vaccine for COVID-19. www.prnewswire.com/news-releases/codagenix-and-serum-institute-of-india-initiate-dosing-in-phase-1-trial-of-covi-vac-a-single-dose-intranasal-live-attenuated-vaccine-for-covid-19-301203130.html

Corbett KS, Flynn B, Foulds KE, et al. (2020). Evaluation of the mRNA-1273 Vaccine against SARS-CoV-2 in Nonhuman Primates. *N Engl J Med* 10.1056/NEJ Moa2024671. doi:10.1056/NEJMoa2024671

Corey L, Mascola JR, Fauci AS, Collins FS (2020). A strategic approach to COVID-19 vaccine R&D. *Science* 368(6494):948–950.

CureVac (2020). Press release on June 18, 2020. CureVac to trial Covid-19 vaccine in Germany and Belgium. www.clinicaltrialsarena.com/news/curevac-covid-19-vaccine-trial/

CureVac Company (2021). Press release on March 23, 2021. CVnCoV Vaccine. www.precisionvaccinations.com/vaccines/cvncov-vaccine

Deccan Herald web desk (2021). New coronavirus strain 70% more infectious: Here's all you need to know. www.deccanherald.com/international/world-news-politics/new-coronavirus-strain-70-more-infectious-heres-all-you-need-to-know-929790.

Deshpande GR, Sapkal GN, Tilekar BN, Yadav PD, Gurav Y, Gaikwad S, Kaushal H, Deshpande KS, Kaduskar O et al. (2020). Neutralizing antibody responses to SARS-CoV-2 in COVID-19 patients. *Indian J Med Res* 152(1&2):82–87.

Folegatti PM, Ewer KJ, Aley PK, et al. (2020). Safety and immunogenicity of the ChAdOx1 nCoV-19 vaccine against SARS-CoV-2: A preliminary report of a phase 1/2, single-blind, randomised controlled trial. [published correction appears in Lancet. 2020 Aug 15;396(10249):466]. *Lancet* 396 (10249):467–478.

Garraud O, Heshmati F, Pozzetto B, Lefrere F, Girot R, Saillol A, Laperche S (2016). Plasma therapy against infectious pathogens, as of yesterday, today and tomorrow. *Transfus Clin Biol* 23(1):39–44.

GlaxoSmithKline (2020). Press article release on June 19, 2020. Glaxo's Covid Vaccine Partnership with Clover Begins Human Tests. www.worldpharmanews.com/gsk/5305-gsk-covid-19-vaccine-development-collaboration-with-clover-biopharmaceuticals-begins-clinical-trials

GlaxoSmithkline (2021). Press release on March 11, 2021. COVID-19 S-Trimer (SCB-2019) Vaccine. www.precisionvaccinations.com/vaccines/covid-19-s-trimer-scb-2019-vaccine

He Y, Zhou Y, Liu S, et al. (2004). Receptor-binding domain of SARS-CoV spike protein induces highly potent neutralizing antibodies: Implication for developing subunit vaccine. *Biochem Biophys Res Commun* 324(2):773–781.

Hoffmann D, Corleis B, Rauch S, Roth N, Mühe N et al. (2021). CVnCoV protects human ACE2 transgenic mice from ancestral B BavPat1 and emerging B.1.351 SARS-CoV-2. https://doi.org/10.1101/2021.03.22.435960.

Hotez PJ, Corry DB, Strych U, Bottazzi ME (2020). COVID-19 vaccines: Neutralizing antibodies and the alum advantage. *Nat Rev Immunol* 1–2.

Huang C, Wang Y, Li X, Ren L, Zhao J, Hu Y, Zhang L, Fan G, Xu J, Gu X, et al. (2020). Clinical features of patients infected with 2019 novel coronavirus in Wuhan, China. *Lancet* 395:497–506

Hui DS, Azhar EI, Madani TA, Ntoumi F, Kock R, et al. (2020). The continuing 2019-nCoV epidemic threat of novel coronaviruses to global health-the latest 2019 novel coronavirus outbreak in Wuhan. *China, Int. J. Infect. Dis* 91:264–266.

Imperial College London (2020). Press statement release on June 15, 2020. Imperial to begin first human trials of new COVID-19 vaccine. www.imperial.ac.uk/news/198314/imperial-begin-first-human-trials-covid-19/

Inovio Pharmaceuticals (2021). Press release on May 12, 2021. INO-4800 COVID-19 Vaccine Dosage and INO-4802 Pan-COVID-19 Vaccine Candidate. www.precisionvaccinations.com/vaccines/ino-4800-covid-19-vaccine

Jackson LA, Anderson EJ, Rouphael NG, et al. (2020). An mRNA Vaccine against SARS-CoV-2 -Preliminary Report [published online ahead of print, July 14, 2020]. *N Engl J Med* NEJMoa2022483.

Johnson & Johnson (2021). Press release on February 27, 2021. Johnson & Johnson COVID-19 vaccine authorized by U.S. FDA for emergency use – first single-shot vaccine in fight against global pandemic. www.jnj.com/johnson-johnson-covid-19-vaccine-authorized-by-u-s-fda-for-emergency-usefirst-single-shot-vaccine-in-fight-against-global-pandemicl

Kai Wu, Angela Choi, Matthew Koch, Sayda Elbashir, LingZhi Ma, et al. (2021). Variant SARS-CoV-2 mRNA vaccines confer broad neutralization as primary or booster series in mice. www.biorxiv.org/content/10.1101/2021.04.13.439482v1

Keech C, Albert G, Cho I, Robertson A, Reed P, Neal S, et al. (2020). Phase 1-2 Trial of a SARS-CoV-2 Recombinant Spike Protein Nanoparticle Vaccine. *N Engl J Med* 383(24):2320–2332.

Kim E, Erdos G, Huang S, et al. (2020). Microneedle array delivered recombinant coronavirus vaccines: Immunogenicity and rapid translational development. *EBioMedicine* 55:102743.

Korber B, Fischer WM, Gnanakaran S, Yoon H, et al. (2020). Tracking Changes in SARS-CoV-2 Spike: Evidence that D614G Increases Infectivity of the COVID-19 Virus. *Cell* 182(4):812–827.e19.

Lamirande EW, DeDiego ML, Roberts A, et al. (2008). A live attenuated severe acute respiratory syndrome coronavirus is immunogenic and efficacious in golden Syrian hamsters. *J Virol* 2(15):7721–7724.

Lara H, Min D, Brenda M, Branda H, Igor S, et al. (2021). Safety and immunogenicity of S-Trimer (SCB-2019), a protein subunit vaccine candidate for COVID-19 in healthy adults: A phase 1, randomised, double-blind, placebo-controlled trial. 397(10275):682–694.

Liu W, Fontanet A, Zhang PH, Zhan L, Xin ZT, Baril L, Tang F, et al. (2006). Two-year prospective study of the humoral immune response of patients with severe acute respiratory syndrome. *J Infect Dis* 193:792–795.

Logunov DY, Dolzhikova IV, Shcheblyakov DV, Tukhvatulin AI, Zubkova OV, et al. (2021). Gam-COVID-Vac Vaccine Trial Group. Safety and efficacy of an rAd26 and rAd5 vector-based heterologous prime-boost COVID-19 vaccine: An interim analysis of a randomised controlled phase 3 trial in Russia. *Lancet* 20;397(10275):671–681.

Logunov DY, Dolzhikova IV, Zubkova OV, et al. (2020). Safety and immunogenicity of an rAd26 and rAd5 vector-based heterologous prime-boost COVID-19 vaccine in two formulations: Two open, non-randomised phase 1/2 studies from Russia. *Lancet* 6736(20):31866–3.

McCallum M, Bassi J, Marco A, Chen A, et al. (2021). SARS-CoV-2 immune evasion by variant B.1.427/B.1.429. *bioRxiv* 1:2021.03.31.437925.

Moderna (2020). press release on November 30, 2020. Moderna Announces Primary Efficacy Analysis in Phase 3 COVE Study for Its COVID-19 Vaccine Candidate and Filing Today with U.S. FDA for Emergency Use Authorization. www.nih.gov/news-events/news-releases/nih-clinical-trialinvestigational-vaccine-covid-19-begins

Mulligan MJ, Lyke KE, Kitchin N, Absalon J, Gurtman A, Lockhart SP, Neuzil K, Raabe V, Bailey R, Swanson KA, Li P, Koury K, Kalina W, Cooper D, et al. (2020). Phase 1/2 Study to Describe the Safety and Immunogenicity of a COVID-19 RNA Vaccine Candidate (BNT162b1) in Adults 18 to 55 Years of Age: Interim Report. *medRxiv* 2020.2006.2030.20142570

Mullard A (2020). COVID-19 vaccine development pipeline gears up. *Lancet* 395(10239):1751–1752.

Novavax. (2021). Press release on May 10, 2021. Novavax Announces Positive Preclinical Data for Combination Influenza and COVID-19 Vaccine Candidate. https://ir.novavax.com/news-releases/news-release-details/novavax-announces-positive-preclinical-data-combination

Ni L, Ye F, Cheng ML, et al. (2020). Detection of SARS-CoV-2-Specific humoral and cellular immunity in COVID-19 convalescent individuals. *Immunity* 52(6):971–977.e3.

Pandey SC, Pande V, Sati D, Upreti S, Samant M (2020). Vaccination strategies to combat novel corona virus SARS-CoV. *Life Sci* 256:117956.

Patel A, Walters J, Reuschel EL, Schultheis E, Parzych E, et al. (2020). Intradermal-delivered DNA vaccine provides anamnestic protection in a rhesus macaque SARS-CoV-2 challenge model. *bioRxiv* 28.225649.

Robbiani DF, Gaebler C, Muecksch F, et al. (2020). Convergent antibody responses to SARS-CoV-2 infection in convalescent individuals. *bioRxiv* 2020.05. 13.092619.

Sanofi and GSK (2020). Press release on September 03, 2020. Sanofi and GSK initiate Phase 1/2 clinical trial of COVID-19 adjuvanted recombinant protein-based vaccine candidate. www.sanofi.com/en/media-room/press-releases/

Sanofi and Translate Bio (2021). Press release on March 12, 2021. Sanofi and Translate Bio Initiate Phase 1/2 Clinical Trial of mRNA COVID-19 Vaccine Candidate. https://investors.translate.bio/news-releases/news-release-details/sanofi-and-translate-bio-initiate-phase-12-clinical-trial-mrna

Sapkal GN, Yadav PD, Ella R, Deshpande GR, Sahay RR, Gupta N, Mohan VK, Abraham P, Bhargava B (2021). Neutralization of UK-variant VUI-202012/01 with COVAXIN vaccinated human serum. *bioRxiv* 2021.01. 26.426986.

Sekimukai H, Iwata-Yoshikawa N, Fukushi S, et al. (2020). Gold nanoparticle-adjuvanted S protein induces a strong antigen-specific IgG response against severe acute respiratory syndrome-related coronavirus infection, but fails to induce protective antibodies and limit eosinophilic infiltration in lungs. *Microbiol Immunol* 64(1):33–51.

Serum Institute of India (2020). Press release March 09, 2020. Serum Institute claims COVID-19 vaccine to be market ready by 2022. www.business today.in/sectors/pharma/ coronavirus-serum-institute-claims-covid-19-vaccine-to-be-market-ready-by-2022/story/397920.

Smith TRF, Patel A, Ramos S, et al. (2020). Immunogenicity of a DNA vaccine candidate for COVID-19. *Nat Commun* 11(1):2601.

Spinney L (2020). When will a coronavirus vaccine be ready? *The Guardian* p. 18.

Tai W, He L, Zhang X, et al. (2020). Characterization of the receptor-binding domain (RBD) of 2019 novel coronavirus: Implication for development of RBD protein as a viral attachment inhibitor and vaccine. *Cell Mol Immunol* 17(6):613–620.

Tebas P, Yang S, Boyer JD, Reuschel EL, Patel A, Christensen-Quick A et al. (2020). Safety and immunogenicity of INO-4800 DNA vaccine against SARS-CoV-2: A preliminary report of an open-label, Phase 1 clinical trial. 100689.

Thomsan Reuter (2021). Press release on April 14, 2021. Brazil's P1 coronavirus variant mutating, may become more dangerous – study. www.reuters.com/article/us-health-coronavirus-brazil-variant-idUKKBN2C11XX).

Thomsan Reuter (2021). Press April 13, 2021. Spain unaware of any delays in J&J COVID vaccine deliveries. Joan FausNathan Allen. www.reuters.com/world/europe/benefits-all-approved-covid-19-vaccines-outweigh-risks-spains-pm-says-2021-04-13/.

University of Queensland (2020). Press release on December 11, 2020. Update on UQ COVID-19 vaccine. www.uq.edu.au/news/article/2020/12/update-uq-covid-19-vaccine

University of Queensland, Australia (2020). Press release on April 09, 2020. International partnership progresses UQ COVID-19 vaccine project. www.uq.edu.au/news/article/2020/04/international-partnership- progresses-uq-covid-19-vaccine-project

U.S. Food and Drug Administration (2020). Press release September 02, 2020. Recommendations for Investigational COVID-19 Convalescent Plasma. www.fda.gov/vaccines-blood-biologics/investigational-new-drug-ind-or-device-exemption-ide-process-cber/recommendations-investigational-covid-19-convalescent-plasma

Vaidyanathan G (2021). Coronavirus variants are spreading in India—what scientists know so far. *Nature* 593: 321–321. https://doi.org/10.1038/ d41586-021-01274-7

Van DN, Lambe T, Spencer A, et al. (2020). ChAdOx1 nCoV-19 vaccination prevents SARS-CoV-2 pneumonia in rhesus macaques. Preprint. *bioRxiv*. 2020; 2020.05.13.093195. doi:10.1101/2020.05.13. 093195

Vaxart company (2020). Press release on September 08, 2020. Vaxart's Oral COVID-19 Vaccine Candidate Induces Potent Systemic and Mucosal Immune Responses in Preclinical Studies. https://investors.vaxart.com/news-releases/news-release-details/vaxarts-oral-covid-19-vaccine-candidate-induces-potent-systemic

Voysey M, Clemens SAC, Madhi SA, Weckx LY, Folegatti PM, Aley PK, et al. (2020). Oxford COVID Vaccine Trial Group. Safety and efficacy of the ChAdOx1 nCoV-19 vaccine (AZD1222) against SARS-CoV-2: An interim analysis of four randomised controlled trials in Brazil, South Africa, and the UK. *Lancet* 8:S0140-6736(20)32661-1.

Wang M, Jiang S, Wang Y (2016). Recent advances in the production of recombinant subunit vaccines in Pichia pastoris. *Bioengineered* 7(3):155–165.

Wang Z, Yuan Z, Matsumoto M, Hengge UR, Chang YF (2005). Immune responses with DNA vaccines encoded different gene fragments of severe acute respiratory syndrome coronavirus in BALB/c mice. *Biochem Biophys Res Commun* 327(1):130–135.

WHO (2020). www.who.int/emergencies/diseases /novel-coronavirus-2019.

Yuki K, Fujiogi M, Koutsogiannaki S (2020). COVID-19 pathophysiology: A review. *Clin Immunol* 215:108427.

Zhou Y, Fu B, Zheng X, et al. (2020). Pathogenic T cells and inflammatory monocytes incite inflammatory storm in severe COVID-19 patients. *Natl Sci Rev* nwaa041.

Zhu FC, Li YH, Guan XH, et al. (2020). Safety, tolerability, and immunogenicity of a recombinant adenovirus type-5 vectored COVID-19 vaccine: A dose-escalation, open-label, non-randomised, first-in-human trial. *Lancet* 395(10240):1845–1854.

Zhu N, Zhang D, Wang W, et al. (2020). A Novel Coronavirus from Patients with Pneumonia in China, 2019. *N Engl J Med* 382(8):727–733.

13 Preclinical Safety Assessment of Vaccines Developed Against COVID-19

Recent Updates

Shubham Adhikary,[1] Harpal S. Buttar,[2] Hardeep Singh Tuli,[3] and Ginpreet Kaur[*1]

[1]Department of Pharmacology, SPP School of Pharmacy & Technology Management, SVKM's NMIMS, Mumbai, Maharashtra, India
[2]Department of Pathology & Laboratory Medicine, University of Ottawa, Faculty of Medicine, Ottawa, Ontario, Canada
[3]Department of Biotechnology, Maharishi Markandeshwar (Deemed to be University), Mullana-Ambala, Haryana, India

CONTENTS

13.1 INTRODUCTION

The Covid-19 pandemic has infected millions of people and still continues to do so without any extenuation due to its high prevalence, long incubation period (Backer et al., 2020), and a paucity of established treatments and/or vaccines (Jean et al., 2020). As per recent updates (January 2023), the total number of cases are **669,582,266 (active and recovered), with the total number of deaths being 6,718,277, and 640,899,774 have recovered** (*Covid Live Update: 150,379,130 Cases and 3,167,228 Deaths from the Coronavirus – Worldometer*, n.d.).

Vaccines by definition are biological agents intended to produce an immune response in the body to a specific antigen derived from an infectious pathogen or disease-causing organism, such as a virus, bacteria, fungi, or parasite. Vaccines have transformed public health safety and prophylaxis, especially after national immunization programs became properly established in the 1960s (Clem,

DOI: 10.1201/9781003358909-13

2011). Vaccines cleverly employ the remarkable ability of the human immune system to respond to the invasion of foreign pathogens and to remember the interactions of pathogenic antigens. In the fifteenth century, both the Chinese and the Turks were trying to induce immunity against smallpox using dried crusts from smallpox lesions by either inhaling the crushed materials or inserting them into the body through small skin cuts. These were the first historical attempts of inoculation. These crude attempts of immunization further led to Edward Jenner's experiments in England where he inoculated a handful of friends and neighbours with cowpox. He later discovered that the cowpox inoculation offered smallpox immunity in humans (Riedel, 2005). This groundbreaking work of Edward Jenner ultimately led to the global eradication of smallpox, as officially declared by the WHO in 1980. These pioneer efforts have led to the discovery of a myriad of vaccines available today. Although all those vaccines were efficacious, their underlying mechanisms to produce immunity at the cellular and molecular levels remained to be ascertained.

The need to expeditiously develop a vaccine against **SARS-CoV-2 or Covid-19** came at a time where there is an explosion of knowledge in the scientific community regarding genomics and proteomics as well as cellular and molecular biology has collectively opened and strengthened a new era of vaccine development (McCullers & Dunn, 2008). Currently, there are more than 300 vaccine candidates in different phases of research, with more than 60 candidates being tested in human clinical trials. According to recent findings, several presently available vaccines offer partial to total protection against upcoming SARS-CoV-2 or Covid-19 strains. With the new variants emerging, new mass vaccination strategies are being developed by healthcare authorities (Dal-Ré et al., 2021; Tregoning et al., 2020; Ye et al., 2020).

The virus mainly affects the respiratory system, although other systems can also get involved in complicated cases. The early case series from Wuhan, China described signs of lower respiratory tract infection such as fever, dry cough, and dyspnea (Huang et al., 2020). Headaches, weakness, dizziness, and diarrhoea were also observed (Yuki et al., 2020). It is now widely believed that the symptoms of Covid-19 are heterogeneous in nature, ranging from minimal symptoms similar to rhinitis to significant hypoxia with ARDS (Shi et al., 2020). The interval between the inception of symptoms and the development of ARDS might be as little as 9 days, implying that this condition is deadly. Despite the high recovery rates, a disturbingly high number of individuals have died from the disease across the world. Current clinical treatments available are largely supportive with no targeted therapy available (Xu et al., 2020). Drugs like lopinavir-ritonavir, remdesivir (Aleissa et al., 2021), hydroxychloroquine (Seligmann, 2021), and azithromycin have been tested in clinical trials (*Efficacy and Safety of Hydroxychloroquine* CP *and Azithromycin for the Treatment of Ambulatory Patients With Mild COVID-19 – Full Text View – ClinicalTrials.Gov*, n.d.; Gautret et al., 2020), but none of them have been proven to be a definite therapy yet (Zhou et al., 2020; Funk et al., 2020). Convalescent plasma treatment has been shown to be useful for a better course of Covid-19 in severe and critically ill patients, due to the fact that a high number of patients require ICU support (Fda & Cber, 2021). The ideal dosage and transfusion period, as well as the safety and effectiveness of CP transfusion, must all be thoroughly explored in well-designed randomized clinical trials (Altuntas et al., 2020; Robbiani et al., 2020). Large-scale immunization plans have been executed by governments and many vaccines are already under clinical trials.

The vaccines being developed to battle the pandemic are using next-generation vaccine platforms, with the main advantage being that they can be made using the genetic sequence alone. If the viral proteins that play a major role in imparting immunity from the virus are known, then the presence of coding sequences of the viral protein(s) is sufficient to initiate vaccine development, rather than relying on the ability to culture the virus. This makes these technologies highly malleable and speeds up the development considerably. For Covid-19, several viral vector, nucleic acid-based vaccines and antigen presenting cells are under development (Perlman & Netland, 2009). Viral vector vaccines are comprised of a recombinant virus that has been subdued to diminish its pathogenicity and has had genes encoding viral antigen(s) cloned using recombinant DNA technology

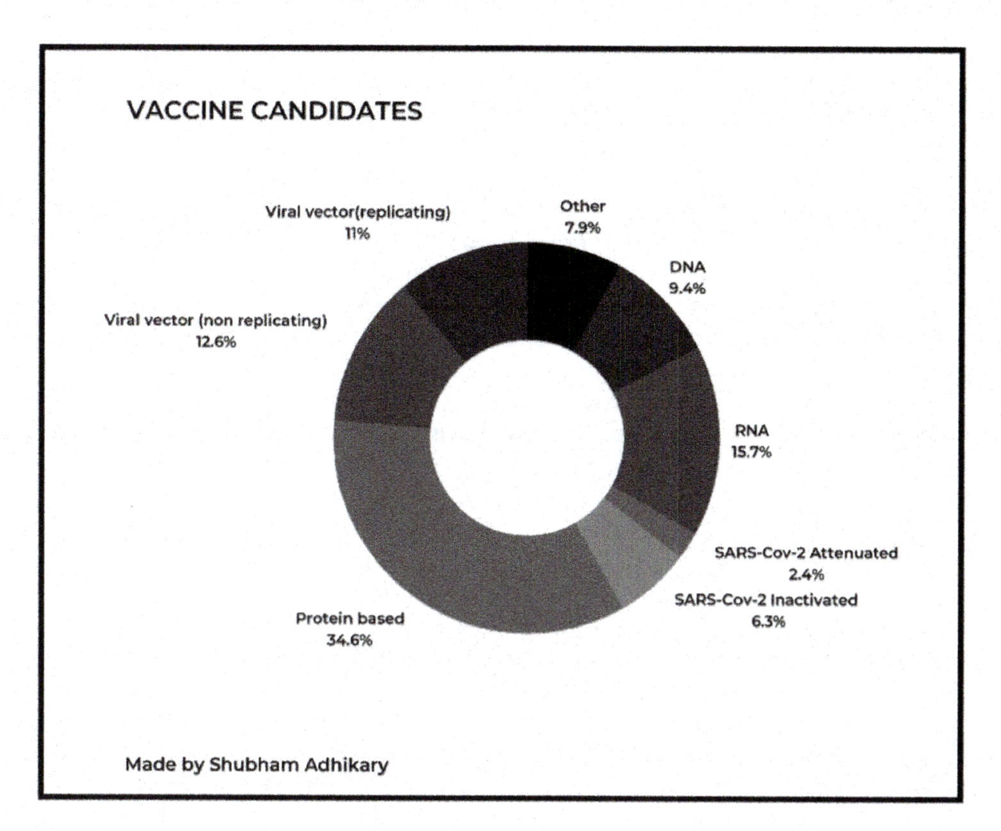

FIGURE 13.1 Diagrammatic representation of different categories of vaccines being developed against SARS-CoV-2 or Covid-19.

(van Riel & de Wit, 2020). There are two types of vector vaccines: Replicating or non-replicating. Nucleic acid-based vaccines, which may be made of DNA or mRNA, can be readily changed as new viruses appear, which is why these were among the first Covid-19 vaccines to enter clinical trials (Ong et al., 2020). A synthetic DNA construct encoding the vaccination antigen is used to create DNA vaccines (Lurie et al., 2020). The aim of this chapter is to highlight the pre-clinical trials conducted for some of the potential vaccines against SARS-CoV-2. It also discusses the various (Thangavel & Bouvier, 2014; V et al., n.d.) used for exploring the potential and their results. In addition, it collates the preclinical development data of various chief vaccine candidates (Figure 13.1), which could function as a platform for future research (Funk et al., 2020).

13.2 MOLECULAR EVOLUTION

Despite the virus's sluggish mutation rate, researchers have catalogued more than 12,000 mutations in SARS-CoV-2 genomes. Compared with HIV, SARS-CoV-2 changes much more slowly as it spreads. The variants of concern have mostly been identified with a mutation in the gene encoding the spike protein, which helps virus particles to penetrate cells. The D614G mutation occurred at the 614th amino acid position of the spike protein, where the amino acid aspartate (D, in biochemical shorthand) was often substituted by glycine (G) due to a copying error that changed a single nucleotide in the virus's 29,903-letter RNA code. This variant became predominant sometime last year and was relatively mild. All vaccines worked against this, which is why the spread and infectivity was controlled. The three new variants that have rapidly become dominant within their countries have raised concerns are B.1.1.7 (also known as VOC-202012/01), 501Y.V2 (B.1.351), and P.1(B.1.1.28.1). The B.1.1.7 variant (23 mutations and 17 amino acid modifications) was sequenced first in the United Kingdom; the 501Y.V2 variant (23 mutations and 17 amino acid changes) was first reported in South Africa; and the P.1 variant (approximately 35 mutations and 17 amino acid changes) was first reported in Brazil. The B.1.1.7 strain had been recorded in 93 countries by February 22, 2021, the 501Y.V2 variant in 45, and the P.1 variant in 21. The N501Y mutation occurs in all three mutants, changing the amino acid asparagine (N) to tyrosine (Y) at position 501 in the spike protein's receptor-binding domain. K417N/T and E484K are two more receptor-binding–domain alterations seen in both the 501Y.V2 and P.1 variants. These alterations improve the receptor-binding domain's affinity for the angiotensin-converting enzyme 2 (ACE2) receptor. B.1.1.7 spreads more efficiently than earlier types of the coronavirus and is now the dominant form of SARS-CoV-2 in many countries. B.1.1.7 also likely causes more severe disease. Given that certain vaccines continue to have protective effects against the B.1.351 variants, it is necessary to continue the rapid and massive vaccination programmes against SARS-CoV-2 along with close monitoring of its genetic mutations.

13.3 PRECLINICAL DATA

The preclinical trials data concerning the evaluation of these vaccines have been retrieved from research papers published in reputable journals. As of June 6, 2021, there are over 100 candidates that have completed their preclinical studies and are in clinical evaluation (*Draft Landscape and Tracker of COVID-19 Candidate Vaccines*, n.d.). The approaches being applied for effective Covid-19 vaccine development (Calina et al., 2020; Vinceti et al., 2021) involve a replacement virus target and sometimes novel vaccine technology platforms and assess vaccine efficacy.

Covid-19-specific animal models have been developed, including ACE2-transgenic mice, hamsters, ferrets, and nonhuman primates (Thanh Le et al., 2020).

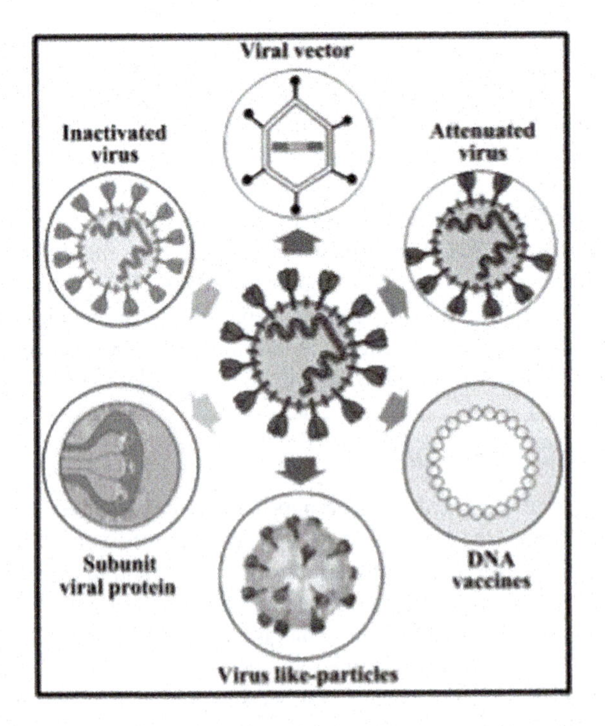

FIGURE 13.2 Strategy types for Covid-19 vaccine development.

13.3.1 University of Oxford/AstraZeneca

The AZD1222 vaccine from Oxford is based on a genetically modified virus (ChAdOx1) that causes the common cold in chimpanzees. It is a nonreplicating viral vector (van Doremalen et al., 2020). A group of macaque monkeys were infected with Covid-19 and treated with the ChAdOx1 vaccination in the study's preclinical phase. The vaccination did not prevent illness and did not prevent the animals from transferring the sickness to other monkeys, according to the findings. The findings did show that the vaccination might help to lessen the severity of the sickness (Dicks et al., 2012).

The vaccine was found to prevent significant lung damage and pneumonia, which has shown to be one of the more prominent reasons for mortality among patients infected with Covid-19 (*Oxford COVID-19 Vaccine Programme Opens for Clinical Trial Recruitment | University of Oxford*, n.d.; Robbiani et al., 2020; Yu et al., 2020). It is worth noting that the virus load put into the Oxford animals was far higher than other competing trials (Folegatti et al., 2020).

13.3.2 Moderna

Moderna's mRNA-1273 vaccine is a new technology so they started with phase I of clinical trials before the completion of preclinical trials (*Moderna's Work on Our COVID-19 Vaccine | Moderna, Inc.*, n.d.). The usual time taken to launch a vaccine is 15–20 years. But the revolutionary mRNA technology has paved the way for a swift vaccine development program. In the preclinical trials, nonhuman primates were given 10 or 100 g of mRNA-1273, a vaccine encoding the SARS-CoV-2 prefusion-stabilized spike protein, or no vaccination. Antibody and T cell responses were measured prior to upper- and lower-airway SARS-CoV-2 exposure (Corbett et al., 2020; *Safety and Immunogenicity Study of 2019-NCoV Vaccine (MRNA-1273) for Prophylaxis of SARS-CoV-2 Infection (COVID-19) – Full Text View – ClinicalTrials.Gov*, n.d.). The cited study demonstrated early suppression of viral replication in the upper and lower respiratory tracts following a high-dose SARS-CoV-2 challenge (about 8105 PFU) (Jackson et al., 2020). The capacity to inhibit viral replication in both the lower and upper airways has significant implications for vaccine-induced SARS-CoV-2 illness and transmission prevention (Anderson et al., 2020).

13.3.3 Sinovac

Sinovac PicoVacc is an inactivated vaccine in which the virus is inactivated by heat or chemical treatment, but the viral surface proteins are not. However, it is possible that it may not be immunogenic, and that repeated doses would be necessary. The mentioned study shows that PiCoVacc is safe in macaques and does not increase infection. Data also demonstrates against SARS-CoV-2 challenge with a 6-mg dose of PiCoVacc in macaques (Kandeil et al., 2021). No pyrexia or decrease in weight was observed after immunization with this vaccine, and the appetite was observed to be normal (Tostanoski et al., 2020). Histopathological examinations of many organs, including the lung, heart, spleen, liver, kidney, and brain revealed that the vaccination did not produce any significant disease in macaques (Gao et al., 2020).

13.3.4 CanSino Biologics (Ad5-nCoV)

It is a genetically engineered vaccine candidate that uses the replication-defective adenovirus type 5 vector to produce SARS-CoV-2 spike protein (Zhu et al., 2020). In preclinical trials, seven out of eight ferrets were protected from having detectable viral copies when challenged with SARS-CoV-2 through nasal dripping 21 days after vaccination, whereas only one out of eight ferrets in the control group was virus-free (*China's CanSino Biologics Reveals Phase 2 Coronavirus Vaccine Results – PMLiVE*, n.d.; Sprangers et al., 2003; Wei C, unpublished).

13.3.5 Sinopharm Vaccine

BBIBP-CorV, Sinopharm's vaccine against Covid-19, is also an inactivated form of the virus. It induced high levels of neutralizing antibodies in six mammalian species, including rats, mice, guinea pigs, rabbits, cynomolgus monkeys, and rhesus macaques, protecting them against SARS-CoV-2 infection (Isakova-Sivak & Rudenko, 2021). BBIBP-CorV vaccination in two doses (2 mg/dose) in rhesus macaques provided extremely effective protection against SARS-CoV-2 without causing any adverse effects or immunopathological aggravation (Xia et al., 2021). The viral load in throat swabs, on the other hand, was much reduced in animals given the low dosage of the vaccine and insignificant in those given the greater dosage (Wang et al., 2020).

13.3.6 Inovio Pharmaceuticals Vaccine

Vaccine INO-4800 is a DNA plasmid vaccine with electroporation that expresses the MERS-CoV spike (S) glycoprotein (Smith et al., 2020; Tebas et al., 2021). In the pre-clinical phase, they confirmed the immunization of mice and guinea pigs with INO-4800 and deduced that it prompted antigen-specific T cell responses, active antibodies that neutralize the SARS-CoV-2 virion and block spike protein binding to the ACE2 receptor, along with biodistribution of SARS-CoV-2 targeting antibodies to the lungs (Mukhopadhyay et al., 2021; Ye et al., 2020).

13.3.7 Bharat Biotech

Covaxin (CTRI/2020/07/026300) is one of India's most promising vaccine candidates against Covid-19. It is an (attenuated) inactivated virus vaccine with Alhydroxiquim-II as adjuvant to illicit stronger immune response (Sapkal et al., 2021).

Its preclinical trials were successfully conducted and showed strong immune response in subject animals (Ganneru et al., 2021; Mohandas et al., 2021), them being Rhesus macaques, mice, rat, and rabbit models (*Bharat Biotech's Covaxin Vaccine Yields Positive Phase I Data*, n.d.; *Covaxin and ZyCoV-D: Recent Update of Covid-19 Vaccine Candidates in India – NHI*, n.d.; Yadav, Ella, et al., 2021). It has successfully cleared clinical trials and is currently in circulation in India, Mongolia, Myanmar, Sri Lanka, and Philippines.

13.3.8 Zydus Cadila

ZyCoV-D (CTRI/2020/07/026352) is another vaccine candidate from India that has shown encouraging results in its preclinical studiesIt is a "plasmid DNA" vaccine, which means it contains a nonreplicating and nonintegrating plasmid encoding the new coronavirus gene, making it safe. The DNA platform is considered to have higher vaccine durability and fewer cold chain requirements, making it easier to store and deliver to remote areas of the country and world. Mice, guinea pig, and rabbit models were used to determine the immunogenicity potential of the vaccine via the intradermal method at 25, 100, and 500μg doses. Proof-of-concept was established based on animal investigations, and preclinical toxicology (PCT) experiments were executed in rat and rabbit models (Dey et al., 2021). The study was also conducted in rhesus macaques, which proved to have protective efficacy against the virus (Yadav, Kumar, et al., 2021). The ZyCoV-D DNA vaccine candidate developed by Cadila Healthcare Limited has recently completed phase I/II clinical trial and is waiting for approval for distribution (see Table 13.1).

13.4 SUMMARY AND FUTURE PERSPECTIVES

Advances in vaccine development and sophisticated manufacturing techniques coupled with open reporting and data sharing among the manufacturers have not only cemented the foundation for

TABLE 13.1
Preclinical Safety Testing of Different Vaccines against SARS-CoV-2 or COVID-19

Manufacturer Name	Platform	Vaccine Candidate	Study Identifier/ Phase	Immunogen	Study Location	Animal Model Chosen
Moderna (Corbett et al., 2020)	mRNA-based	mRNA1237	NCT04283461	S protein	United States	Mice and rhesus macaques
Innovio Pharmaceuticals (Smith et al., 2020)	DNA-based	INO-4800	NCT04336410	S protein	United States	Mice and guinea pig
Sinovac (Gao et al., 2020)	Inactivated (+Alum Adjuvant)	PiCoVacc	NCT04352608	Whole virus	China	Rats, mice, and rhesus macaques
Cansino Biologics (Zhu et al., 2020)	Adenovirus viral vector	Ad5-nCoV	NCT04341389	S protein	China	Mice and monkeys
University of Oxford/ AstraZeneca (van Doremalen et al., 2020)	Non-replicating viral vector	ChAdOx1 nCov-19 (AZD1222)	NCT04324606	S protein	United Kingdom	Macaque monkeys
Sinopharm (Xia et al., 2021)	Inactivated	BBIBP- CorV	ChiCTR2000034780	Whole Virus	China	Cynomolgus monkeys, rabbits, guinea pigs, rats, and mice
Bharat Biotech (Sapkal et al., 2021)	Inactivated (Whole - Virion Inactivated)	Covaxin	NCT04471519	Whole Virus	India	Mice and hamsters
Zydus Cadila (Dey et al., 2021)	DNA (DNA Plasmid Vaccine)	ZycovD	CTRI/2020/07/026352	Plasmid DNA	India	Rats, mice, guinea pigs, and rabbits

making different types of vaccines, but have also expedited the development of innovative vaccine making technologies that have made a positive impact during the Covid-19 pandemic. This chapter highlighted the preclinical assessments of eight promising vaccine candidates, first of which was the Moderna vaccine (mRNA-1273), followed by the AstraZeneca/Oxford University's AZD1222. Moderna used both mice and rhesus macaque models to evaluate the safety and efficacy of its vaccine, whereas Oxford University only used the macaque model. Overall, both vaccines have shown relatively strong immune response among the humans. For ChAdOx1 nCoV-19, five out of six lung lobes in the vaccinated group had detectable viral load, protecting the participants from pneumonia (Jain et al., 2020).

Bharat Biotech's Covaxin and Cadilla's ZyCoV-D are inactivated DNA plasmid vaccines, and both candidates have originated from India. Mice, rats, guinea pigs, and rabbits were used for their preclinical assessment, and both have shown positive results in humans (Su et al., 2021). China, where the pandemic first hit, has multiple candidates that show promise in formulating a successful vaccine – one of them being PicoVacc, manufactured by Sinovac. It used rats, mice, and macaque models for preclinical testing. Immunization with two different doses (3 or 6 mg/dose) provided partial or complete protection in macaques against SARS-CoV-2 challenge, respectively, without any antibody-dependent enhancement of infection. Another Chinese vaccine, CanSino Biologics Ad5-nCoV, showed good results and humoral immunization in nonhuman primates (Haynes et al., 2020). At 7 days following inoculation with BBIBP-CorV, no macaques in the low- or high-dose groups had detectable viral load in any lung lobe. Nonetheless, it provided reasonable protection and averted viral interstitial pneumonia in all vaccinated macaques (Munster et al., 2020). INOVIO's DNA Vaccine INO-4800 demonstrated robust neutralizing antibody and T cell immune responses in the chosen animal models. Within the first 40 days of initial genomic reports of the virus, the first vaccine (mRNA-1273) entered clinical development phase, and as of June 6, 2021, there were roughly 102 candidates in clinical trials (*Draft Landscape and Tracker of COVID-19 Candidate Vaccines*, n.d.), many in phase II/III. Although therapeutic reality is far from reality, the combined efforts of academic laboratories and the pharmaceutical industry bodes a positive outcome (Choudhary et al., 2021). As cataclysmic as Covid-19 is, it will serve as an impetus for the biomedical and science community, funding bodies, and stakeholders to devote more streamlined efforts towards development of platform technologies that prepare us better for future pandemics.

Conflict of interest: The authors declare no conflict of interest.

REFERENCES

Aleissa, M. M., Silverman, E. A., Paredes Acosta, L. M., Nutt, C. T., Richterman, A., & Marty, F. M. (2021). New perspectives on antimicrobial agents: Remdesivir treatment for COVID-19. Antimicrobial Agents and Chemotherapy, 65(1). https://doi.org/ 10.1128/AAC. 01814-20

Altuntas, F., Ata, N., Yigenoglu, T. N., Bascı, S., Dal, M. S., Korkmaz, S., Namdaroglu, S., Basturk, A., Hacıbekiroglu, T., Dogu, M. H., Berber, İ., Dal, K., Kınık, K., Haznedaroglu, İ., Yılmaz, F. M., Kılıç, İ., Demircioğlu, S., Yosunkaya, A., Erkurt, M. A., ... Celik, O. (2020). Convalescent plasma therapy in patients with COVID-19. Transfusion and Apheresis Science, 60(1), 102955. https://doi.org/10.1016/j.transci.2020.102955

Anderson, E. J., Rouphael, N. G., Widge, A. T., Jackson, L. A., Roberts, P. C., Makhene, M., Chappell, J. D., Denison, M. R., Stevens, L. J., Pruijssers, A. J., McDermott, A. B., Flach, B., Lin, B. C., Doria-Rose, N. A., O'Dell, S., Schmidt, S. D., Corbett, K. S., Swanson, P. A., Padilla, M., ... Beigel, J. H. (2020). Safety and Immunogenicity of SARS-CoV-2 mRNA-1273 Vaccine in Older Adults. New England Journal of Medicine, 383(25), 2427–2438. https://doi.org/10. 1056/nejmoa2028436

Backer, J. A., Klinkenberg, D., & Wallinga, J. (2020). Incubation period of 2019 novel coronavirus (2019-nCoV) infections among travellers from Wuhan, China, 20–28 January 2020. In Eurosurveillance, 25(5), 20–28. European Centre for Disease Prevention and Control (ECDC). https://doi.org/10. 2807/1560-7917.ES.2020.25.5.2000062

Bharat Biotech's Covaxin vaccine yields positive Phase I data. (n.d.). Retrieved 29 April 2021, from www.clin icaltrialsarena. com/news/bharat-biotech-covaxin-data/

Calina, D., Docea, A. O., Petrakis, D., Egorov, A. M., Ishmukhametov, A. A., Gabibov, A. G., Shtilman, M. I., Kostoff, R., Carvalho, F., Vinceti, M., Spandidos, D. A., & Tsatsakis, A. (2020). Towards effective COVID-19 vaccines: Updates, perspectives and challenges (Review). In International Journal of Molecular Medicine, 46(1), 3–16. NLM (Medline). https://doi.org/10.3892/ijmm.2020.4596

China's CanSino Biologics reveals phase 2 coronavirus vaccine results – PMLiVE. (n.d.). Retrieved 29 April 2021, from www.pmlive.com/pharma_news/chinas_cansino_biologics_reveals_phase_2_coronavirus_vaccine_results_1345203

Choudhary, H. B., Sirvi, I. H., Rajendra Bamb, Y., Bamb, R., & Rajkumarpatekar, R. (2021). COVID-19 Vaccines: Systematic review. Journal of Advanced Research and Reviews, 2021(01), 143–155. https://doi.org/10.30574/wjarr.2021. 10.1.0118

Clem, A. S. (2011). Fundamentals of vaccine immunology. Journal of Global Infectious Diseases, 3(1), 73–78. https://doi.org/10. 4103/ 0974-777X.77299

Corbett, K. S., Flynn, B., Foulds, K. E., Francica, J. R., Boyoglu-Barnum, S., Werner, A. P., Flach, B., O'Connell, S., Bock, K. W., Minai, M., Nagata, B. M., Andersen, H., Martinez, D. R., Noe, A. T., Douek, N., Donaldson, M. M., Nji, N. N., Alvarado, G. S., Edwards, D. K., … Graham, B. S. (2020). Evaluation of the mRNA-1273 Vaccine against SARS-CoV-2 in Nonhuman Primates. New England Journal of Medicine, 383(16), 1544–1555. https://doi.org/10. 1056/nejmoa2024671

Covaxin and ZyCoV-D: Recent Update of Covid-19 Vaccine Candidates in India – NHI. (n.d.). Retrieved 29 April 2021, from https://neucradhealth.in/language/en/ covaxin-and-zycov-d-recent-update-of-covid-19-vaccine-candidates-in-india/

COVID Live Update: 150,379,130 Cases and 3,167,228 Deaths from the Coronavirus – Worldometer. (n.d.). Retrieved 29 April 2021, from www.worldo meters.info/coronavirus/

Dal-Ré, R., Bekker, L.-G., Gluud, C., Holm, S., Jha, V., Poland, G. A., Rosendaal, F. R., Schwarzer-Daum, B., Sevene, E., Tinto, H., Voo, T. C., & Sreeharan, N. (2021). Ongoing and future COVID-19 vaccine clinical trials: challenges and opportunities. The Lancet Infectious Diseases. https://doi.org/ 10.1016/S1473-3099(21)00263-2

Dey, A., Rajanathan, C. T. M., Chandra, H., Pericherla, H. P. R., Kumar, S., Choonia, H. S., Bajpai, M., Singh, A. K., Sinha, A., Saini, G., Dalal, P., Vandriwala, S., Raheem, M. A., Divate, R. D., Navlani, N. L., Sharma, V., Parikh, A., Prasath, S., Rao, S., & Maithal, K. (2021). Immunogenic potential of DNA vaccine candidate, ZyCoV-D against SARS-CoV-2 in animal models. In bioRxiv (p. 2021.01.26.428240). Cold Spring Harbor Laboratory. https://doi.org/10.1101/2021.01.26. 428240

Dicks, M. D. J., Spencer, A. J., Edwards, N. J., Wadell, G., Bojang, K., Gilbert, S. C., Hill, A. V. S., & Cottingham, M. G. (2012). A novel chimpanzee adenovirus vector with low human seroprevalence: Improved systems for vector derivation and comparative immunogenicity. PLoS ONE, 7(7), e40385. https://doi.org/10.1371/journal.pone.0040385

Draft landscape and tracker of COVID-19 candidate vaccines. (n.d.).

Efficacy and Safety of Hydroxychloroquine and Azithromycin for the Treatment of Ambulatory Patients With Mild COVID-19 – Full Text View – ClinicalTrials.gov. (n.d.). Retrieved 29 April 2021, from https://cli nicaltrials.gov/ct2/show/NCT04348474

Fda, & Cber. (2021). Investigational COVID-19 Convalescent Plasma Guidance for Industry Preface Public Comment.

Folegatti, P. M., Ewer, K. J., Aley, P. K., Angus, B., Becker, S., Belij-Rammerstorfer, S., Bellamy, D., Bibi, S., Bittaye, M., Clutterbuck, E. A., Dold, C., Faust, S. N., Finn, A., Flaxman, A. L., Hallis, B., Heath, P., Jenkin, D., Lazarus, R., Makinson, R., … Yau, Y. (2020). Safety and immunogenicity of the ChAdOx1 nCoV-19 vaccine against SARS-CoV-2: a preliminary report of a phase 1/2, single-blind, randomised controlled trial. The Lancet, 396(10249), 467–478. https://doi.org/10.1016/S0140-6736(20)31604-4

Funk, C. D., Laferrière, C., & Ardakani, A. (2020). A Snapshot of the Global Race for Vaccines Targeting SARS-CoV-2 and the COVID-19 Pandemic. In Frontiers in Pharmacology (Vol. 11). Frontiers Media S.A. https://doi.org/10.3389/fphar.2020.00 937

Ganneru, B., Jogdand, H., Daram, V. K., Das, D., Molugu, N. R., Prasad, S. D., Kannappa, S. V., Ella, K. M., Ravikrishnan, R., Awasthi, A., Jose, J., Rao, P., Kumar, D., Ella, R., Abraham, P., Yadav, P. D., Sapkal, G. N., Shete-Aich, A., Desphande, G., … Vadrevu, K. M. (2021). Th1 skewed immune response of whole virion inactivated SARS CoV 2 vaccine and its safety evaluation. IScience, 24(4), 102298. https://doi.org/10.1016/j.isci.2021.102 298

Gao, Q., Bao, L., Mao, H., Wang, L., Xu, K., Yang, M., Li, Y., Zhu, L., Wang, N., Lv, Z., Gao, H., Ge, X., Kan, B., Hu, Y., Liu, J., Cai, F., Jiang, D., Yin, Y., Qin, C., … Qin, C. (2020). Development of an inactivated vaccine candidate for SARS-CoV-2. Science, 369(6499), 77–81. https://doi.org/10.1126/science.abc1932

Gautret, P., Lagier, J. C., Parola, P., Hoang, V. T., Meddeb, L., Mailhe, M., Doudier, B., Courjon, J., Giordanengo, V., Vieira, V. E., Tissot Dupont, H., Honoré, S., Colson, P., Chabrière, E., La Scola, B., Rolain, J. M., Brouqui, P., & Raoult, D. (2020). Hydroxychloroquine and azithromycin as a treatment of COVID-19: results of an open-label non-randomized clinical trial. International Journal of Antimicrobial Agents, 56(1), 105949. https://doi.org/10.1016/j.ijantimicag.2020.105949

Haynes, B. F., Corey, L., Fernandes, P., Gilbert, P. B., Hotez, P. J., Rao, S., Santos, M. R., Schuitemaker, H., Watson, M., & Arvin, A. (2020). Prospects for a safe COVID-19 vaccine. In Science Translational Medicine, 12(568), 948. American Association for the Advancement of Science. https://doi.org/10.1126/scitranslmed.abe0948

Huang, C., Wang, Y., Li, X., Ren, L., Zhao, J., Hu, Y., Zhang, L., Fan, G., Xu, J., Gu, X., Cheng, Z., Yu, T., Xia, J., Wei, Y., Wu, W., Xie, X., Yin, W., Li, H., Liu, M., … Cao, B. (2020). Clinical features of patients infected with 2019 novel coronavirus in Wuhan, China. The Lancet, 395(10223), 497–506. https://doi.org/ 10.1016/S0140-6736(20)30183-5

Isakova-Sivak, I., & Rudenko, L. (2021). A promising inactivated whole-virion SARS-CoV-2 vaccine. In The Lancet Infectious Diseases, 21(1), 2–3. Lancet Publishing Group. https://doi.org/10.1016/S1473-3099(20)30832-X

Jackson, L. A., Anderson, E. J., Rouphael, N. G., Roberts, P. C., Makhene, M., Coler, R. N., McCullough, M. P., Chappell, J. D., Denison, M. R., Stevens, L. J., Pruijssers, A. J., McDermott, A., Flach, B., Doria-Rose, N. A., Corbett, K. S., Morabito, K. M., O'Dell, S., Schmidt, S. D., Swanson, P. A., … Beigel, J. H. (2020). An mRNA Vaccine against SARS-CoV-2 – Preliminary Report. New England Journal of Medicine, 383(20), 1920–1931. https://doi.org/10.1056/ nejmoa2022483

Jain, S., Batra, H., Yadav, P., & Chand, S. (2020). Covid-19 vaccines currently under preclinical and clinical studies, and associated antiviral immune response. In Vaccines, 8(4), 1–16. MDPI AG. https://doi.org/10.3390/vaccines8040649

Jean, S. S., Lee, P. I., & Hsueh, P. R. (2020). Treatment options for COVID-19: the reality and challenges. In Journal of Microbiology, Immunology and Infection, 53(3), 436–443. Elsevier Ltd. https://doi.org/10.1016/j.jmii.2020. 03.034

Kandeil, A., Mostafa, A., Hegazy, R. R., El-Shesheny, R., El Taweel, A., Gomaa, M. R., Shehata, M., Elbaset, M. A., Kayed, A. E., Mahmoud, S. H., Moatasim, Y., Kutkat, O., Yassen, N. N., Shabana, M. E., GabAllah, M., Kamel, M. N., Abo Shama, N. M., El Sayes, M., Ahmed, A. N., … Ali, M. A. (2021). Immunogenicity and Safety of an Inactivated SARS-CoV-2 Vaccine: Preclinical Studies. Vaccines, 9(3), 214. https://doi.org/10.3390/vaccines9030214

Lurie, N., Saville, M., Hatchett, R., & Halton, J. (2020). Developing Covid-19 vaccines at pandemic speed. New England Journal of Medicine, 382(21), 1969–1973. https://doi.org/10.1056/nejmp 2005630

McCullers, J. A., & Dunn, J. D. (2008). Advances in vaccine technology and their impact on managed care. In P and T, 33(1), 35. MediMedia, USA.

Moderna's Work on our COVID-19 Vaccine | Moderna, Inc. (n.d.). Retrieved 29 April 2021, from www.modernatx.com/ modernas-work-potential-vaccine-against-covid-19

Mohandas, S., Yadav, P. D., Shete-Aich, A., Abraham, P., Vadrevu, K. M., Sapkal, G., Mote, C., Nyayanit, D., Gupta, N., Srinivas, V. K., Kadam, M., Kumar, A., Majumdar, T., Jain, R., Deshpande, G., Patil, S., Sarkale, P., Patil, D., Ella, R., … Bhargava, B. (2021). Immunogenicity and protective efficacy of BBV152, whole virion inactivated SARS-CoV-2 vaccine candidates in the Syrian hamster model. IScience, 24(2), 102054. https://doi.org/10.1016/j.isci.2021.102054

Mukhopadhyay, L., Yadav, P., Gupta, N., Mohandas, S., Patil, D., Shete-Aich, A., Panda, S., & Bhargava, B. (2021). Comparison of the immunogenicity & protective efficacy of various SARS-CoV-2 vaccine candidates in non-human primates. In Indian Journal of Medical Research, 153(1) 93–114). Wolters Kluwer Medknow Publications. https://doi.org/10.4103/ijmr. IJMR_4431_20

Munster, V. J., Feldmann, F., Williamson, B. N., van Doremalen, N., Pérez-Pérez, L., Schulz, J., Meade-White, K., Okumura, A., Callison, J., Brumbaugh, B., Avanzato, V. A., Rosenke, R., Hanley, P. W., Saturday, G.,

Scott, D., Fischer, E. R., & de Wit, E. (2020). Respiratory disease in rhesus macaques inoculated with SARS-CoV-2. Nature, 585(7824), 268–272. https://doi.org/ 10.1038/s41586-020-2324-7

Ong, E., Wong, M. U., Huffman, A., & He, Y. (2020). COVID-19 coronavirus vaccine design using reverse vaccinology and machine learning. In bioRxiv. bioRxiv. https://doi.org/ 10.1101/2020.03.20.000 141

Oxford COVID-19 vaccine programme opens for clinical trial recruitment | University of Oxford. (n.d.). Retrieved 29 April 2021, from www.ox.ac.uk/news/2020-03-27-oxford-covid-19-vaccine-programme-opens-clinical-trial-recruitment

Perlman, S., & Netland, J. (2009). Coronaviruses post-SARS: update on replication and pathogenesis. In Nature Reviews Microbiology, 7(6), 439–450. https://doi.org/10.1038/ nrmicro2147

Riedel, S. (2005). Edward Jenner and the History of Smallpox and Vaccination. Baylor University Medical Center Proceedings, 18(1), 21–25. https://doi.org/10.1080/08998280.2005.11928 028

Robbiani, D. F., Gaebler, C., Muecksch, F., Lorenzi, J. C. C., Wang, Z., Cho, A., Agudelo, M., Barnes, C. O., Gazumyan, A., Finkin, S., Hägglöf, T., Oliveira, T. Y., Viant, C., Hurley, A., Hoffmann, H. H., Millard, K. G., Kost, R. G., Cipolla, M., Gordon, K., ... Nussenzweig, M. C. (2020). Convergent antibody responses to SARS-CoV-2 in convalescent individuals. Nature, 584(7821), 437–442. https://doi.org/ 10.1038/s41586-020-2456-9

Safety and Immunogenicity Study of 2019-nCoV Vaccine (mRNA-1273) for Prophylaxis of SARS-CoV-2 Infection (COVID-19) – Full Text View – ClinicalTrials.gov. (n.d.). Retrieved 29 April 2021, from https://clinicaltrials.gov/ct2/show/NCT04283461

Sapkal, G. N., Yadav, P. D., Ella, R., Deshpande, G. R., Sahay, R. R., Gupta, N., Vadrevu, K. M., Abraham, P., Panda, S., & Bhargava, B. (2021). Inactivated COVID-19 vaccine BBV152/COVAXIN effectively neutralizes recently emerged B.1.1.7 variant of SARS-CoV-2. Journal of Travel Medicine, 28(4). https:// doi.org/10.1093/jtm/taab051

Seligmann, H. (2021). Balanced evaluation of preliminary data on a candidate COVID-19 hydroxychloroquine treatment. In International Journal of Antimicrobial Agents, 57(3). Elsevier B.V. https://doi.org/10.1016/ j.ijantimicag.2021.106292

Shi, H., Han, X., Jiang, N., Cao, Y., Alwalid, O., Gu, J., Fan, Y., & Zheng, C. (2020). Radiological findings from 81 patients with COVID-19 pneumonia in Wuhan, China: a descriptive study. The Lancet Infectious Diseases, 20(4), 425–434. https://doi.org/ 10.1016/S1473-3099(20)30086-4

Smith, T. R. F., Patel, A., Ramos, S., Elwood, D., Zhu, X., Yan, J., Gary, E. N., Walker, S. N., Schultheis, K., Purwar, M., Xu, Z., Walters, J., Bhojnagarwala, P., Yang, M., Chokkalingam, N., Pezzoli, P., Parzych, E., Reuschel, E. L., Doan, A., ... Broderick, K. E. (2020). Immunogenicity of a DNA vaccine candidate for COVID-19. Nature Communications, 11(1), 1–13. https://doi.org/ 10.1038/s41467-020-16505-0

Sprangers, M. C., Lakhai, W., Koudstaal, W., Verhoeven, M., Koel, B. F., Vogels, R., Goudsmit, J., Havenga, M. J. E., & Kostense, S. (2003). Quantifying Adenovirus-Neutralizing Antibodies by Luciferase Transgene Detection: Addressing Preexisting Immunity to Vaccine and Gene Therapy Vectors. Journal of Clinical Microbiology, 41(11), 5046–5052. https://doi.org/10.1128/JCM.41.11.5046-5052.2003

Su, S., Du, L., & Jiang, S. (2021). Learning from the past: development of safe and effective COVID-19 vaccines. In Nature Reviews Microbiology, 19(3), 211–219. Nature Research. https://doi.org/10. 1038/ s41579-020-00462-y

Tebas, P., Yang, S. P., Boyer, J. D., Reuschel, E. L., Patel, A., Christensen-Quick, A., Andrade, V. M., Morrow, M. P., Kraynyak, K., Agnes, J., Purwar, M., Sylvester, A., Pawlicki, J., Gillespie, E., Maricic, I., Zaidi, F. I., Kim, K. Y., Dia, Y., Frase, D., ... Humeau, L. M. (2021). Safety and immunogenicity of INO-4800 DNA vaccine against SARS-CoV-2: a preliminary report of an open-label, Phase 1 clinical trial. EClinicalMedicine, 31, 100689. https://doi.org/10. 1016/j.eclinm.2020.100689

Thangavel, R. R., & Bouvier, N. M. (2014). Animal models for influenza virus pathogenesis, transmission, and immunology. In Journal of Immunological Methods, 410, 60–79. Elsevier. https://doi.org/10.1016/ j.jim.2014.03.023

Thanh Le, T., Andreadakis, Z., Kumar, A., Gómez Román, R., Tollefsen, S., Saville, M., & Mayhew, S. (2020). The COVID-19 vaccine development landscape. In Nature reviews. Drug discovery, 19(5), 305–306. NLM (Medline). https://doi.org/10.1038/ d41573-020-00073-5

Tostanoski, L. H., Wegmann, F., Martinot, A. J., Loos, C., McMahan, K., Mercado, N. B., Yu, J., Chan, C. N., Bondoc, S., Starke, C. E., Nekorchuk, M., Busman-Sahay, K., Piedra-Mora, C., Wrijil, L. M., Ducat, S., Custers, J., Atyeo, C., Fischinger, S., Burke, J. S., ... Barouch, D. H. (2020). Ad26 vaccine protects against SARS-CoV-2 severe clinical disease in hamsters. Nature Medicine, 26(11), 1694–1700. https://doi.org/10.1038/s41591-020-1070-6

Tregoning, J. S., Brown, E. S., Cheeseman, H. M., Flight, K. E., Higham, S. L., Lemm, N. M., Pierce, B. F., Stirling, D. C., Wang, Z., & Pollock, K. M. (2020). Vaccines for COVID-19. In Clinical and Experimental Immunology, 202(2), 162–192. Blackwell Publishing Ltd. https://doi.org/10.1111 /cei.13517

V, V. B. B., Canada, H., England, N., Biosafety, R., Food, U. S., & Agency, P. H. (n.d.). Mapping of animal models capacity to accelerate COVID-19 vaccines and therapeutics development Purpose of the document. This document is primarily intended for COVID-19 vaccine and/or therapeutic developers.

van Doremalen, N., Lambe, T., Spencer, A., Belij-Rammerstorfer, S., Purushotham, J. N., Port, J. R., Avanzato, V. A., Bushmaker, T., Flaxman, A., Ulaszewska, M., Feldmann, F., Allen, E. R., Sharpe, H., Schulz, J., Holbrook, M., Okumura, A., Meade-White, K., Pérez-Pérez, L., Edwards, N. J., … Munster, V. J. (2020). ChAdOx1 nCoV-19 vaccine prevents SARS-CoV-2 pneumonia in rhesus macaques. Nature, 586(7830), 578–582. https://doi.org/10.1038/s41586-020-2608-y

van Riel, D., & de Wit, E. (2020). Next-generation vaccine platforms for COVID-19. In Nature Materials (Vol. 19, Issue 8, pp. 810–812). Nature Research. https://doi.org/10.1038/s41563-020-0746-0

Vinceti, M., Filippini, T., Rothman, K. J., Di Federico, S., & Orsini, N. (2021). SARS-CoV-2 infection incidence during the first and second COVID-19 waves in Italy. Environmental Research, 197, 111097. https://doi.org/10.1016/j.envres.2021.111097

Wang, H., Zhang, Y., Huang, B., Deng, W., Quan, Y., Wang, W., Xu, W., Zhao, Y., Li, N., Zhang, J., Liang, H., Bao, L., Xu, Y., Ding, L., Zhou, W., Gao, H., Liu, J., Niu, P., Zhao, L., … Yang, X. (2020). Development of an Inactivated Vaccine Candidate, BBIBP-CorV, with Potent Protection against SARS-CoV-2. Cell, 182(3), 713–721.e9. https://doi.org/10.1016/j.cell.2020.06.008

Xia, S., Zhang, Y., Wang, Y., Wang, H., Yang, Y., Gao, G. F., Tan, W., Wu, G., Xu, M., Lou, Z., Huang, W., Xu, W., Huang, B., Wang, H., Wang, W., Zhang, W., Li, N., Xie, Z., Ding, L., … Yang, X. (2021). Safety and immunogenicity of an inactivated SARS-CoV-2 vaccine, BBIBP-CorV: a randomised, double-blind, placebo-controlled, phase 1/2 trial. The Lancet Infectious Diseases, 21(1), 39–51. https://doi.org/10.1016/S1473-3099(20)30831-8

Xu, X., Chen, P., Wang, J., Feng, J., Zhou, H., Li, X., Zhong, W., & Hao, P. (2020). Evolution of the novel coronavirus from the ongoing Wuhan outbreak and modeling of its spike protein for risk of human transmission. In Science China Life Sciences, 63(3), 457–460). Science in China Press. https://doi.org/10.1007/s11427-020-1637-5

Yadav, P. D., Ella, R., Kumar, S., Patil, D. R., Mohandas, S., Shete, A. M., Vadrevu, K. M., Bhati, G., Sapkal, G., Kaushal, H., Patil, S., Jain, R., Deshpande, G., Gupta, N., Agarwal, K., Gokhale, M., Mathapati, B., Metkari, S., Mote, C.,… Bhargava, B. (2021). Immunogenicity and protective efficacy of inactivated SARS-CoV-2 vaccine candidate, BBV152 in rhesus macaques. Nature Communications, 12(1), 1–11. https://doi.org/10.1038/s41467-021-21639-w

Yadav, P. D., Kumar, S., Agarwal, K., Jain, M., Patil, D. R., Maithal, K., Mathapati, B., Giri, S., Mohandas, S., Shete, A., Sapkal, G., Patil, D. Y., Dey, A., Chandra, H., Deshpande, G., Gupta, N., Nyayanit, D., Kaushal, H., Sahay, R., … Abraham, P. (2021). Assessment of immunogenicity and protective efficacy of ZyCoV-D DNA vaccine candidates in Rhesus macaques against SARS-CoV-2 infection. BioRxiv, 2021.02.02.429480. https://doi.org/10.1101/2021.02.02.429480

Ye, T., Zhong, Z., García-Sastre, A., Schotsaert, M., & De Geest, B. G. (2020). Current Status of COVID-19 (Pre)Clinical Vaccine Development. Angewandte Chemie International Edition, 59(43), 18885–18897. https://doi.org/10.1002/anie.202008319

Yu, J., Tostanosk, L. H., Peter, L., Mercad, N. B., McMahan, K., Mahrokhia, S. H., Nkolol, J. P., Liu, J., Li, Z., Chandrashekar, A., Martine, D. R., Loos, C., Atyeo, C., Fischinger, S., Burk, J. S., Slei, M. D., Chen, Y., Zuiani, A., Lelis, F. J. N., … Barou, D. H. (2020). DNA vaccine protection against SARS-CoV-2 in rhesus macaques. Science, 369(6505), 806–811. https://doi.org/10.1126/science.abc6284

Yuki, K., Fujiogi, M., & Koutsogiannaki, S. (2020). COVID-19 pathophysiology: a review. In Clinical Immunology, 215, 108427. Academic Press Inc. https://doi.org/10.1016/j.clim.2020.108427

Zhou, F., Yu, T., Du, R., Fan, G., Liu, Y., Liu, Z., Xiang, J., Wang, Y., Song, B., Gu, X., Guan, L., Wei, Y., Li, H., Wu, X., Xu, J., Tu, S., Zhang, Y., Chen, H., & Cao, B. (2020). Clinical course and risk factors for mortality of adult inpatients with COVID-19 in Wuhan, China: a retrospective cohort study. The Lancet, 395(10229), 1054–1062. https://doi.org/10.1016/S0140-6736 (20)30566-3

Zhu, F. C., Li, Y. H., Guan, X. H., Hou, L. H., Wang, W. J., Li, J. X., Wu, S. P., Wang, B. Sen, Wang, Z., Wang, L., Jia, S. Y., Jiang, H. D., Wang, L., Jiang, T., Hu, Y., Gou, J. B., Xu, S. B., Xu, J. J., Wang, X. W., … Chen, W. (2020). Safety, tolerability, and immunogenicity of a recombinant adenovirus type-5 vectored COVID-19 vaccine: a dose-escalation, open-label, non-randomised, first-in-human trial. The Lancet, 395(10240), 1845–1854. https://doi.org/10.1016/S0140-6736(20)31208-3

14 Mucormycosis in Covid-19 Patients

*Guneet Kaur[1], Samander Kaushik[2], Sulochana Kaushik[3], Sandeep Singh[4], Gagandeep Singh[5], Immaculata Xess[5], and Pankaj Seth[1]**

[1]National Brain Research Centre, Manesar, Gurgaon, Haryana, India
[2]Centre for Biotechnology, Maharshi Dayanand University, Rohtak, Haryana, India
[3]Department of Genetics, Maharshi Dayanand University, Rohtak, Haryana, India
[4]Department of Biochemistry, Maharshi Dayanand University, Rohtak, Haryana, India
[5]Department of Microbiology, All India Institute of Medical Sciences, New Delhi, India
*Email: pseth.nbrc@gov.in

CONTENTS

14.1 INTRODUCTION

The emergence of new viral infections has greatly impacted the nation's population, health sector, and economy. In the past few decades, viruses like Human Immunodeficiency Virus-1 (HIV-1), Ebola Hemorrhagic Fever (EHF), Marburg virus (MARV), Zika virus (ZIKV), Nipah virus (NiV), SARS-CoV, MERS-CoV, and now SARS-CoV2 causing Covid-19 disease have resulted in devastating outcomes. The most recent and ongoing Covid-19 pandemic has proven to be the world's worst public health crisis. Worldwide, this virus has spread across over 220 countries and has

DOI: 10.1201/9781003358909-14

resulted in 640,429,188 confirmed cases and accounted for 6,615,648 fatalities, as of November 14, 2022 (Worldometer, 2022). The Ministry of Health and Family Welfare of India has so far reported 30,133,417 laboratory-confirmed cases, with 3,93,338 fatalities in the country (Covid19India, 2020). Compared to mortality rates of SARS-CoV (10%) and MERS-CoV (37.1%), SARS-CoV2 has a lower mortality rate of 2.08%. Although the mortality rate in India seems about half as that of the rest of the world, India being the second most populated country in the world, the number of deaths is high, and the death reporting criteria remains debatable. The current wave of the virus has amplified the number of cases across India, largely affecting rural areas since 2020. This has resulted in a large number of patients being admitted to different Covid-19 treatment centres, where they are subjected to symptomatic treatments with supportive treatment. Many deaths have been attributed to the occurrence of secondary bacterial and/or fungal infections in such patients. In the majority of cases, fungal infections are more common, greatly attributing to increased morbidity and mortality (Martin et al., 2003). Reports suggest that fungal infections affect diseases that impact immune systems as is the case of HIV-AIDS where about 90% of patients get fungal infections during the course of their treatment (Diamond, 1991). Of such infections, mucormycosis has emerged as the most common opportunistic infection in Covid-19 patients. During the second wave of Covid-19 (April–June 2021), a great number of mucormycosis cases have been reported.

Since Covid patients already have a weakened immune system, other predisposing factors such as uncontrolled diabetes, HIV-AIDS, neutropenia, and long-term steroid use contribute to disease manifestation. This chapter focuses on the epidemiology, clinical manifestations, predisposing risk factors, diagnosis, treatment, and management of mucormycosis associated with Covid-19.

14.2 MUCORMYCOSIS-ETIOLOGIC AGENT

The term "mucormycosis" was coined by American pathologist R.D. Baker to denote fungal disease "mycosis" caused by members of the order Mucaroles (Kwon-Chung, 2012). Fungal pathogens exhibit high diversity, which led to re-analysis of its classification on the basis of molecular phylogeny. Earlier, Mucaroles belonged to the phylum Zygomycota. As per current classification, they are classified as phylum Glomeromycota, which are further subdivided into Mucoromycotina, Entomophthoromycotina, Kickxellales, and Zoopagomycotina (Hibbett et al., 2007; Binder, Maurer and Lass-Flörl, 2014) (as shown in Figure 14.1). The order Mucaroles is comprised of many pathogenic genera including *Rhizopus*, *Mucor*, *Rhizomucor*, *Lichtheimia*, Apophysomyces, *Saksenaea*, *Cunninghamella*, *Sycephalastrum*, etc. (Kwon-Chung, 2012).

Mucormycosis is a rare but highly invasive fungal infection. These pathogens are found worldwide, and commonly feed onto decaying organic matter (such as fruit and vegetable waste), animal faeces, and forest soils. Infection caused by Mucaroles is characterized by its rapid progression, which predominantly affects immunocompromised hosts. Among all, *Rhizopus arrhizus* is the most common species in cultured confirmed cases worldwide (Jeong et al., 2019).

14.3 EPIDEMIOLOGY AND DISEASE MANIFESTATIONS

There is a high incidence of fungal infections in patients with uncontrolled diabetes, especially with diabetes ketoacidosis (DKA), which is highly prevalent in the Asian population. In India, high incidence of diabetes-related mucormycosis cases have been reported in northern (67%), western (55.6%), and southern (22%) populations (Patel et al., 2017; Prakash and Chakrabarti, 2019). Other predisposing conditions involve haematological manifestations (HM) and severe neutropenia (Binder, Maurer and Lass-Flörl, 2014). In India, HM patients are at 1–9% risk of developing mucormycosis (Skiada, Pavleas and Drogari-Apiranthitou, 2020). Mononuclear and polymorphonuclear phagocytes in a healthy human generally inhibit the germinating spores, and patients with decreased amounts of phagocytes are found to be at a higher risk of developing the disease. Other

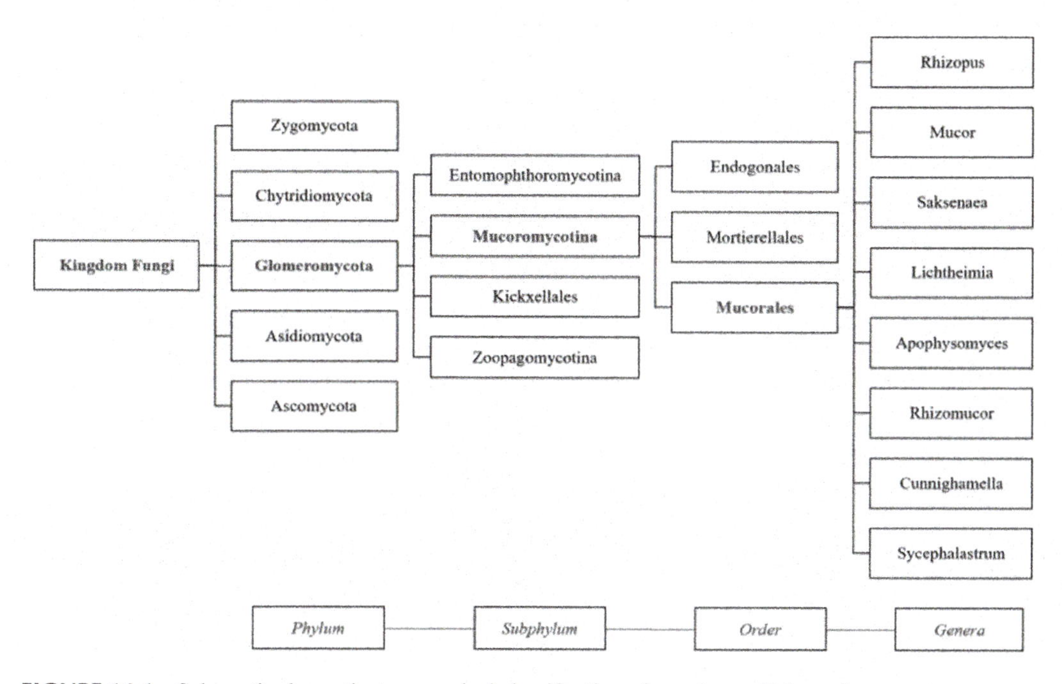

FIGURE 14.1 Schematic shows the taxonomical classification of members of Mucorales.

underlying causes involve patients who underwent organ transplants (2.6–11%) (Roden et al., 2005; Jeong et al., 2019), extended hospital stays (9%) (Chakrabarti et al., 2009), trauma (7.5–22%), or patients treated with high doses of corticosteroids.

The above listed factors greatly increase the incidence of fungal coinfections. High glucose level due to history of diabetes or prolonged use of steroids provides the ideal environment for germination and propagation of fungal spores (Gupta, Sharma and Chakrabarti, 2021). Other conditions include low oxygen levels and high iron load in patients. Since a majority of Covid-19 treatment involves use of corticosteroids, mechanical ventilation, and monoclonal antibodies and antibiotics, it can lead to immunosuppression in patients. This results in serious complications including exposure of Covid-19 patients to opportunistic infections such as mucormycosis.

During the current wave of Covid-19, many patients have blurred vision, chest pain, and difficulty breathing. Upon testing, such symptoms were diagnosed as a secondary infection of mucormycosis. Although it remains a rare manifestation in immunocompetent people, it can quickly turn fatal for immunocompromised patients if left untreated. Mucormycosis is today commonly termed as black fungus, because of its characteristic appearance of blackened necrotic tissue around the affected area.

Around 51,775 confirmed cases of black fungus have been reported by the Government of India as of November 29, 2021. Of all states, Maharashtra, Gujarat, and Andhra Pradesh have been the worst hit states accounting for 43.9 % of all cases in the country. Table 14.1 depicts the state-wise distribution of confirmed mucormycosis cases across India according to the recent numbers (November, 2021).

14.4 CLASSIFICATION OF MUCORMYCOSIS BASED ON ANATOMIC SITE OF INFECTION

The disease is further classified according to the predominant anatomical site of infection—rhino-orbital-cerebral (ROCM), pulmonary, cutaneous, gastrointestinal, or disseminated (Table 14.2). The entry of fungal spores usually happens through the respiratory tract, injured or macerated skin

TABLE 14.1
State-wise Distribution of Mucormycosis Cases

State	No. of Confirmed Cases
Gujarat	7257
Maharashtra	10366
Andhra Pradesh	5107
Madhya Pradesh	2381
Telangana	2638
Uttar Pradesh	2603
Central Institutions	592
Rajasthan	3711
Karnataka	4025
Haryana	1792
Tamil Nadu	5007
Bihar	842
Delhi	2069
Punjab	691
Uttarakhand	590
Chattisgarh	429
Kerala	148
Jharkhand	132
Odisha	227
Goa	33
Chandigarh	739
Dadar and Nagar Haveli	6
Jammu & Kashmir	58
Himachal Pradesh	38
Puducherry	225
West Bengal	328
Tripura	1

TABLE 14.2
Relationship between Predisposing Risk Factors and Predominant Site of Infection

Predisposing Factors	Anatomic Site of Infection
Diabetes Mellitus	Rhino-orbito-cerebral, cutaneous
Haematological Malignancies Neutropenia	Pulmonary, cutaneous, sino-orbital, or disseminated
Corticosteroids	Pulmonary, rhino-cerebral, or disseminated
Organ Transplants	Cutaneous, pulmonary, rhino-orbito-cerebral, or disseminated
Malnutrition	Gastrointestinal, disseminated
Trauma/skin wounds Injection/Ccatheter site	Cutaneous

or through catheter/injection site, or it may enter via ingestion of contaminated food (Spellberg, Edwards and Ibrahim, 2005; Binder, Maurer, and Lass-Flörl, 2014).

14.4.1 Rhino-Orbital-Cerebral Mucormycosis (ROCM)

This is the most common form of the disease, and is defined by infection of paranasal sinuses and possible infection extended to the brain. About 70% of ROCM cases are found in patients with uncontrolled diabetes. Symptoms involve unilateral facial swelling, sinus pain, fever, and headache. Nasal ulceration occurs in a few cases. Disease progression results in tissue necrosis around nasal mucosa resulting in black eschar around the infected area. Fungal invasion of orbital blood vessels results in thrombosis, resulting in periorbital swelling, superior orbital fissure syndrome, diplopia, conjunctival suffusion, and cavernous sinus thrombosis (CST). In mucormycosis, CST usually results in blurred vision or even complete loss of eyesight (Dinakaran et al., 2020). Patients also present with proptosis or chemosis because of involvement of orbital cavity (as depicted in Figure 14.2).

Progression of ROCM from orbital cavity to the brain regions occurs through ophthalmic artery, superior orbital fissure (causing sinus thrombosis), and cribriform plate (via perivascular and perineural channels) leading to cranial neuropathies, cerebral infarctions, and unconsciousness (Wali, Balkhair and Al-Mujaini, 2012). Clinical findings done using computerized tomography (CT) scanning of head and/or sinuses comprise mucosal thickening, sinusitis, and bone erosion. Mortality in confirmed cases of ROCM is very high and approximately 80–90% (Wali, Balkhair and Al-Mujaini, 2012; Binder, Maurer and Lass-Flörl, 2014).

The states of Gujarat and Maharashtra have reported the highest number of ROCM cases, 22% and 21%, respectively. Furthermore, 78% of the patients had history of diabetes, and 57% had been on oxygen support due to Covid-19 and 87% were getting treatment of corticosteroids. This

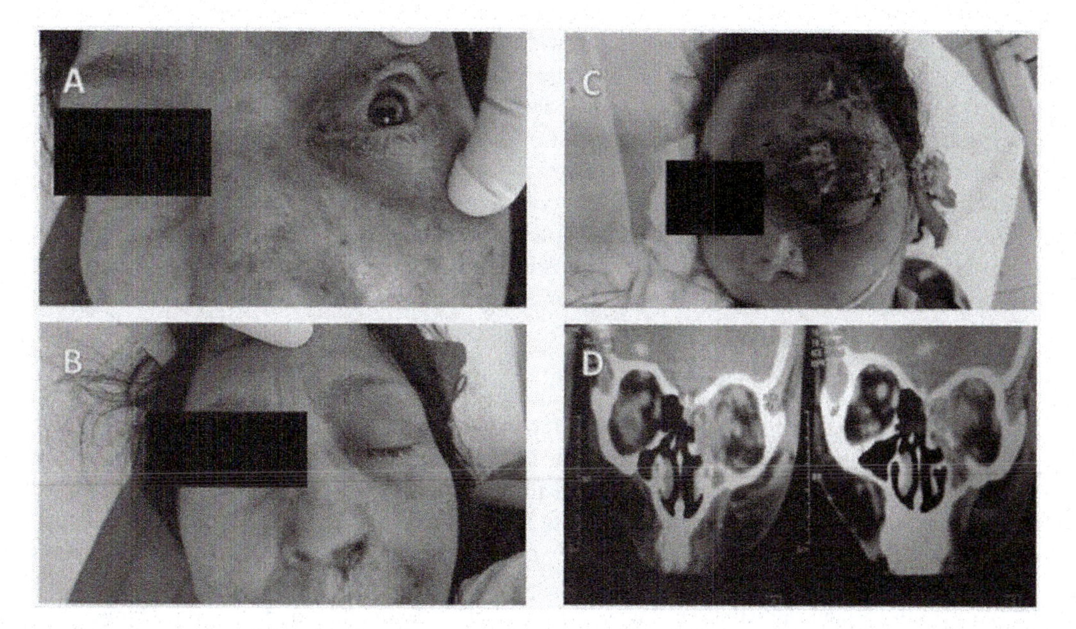

FIGURE 14.2 A–C depicts complete ptosis with no eye movement, complete vision loss, and inflammation around cheek and nose with nasal discharge. D shows CT scan of another patient exhibiting extensive rhino-orbital disease.

highlights how high-dose steroid treatment and uncontrolled diabetes accelerates the development of Covid-19-associated mucormycosis (Sen et al., 2021).

14.4.2 Pulmonary Mucormycosis

This is the second most common form of mucormycosis and has a high mortality rate, up to 70%. It has high incidence in leukemic patients and such patients have severe neutropenia. Clinical manifestations involve fever, cough, pleuritic chest pain, and dyspnoea. Chest CT shows characteristic appearance of multiple nodules, lobar consolidation, and cavitation. If pathogen invades into blood vessels, it leads to tissue necrosis resulting in major haemoptysis, which may also disseminate into the systemic route (Lin, Moua and Limper, 2017). Hematogenous dissemination eventually results in infection of contralateral lung and other organs. Lung biopsy is the most common method to diagnose pulmonary infection.

Mucormycosis infection generally causes damage to bronchial alveolar macrophages as well as T cells (Pasero et al., 2020). In a brief study, it was observed that Mucorales-specific CD4+ and CD8+ T cells were active in haematological patients. This further resulted in production of IL-4, IL-17, IL-10, and IFN-γ. The authors concluded that Mucorales-specific-T cells could be useful diagnostic markers (Potenza et al., 2016).

14.4.3 Cutaneous Mucormycosis

It is frequently observed in patients with uncontrolled diabetes, which is a major health concern in developing countries like India. It can be classified into primary and secondary infection. Primary cutaneous infection arises via direct inoculation and may also arise at injection/catheter sites, along with use of contaminated surgical dressing material. It is also described as hospital-associated mucormycosis because of the use of contaminated tapes and ostomy bags. The secondary form usually occurs as a result of dissemination from other anatomic sites, commonly from rhino-cerebral infection. This form of mucormycosis also occurs in patients with no history of predisposing conditions (Castrejón-pérez, Miranda, and Welsh, no date). In a brief study, it was found that about 50% of cases have no underlying factor, whereas in 40% of cases, patients were found to be immunocompetent (Roden et al., 2005; Skiada and Petrikkos, 2009). A few cases have been reported in which patients developed Covid-19-associated cutaneous mucormycosis, those with a history of predisposed risk factors such as heart transplant or diabetes. *Rhizopus arrhizus* commonly causes cutaneous infections. It was mainly found to be a secondary infection, disseminated from ROCM ('Insult to injury Covid-19-associated mucormycosis', no date; Khatri et al., 2021).

Arms and legs are commonly affected areas observed in skin infection. Others include scalp, neck, face, back, and abdominal area. Clinical presentations involve formation of eschar around necrotic tissue, scaly plaques, tender nodules, and ulcers (Castrejón-pérez, Miranda, and Welsh, no date).

14.4.4 Gastrointestinal Mucormycosis

Gastrointestinal (GI) manifestation is relatively rare since it is poorly diagnosed due to non-specific clinical presentations. It has low incidence of 5–13% compared to other forms of mucormycosis. Delay in diagnosis or no diagnosis contributes to its high mortality rate (85%). It is caused by ingesting contaminated food, fermented products such as milk, porridge, and dried bread. The disease has high prevalence in premature neonates and adults and children with malnutrition (Kaur et al., 2018). Rare cases are also found in immune-compromised patients.

It commonly infects intestines followed by stomach. CT imaging findings show bowel wall thickening and/or lack of wall enhancement, pneumatosis, bowel perforation, and acute ischemia (Ghuman et al., 2021). Common symptoms involve nausea, vomiting, and abdominal pain (Binder, Maurer and Lass-Flörl, 2014).

With the dawn of the second wave of Covid-19, cases of GI invasive mucormycosis (IM) have started appearing. A few have reported patients with mesenteric thrombosis as a cause of IM along with symptoms such as abdominal distension and constipation (Jain et al., 2021). Another case study revealed occurrence of gastric ulcers and presence of necrotic debris after doing esophagogastroduodenoscopy (Monte Junior et al., 2020).

14.4.5 Disseminated Mucormycosis

Disseminated mucormycosis is another rare form of disease involving two or more organ systems. It may originate from any site of infection. The highest incidence of dissemination occurs in pulmonary mucormycosis. The most prevalent symptom is usually fever, followed by hemoptysis and chest pain (Soliman et al., 2019). The most common site of dissemination is the brain, but metastatic lesions may occur in other organs. Brain infection differs from rhinocerebral mucormycosis as it results in infarction and abscess formation. Patients develop focal neurological deficits or comatose conditions. The disseminated form has a high mortality rate (> 90%) (Spellberg, Edwards, and Ibrahim, 2005).

One case study reported that a patient with severe Covid-19 infection developed disseminated mucormycosis in the lungs and brain, which was revealed upon postmortem studies (Mehta and Pandey, 2020). Another documented case described an immunocompetent patient with Covid-19 pneumonia and multiorgan failure upon autoptic identification (Krishna et al., 2021).

14.5 DIAGNOSIS

Early diagnosis of mucormycosis proves crucial towards understanding the prognosis and giving the correct treatment timely. Sample collection is an important measure. Usually tissue biopsies prove to provide better insights of the spread of the disease. Respiratory samples such as bronchoalveolar lavage (BAL) and induced sputum are taken for diagnosing pulmonary mucormycosis. CT-guided invasive lung biopsy also helps in definitive diagnosis. Tissue samples are transported in a saline solution in a sterile container, at room temperature, since Mucaroles lose viability at lower temperatures. The following diagnostic measures are done routinely.

14.5.1 Direct Microscopy

KOH or KOH-calcofluor mount preparation is done to visualize broad (6–25 μm) ribbon-like, thin-walled aseptate/pauci-septate hyphae of Mucaroles (Figure 14.3).

14.5.2 Histopathology

Histopathological stains like hematoxylin and eosin, Grocott–Gomori methenamine silver, and periodic acid–Schiff are used routinely to visualize the characteristic hyphal element of mucaroles in infected tissues.

14.5.3 Culture and Identification

Fungal cultures are done on Sabouraud Dextrose Agar (SDA), which gives a cotton-like white or greyish black colony (Figure 14.4). Identification is then done by morphology on lactophenol cotton mount, thermotolerance studies, by MALDI-TOF, or other molecular methods.

Molecular methods involve nucleic acid amplification from clinical samples targeting the internal transcribed spacer (ITS) regions of fungal ribosomal small subunit.

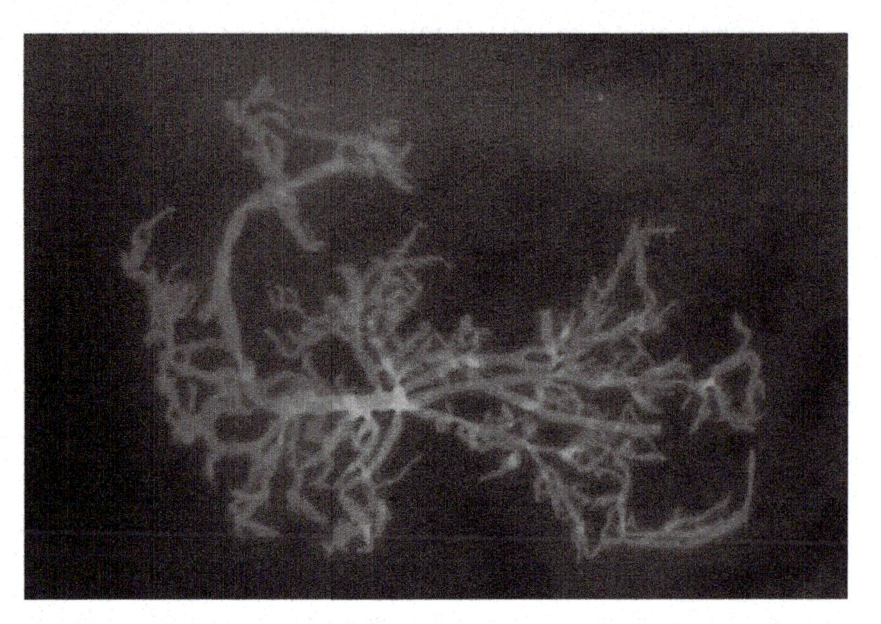

FIGURE 14.3 KOH-calcofluor mount showing broad aseptate hyphae with right angle branching and ribbonlike folding suggestive of Mucorales (40x).

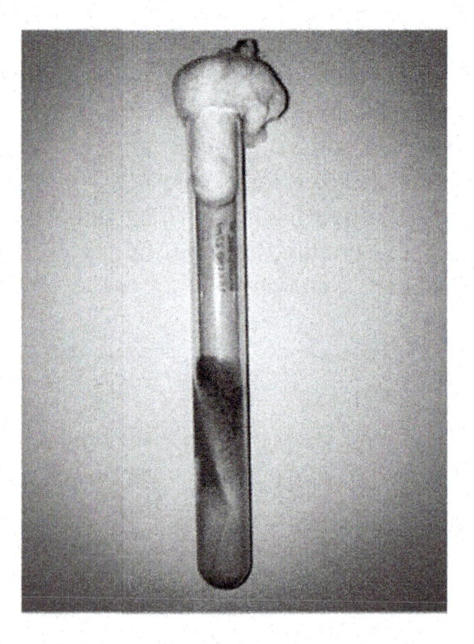

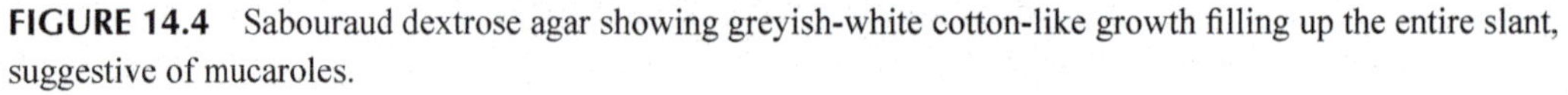

FIGURE 14.4 Sabouraud dextrose agar showing greyish-white cotton-like growth filling up the entire slant, suggestive of mucaroles.

14.6 TREATMENT

The three basic principles of treatment for mucormycosis cases include surgical debridement, correction of the immunosuppressive/metabolic disease, and appropriate antifungal therapy. The first drug of choice for mucormycosis is liposomal amphotericin B; 3-10mg/kg/day, following posaconazole and isavuconazole. For refractory diseases, intolerance or toxicity cases or prerenal compromised cases, azoles may be preferred upfront as salvage therapy (Table 14.3).

TABLE 14.3
Widely Used Antifungal Treatment Options

Drug	Dose
First-line Antifungal Therapy	
Liposomal Amphotericin B (L-AmB)	5 mg/kg/day; diluted in 5% dextrose and given over 2–3 hours as infusion, higher dose 10 mg/kg/day may be administered in the case of brain involvement.
Amphotericin deoxycholate (D-AmB)	Only given if cost and availability of L-AmB is an issue; 1mg/kg/day in 5% dextrose, slow infusion for 6–8 hours. Requires premedication to avoid infusion reaction. Renal functions and potassium levels are monitored routinely.
Second-Line/ Salvage therapy	
Posaconazole	300 mg twice on day 1, followed by 300 mg once a day. Interacting drugs to be avoided.
Isavuconazole	200 mg thrice a day for two days, followed by 200 mg once a day.

14.7 ROLE OF MEDICINAL PLANTS IN FUNGAL INFECTIONS

SARS-CoV-1 and SARS-CoV-2 are two separate viruses that belong to the same family and have comparable prevalence, biology, and clinical features (Peeri et al., 2020). The incidence of fungal comorbidity is 14.8–27% in SARS-CoV-1 patients and up to 21.9–33% is observed in very sick patients (Zhang et al., 2003; Yin et al., 2004). For SARS-CoV-1 patients, fungal comorbidity proved to be one of the leading causes (25–73.7%) of their mortality (Li and Pan, 2003). Experts are now aware of fungal coinfections in the SARS-CoV-2 epidemic. Clinicians are increasingly concentrating on early detection and treatment in critically ill or immune-compromised patients (Xu et al., 2020). Mucormycosis co-infection may prove to be a significant health danger and a life-threatening condition, particularly in immunocompromised individuals. In immune-compromised hosts, early, targeted, and cost-effective antifungal treatment is critical. The treatments of fungal infections are difficult as compared to bacteria. In terms of cellular organization, fungus is similar to higher animals and significantly different from prokaryotes. The cell wall and ribosome of fungi are different from bacterial, and the mode of actions of antibiotics on cell walls and ribosomes is ineffective for fungi. The fungal cell wall is made of chitin, which is different from the cell wall composition of a bacterial cell, making its treatment challenging. There are a few antifungal drugs like amphotericin B, azoles, echinocandins, terbinafine, and flucytosine that are regularly used in the treatment of fungal infections. Amphotericin B is one of the most widely used antifungal agents; however, it is expensive and has multiple side effects. Fluconazole is considered one of the safest antifungal drugs. Unfortunately, fluconazole is fungistatic and is used for treatment of only yeast infections and fungi may develop resistance, making their utility limited (Andriole, 2000; Balkis et al., 2002; Ishida et al., 2006). Voriconazole is the first line of treatment against Aspergillosis. Posaconazole and Isavuconazole are effective against a wide spectrum of fungal infections, but are very expensive and may not be available everywhere. Several other studies indicate that azole medication may also increase the resistance in the *Aspergillus* spp. and dermatophytes (Balkis et al., 2002; Howard et al., 2006; Santos and Hamdan, 2007).

Antifungal drug resistance, high cost, severe toxicity issues, and limited activity push researchers to look toward the development of a newer antimycotic agent, which is economical, with a wider spectrum and fewer dose-limiting adverse responses. Additionally, fungi grow more slowly than bacteria and require more time for their specific and accurate diagnosis. Sometimes, delay in laboratory diagnosis of fungal infection may hamper their treatment.

Medicinal plants are the best source of a variety of medications, according to the World Health Organization. Traditional remedies based on medicinal plant products are used by about 80% of the population in poor nations (Nascimento et al., 2000). Since ancient times, medicinal plant products have been widely utilized to cure a variety of ailments including diabetes, hypertension, peptic ulcer disease, microbial infection, sexual problems, and many more. Various *in vitro* and *in vivo* studies suggest that medicinal plants might play a key role in treating viral illnesses such as Herpes Simplex Viruses, Dengue Fever, and Chikungunya (Vachirayonstien et al., 2010; Kaushik et al., 2018, 2021; Kaushik, Jangra, et al., 2020; Kaushik, Kaushik, et al., 2020; Sharma et al., 2021). Antimalaria and antiprotozoan activity has been found in medicinal plant extracts or green manufactured nanoparticles (Dutta et al., 2017; Kumar et al., 2020). The development of phytochemical and phytopharmacological analytic sciences has allowed us to determine the precise composition and biological activity of a variety of therapeutic plant products.

Effective antifungals are the need of the hour. Antifungal agents are either chemically synthesized or isolated from medicinal plants. The toxic nature of chemical-based antifungals makes them inferior to plant-originated antifungal agents. Due to the high morbidity and mortality of fungal infections, there is a great need for novel antifungal agents. The natural product-based treatment should be less expensive, broad-spectrum in action, and with fewer adverse effects than chemically synthesized drugs. Several *in vitro* and *in vivo* investigations involving plant components utilized in ethnomedicine have demonstrated potential antifungal effectiveness with no side effects, particularly when plant essential oils are employed (Mondello et al., 2003; Soković et al., 2006; Park et al., 2007; Bansod and Rai, 2008; Bajpai, Yoon and Kang, 2009). The saponins, tannins, alkaloids, anthraquinone, and phenols are found in the stem bark and oil-extract formulation of *Polyscias fulva*. The capacity of these phytocomponents to protect host cells from reactive oxygen species and free radical-induced damage is well recognized. The crude dichloromethane methanol (1:1 v/v) extract from the stem bark of *Polyscias fulva* has antidermatophyte properties. The extract-oil formulation at 5% may constitute an alternative means to reduce the fungal infections caused by dermatophytes (Njateng et al., 2013). Essential oil extract from the aerial parts (umbels and mature seeds) of *Ferulago capillaries* contains limonene and α-pinene compounds. Essential oil inhibited germ tube formation in *C. albicans*. It also displays a broad fungicidal activity by inducing oxidative stress, which affects enzymes activity and the membrane potential of mitochondria. The antifungal activity of *Ferulago capillaries* essential oil against *Candida, Cryptococcus, Aspergillus*, and *dermatophyte* species suggests its therapeutic potential on superficial infections of these fungi (Pinto et al., 2013). The crude extract and essential oil from the *Moringa oleifera Lam* contain pentacosane and hexacosane. These extracts of *M. oleifera* inhibit the growth of *T. rubrum, T. mentagrophytes, E. floccosum,* and *M. canis* fungi through the cell lysis mechanisms (Chuang et al., 2007). Essential oil extracts from the bulbs of *Allium sativum* (Di-2-propenyl trisulfide and di-2-propenyl disulfide), bulbs of *A. cepa* (dipropyl disulfide), and whole plant of *A. fistulosum* (dipropyl disulfide, trans-propenyl propyl disulfide, and dimethyl trisulfide) have fungistatic activity against *T. erinacei, T. rubrum*, and *T. soudanense.* On Sabouraud agar culture plates, dosage-dependent inhibitory effects of essential oils of *A. sativum, A. cepa,* and *A. fistulosum* against *Trichophyton* species have been observed. In the checkerboard titre test and disk diffusion test, these oils demonstrated considerable synergistic antifungal action when coupled with ketoconazole. *Allium spp.* essential oil extracts contain fungistatic properties, making them potent antifungal agents (Pyun and Shin, 2006). *Curcuma longa* is a well-known medicinal herb in India. Turmerone, atlantone, and zingiberone are found in the paste of *C. longa* rhizome, which stimulates our immune system and aids in quick recovery from *T. mentagrophytes, T. rubrum, E. floccosum,* and *M. gypseum* (Jankasem, Wuthi-Udomlert and Gritsanapan, 2013). The essential oil of *Eugenia cariophyllata*, which contains eugenol, inhibits spore germination and mycelial development of *C. albicans, C. tropicalis, C. krusei, Trichophyton rubrum, T. mentagrophytes*, and *Geotrichum candidum*. Essential oils and eugenol from *E. cariophyllata* are effective against mold and yeast strains identified from onychomycosis (Gayoso et al., 2005). Essential oils and methanol extracts from the aerial portions of *Salvia cryptantha* and *Salvia multicaulis* include the antifungal compounds pinene, eucalyptol, camphor, camphene, and borneol, which inhibit the growth of *C. albicans* and *C. krusei* (Tepe et al., 2004). Trans-geraniol, b-elemene, E-citral, and linalool are found in *Cymbopogon martini* extracts, whereas m-cymene, myrtenol, a-terpene, caryophyllene oxide, and 2,3-epoxy carvone *Chenopodium ambrosioides* have the capability to suppress the development of *Microsporum gypseum, Trichophyton rubrum,* and other filamentous fungi. The extract of *Cymbopogon martini* contains trans-geraniol, b-element, E-citral, and linalool while *Chenopodium ambrosioide* (m-cymene, myrtenol, a-terpene, caryophyllene oxide and 2,3-epoxy carvone) inhibits growth of *Microsporumgypseum*, *Trichophyton rubrum,* and other filamentous fungi. The findings support the use of essential oils in the treatment of dermatophyte diseases on a scientific validation level (Prasad et al., 2010). The phytoconstituent of *Leptadenia lancifolia* exhibits antifungal activity against several fungal illnesses. It can be used to treat mycotic infection, which is a very frequent health problem all over the world, especially in developing countries. To treat ringworm, the sap from the leaves is used topically (Doughari and Obidah, 2008). *Syzygium jambolanum* seeds were reported to be highly efficient against *Candida albicans, Aspergillus flavus, Aspergillus fumigatus, Aspergillus niger, Bacillus subtilis*, and *Staphylococcus aureus* in a study evaluating their antibacterial and antifungal properties (Chandrasekaran and Venkatesalu, 2004). Anthraquinones, the main components of *Saprosma fragrans*, showed strong antifungal activity. In research examining the antibacterial and antifungal characteristics of *Syzygium jambolanum* seeds, it was shown that they were extremely effective against *C. albicans, A. flavus, A. fumigatus, A. niger, Bacillus subtilis,* and *Staphylococcus aureus* (Chandrasekaran and Venkatesalu, 2004). *Saprosma fragrans* contains anthraquinones, which have potent antifungal properties. The antifungal action is mostly attributed to 3, 4-dihydroxy methoxy anthraquinone-2-carboxaldehyde (Singh et al., 2006). Both *C. albicans* and *Cryptococcus neoformans* are vulnerable to the antifungal properties of *Euphorbia fusiformis*. Rheumatism, gout, paralysis, arthritis, inflammation, and bacterial infections have all been treated with it in herbal medicine for a long time. *C. albicans* was also more resistant to the combined extract formulations than *C. neoformans*. The n-hexane extracts of *Datura metel* have an antifungal efficacy against *Ascochyta rabiei* (Shafique and Shafique, 2008).

14.8 CONCLUDING REMARKS

Although the occurrence of mucormycosis has been known for decades, the recent Covid-19 pandemic has placed this fungal infection into the limelight, due to its higher occurrence, morbidity, and mortality. A significant increase in the incidence of mucormycosis in Covid-19 patients during the second wave in India has left scientists and clinicians grappling for effective treatment and searching for newer antifungals that are affordable, less nephrotoxic, and easily available in abundance during such times of crisis. This global pandemic has highlighted the need and necessitated efforts towards more precise and quicker diagnosis of this disease for timely intervention to contain the spread of this fungus to vital organs like pulmonary and nervous system tissues. In current times, considering the fewer side effects of liposomal preparations of amphotericin B, it is the drug of choice for treatment of mucormycosis; however, its high cost makes it unaffordable for many. Several medicinal plants offer significant antifungal properties, are affordable, and have minimal side effects. Due to the fact that several natural plant products have previously been tested against various fungal diseases, they may have a role as adjunct therapy to current pharmacological products and need to be explored in detail. Perhaps the sudden surge of mucormycosis as an opportunistic infection in Covid-19 pandemic is an eye opener for future diseases that may affect populations in the future as emerging pathogens may lead to more pandemics.

REFERENCES

Andriole, V. T. (2000) 'Current and future antifungal therapy: new targets for antifungal therapy', *International Journal of Antimicrobial Agents*, 16(3), pp. 317–321. doi: 10.1016/s0924-8579(00)00258-2

Bajpai, V. K., Yoon, J. I. and Kang, S. C. (2009) 'Antifungal potential of essential oil and various organic extracts of Nandina domestica Thunb. against skin infectious fungal pathogens', *Applied Microbiology and Biotechnology*, 83(6), pp. 1127–1133.

Balkis, M. M. et al. (2002) 'Mechanisms of fungal resistance: An overview', *Drugs*, 62(7), pp. 1025–1040. doi: 10.2165/000 03495-200262070-00004

Bansod, S. and Rai, M. (2008) 'Antifungal activity of essential oils from Indian medicinal plants against human pathogenic Aspergillus fumigatus and A. niger', *World Journal of Medical Sciences*, 3(2), pp. 81–88.

Binder, U., Maurer, E. and Lass-Flörl, C. (2014) 'Mucormycosis – from the pathogens to the disease', *Clinical Microbiology and Infection*, 20(6), pp. 60–66. doi: 10.1111/1469-0691.12566

Castrejón-pérez, A. D., Miranda, I. and Welsh, O. (no date) 'Cutaneous mucormycosis', pp. 304–311.

Chakrabarti, A. et al. (2009) 'Invasive zygomycosis in India: experience in a tertiary care hospital', *Postgraduate Medical Journal*, 85(1009), pp. 573 LP—581. doi: 10.1136/pgmj.2008.076463

Chandrasekaran, M. and Venkatesalu, V. (2004) 'Antibacterial and antifungal activity of Syzygium jambolanum seeds', *Journal of ethnopharmacology*, 91(1), pp. 105–108. doi: 10.1016/j.jep.2003.12.012

Chuang, P.-H. et al. (2007) 'Anti-fungal activity of crude extracts and essential oil of Moringa oleifera Lam', *Bioresource Technology*, 98(1), pp. 232–236.

Covid19India (2020) 'Coronavirus Outbreak in India – covid19india.org', pp. 3–7. Available at: www.covid 19india.org/

Diamond, R. D. (1991) 'The growing problem of mycoses in patients infected with the human immunodeficiency virus', *Reviews of Infectious Diseases*, 13(3), pp. 480–486. doi: 10.1093/clinids/13.3.480

Dinakaran, D. et al. (2020) 'Neuropsychiatric aspects of COVID-19 pandemic: A selective review', *Asian Journal of Psychiatry*, 53, p. 102188. doi: 10.1016/j.ajp.2020.102188

Doughari, J. H. and Obidah, J. S. (2008) 'Antibacterial potentials of stem bark extracts of Leptadenia lancifolia against some pathogenic bacteria', *Pharmacologyonline*, 3, pp. 172–180.

Dutta, P. P. et al. (2017) 'Antimalarial silver and gold nanoparticles: Green synthesis, characterization and in vitro study', *Biomedicine & Pharmacotherapy*, 91, pp. 567–580. doi: https://doi.org/10.1016/j.bio pha.2017.04.032

Gayoso, C. W. et al. (2005) 'Sensitivity of fungi isolated from onychomycosis to Eugenia cariophyllata essential oil and eugenol', *Fitoterapia*, 76(2), pp. 247–249.

Ghuman, S. S. et al. (2021) 'CT appearance of gastrointestinal tract mucormycosis', *Abdominal Radiology (New York)*, 46(5), pp. 1837–1845. doi: 10.1007/s00261-020-02854-3

Gupta, A., Sharma, A. and Chakrabarti, A. (2021) 'The emergence of post-COVID-19 mucormycosis in India: Can we prevent it?', *Indian Journal of Ophthalmology*, 69(7). Available at: https://journals.lww.com/ijo/Fulltext/2021/07000/The_emergence_of_post_COVID_19_mucormycosis_in.2.aspx.

Hibbett, D. S. et al. (2007) 'A higher-level phylogenetic classification of the Fungi', *Mycological Research*, 111(5), pp. 509–547. doi: 10.1016/j.mycres.2007.03.004

Howard, S. J. et al. (2006) 'Multi-azole resistance in Aspergillus fumigatus', *International Journal of Anti-Microbial Agents*, 28(5), pp. 450–453. doi: 10.1016/j.ijantimicag.2006.08.017

'Insult to injury_ COVID-19-associated mucormycosis' (no date).

Ishida, K. et al. (2006) 'Influence of tannins from Stryphnodendron adstringens on growth and virulence factors of Candida albicans', *Journal of Antimicrobial Chemotherapy*, 58(5), pp. 942–949.

Jain, M. et al. (2021) 'Post-COVID-19 gastrointestinal invasive mucormycosis', *The Indian Journal of Surgery*, pp. 1–3. doi: 10.1007/s12262-021-03007-6

Jankasem, M., Wuthi-Udomlert, M. and Gritsanapan, W. (2013) 'Antidermatophytic properties of ar-turmerone, turmeric oil, and curcuma longa preparations', *ISRN Dermatology*, 2013, p. 250597. doi: 10.1155/2013/ 250597

Jeong, W. et al. (2019) 'The epidemiology and clinical manifestations of mucormycosis: a systematic review and meta-analysis of case reports', *Clinical Microbiology and Infection*, 25(1), pp. 26–34. doi: https:// doi.org/10.1016/j.cmi.2018.07.011

Kaur, H. et al. (2018) 'Gastrointestinal mucormycosis in apparently immunocompetent hosts-A review', *Mycoses*, 61(12), pp. 898–908. doi: 10.1111/myc.12798

Kaushik, Sulochana et al. (2018) 'Antiviral and therapeutic uses of medicinal plants and their derivatives against dengue viruses', *Pharmacognosy Reviews*, 12(24).

Kaushik, Sulochana, Jangra, G., et al. (2020) 'Anti-viral activity of Zingiber officinale (Ginger) ingredients against the Chikungunya virus', *VirusDisease*, 31(3), pp. 270–276. doi: 10.1007/s13337-020-00584-0

Kaushik, Sulochana, Kaushik, Samander, et al. (2020) 'In-vitro and in silico activity of Cyamopsis tetragonoloba (Gaur) L. supercritical extract against the dengue-2 virus', *VirusDisease*, 31(4), pp. 470–478. doi: 10.1007/s13337-020-00624-9

Kaushik, Sulochana et al. (2021) 'Identification and characterization of new potent inhibitors of dengue virus NS5 proteinase from Andrographis paniculata supercritical extracts on in animal cell culture and in silico approaches', *Journal of Ethnopharmacology*, 267, p. 113541. doi: https://doi.org/ 10.1016/ j.jep.2020.113541.

Khatri, A. et al. (2021) 'Mucormycosis after Coronavirus disease 2019 infection in a heart transplant recipient – Case report and review of literature', *Journal de mycologie medicale*, p. 101125. doi: 10.1016/ j.mycmed.2021.101125

Krishna, V. et al. (2021) 'Autoptic identification of disseminated mucormycosis in a young male presenting with cerebrovascular event, multi-organ dysfunction and COVID-19 infection', *IDCases*, p. e01172. doi: 10.1016/j. idcr.2021.e01172

Kumar, R. et al. (2020) 'Green synthesized Allium cepa nanoparticles with enhanced antiprotozoal activities for E. gingivalis', *Chemical Biology Letters*, 7(4 SE-Articles), pp. 247–250. Available at: https://pubs. thesciencein.org/journal/index.php /cbl/article/view/246

Kwon-Chung, K. J. (2012) 'Taxonomy of fungi causing mucormycosis and entomophthoramycosis (zygomycosis) and nomenclature of the disease: Molecular mycologic perspectives', *Clinical Infectious Diseases*. Oxford Academic, pp. S8–S15. doi: 10.1093/cid/cir864

Li, C. S. and Pan, S. F. (2003) 'Analysis and causation discussion of 185 severe acute respiratory syndrome dead cases', *Zhongguo wei zhong bing ji jiu yi xue= Chinese critical care medicine= Zhongguo weizhongbing jijiuyixue*, 15(10), pp. 582–584.

Lin, E., Moua, T. and Limper, A. H. (2017) 'Pulmonary mucormycosis: clinical features and outcomes', *Infection*, 45(4), pp. 443–448. doi: 10.1007/s15010-017-0991-6

Martin, G. S. et al. (2003) 'The Epidemiology of Sepsis in the United States from 1979 through 2000', *New England Journal of Medicine*, 348(16), pp. 1546–1554. doi: 10.1056/ nejmoa022139

Mehta, S. and Pandey, A. (2020) 'Rhino-orbital mucormycosis associated with COVID-19', *Cureus*, 12(9), pp. 10–14. doi: 10.7759/cureus.10726

Mondello, F. et al. (2003) 'In vitro and in vivo activity of tea tree oil against azole-susceptible and -resistant human pathogenic yeasts', *The Journal of Antimicrobial Chemotherapy*, 51(5), pp. 1223–1229. doi: 10.1093/jac/dkg202

Monte Junior, E. S. do et al. (2020) 'Rare and fatal gastrointestinal mucormycosis (zygomycosis) in a COVID-19 patient: A case report', *Clinical Endoscopy*, pp. 746–749. doi: 10.5946/ce.2020.180

Nascimento, G. G. F. et al. (2000) 'Antibacterial activity of plant extracts and phytochemicals on antibiotic-resistant bacteria', *Brazilian Journal of Microbiology*, 31, pp. 247–256.

Njateng, G. S. S. et al. (2013) 'In vitro and in vivo antidermatophytic activity of the dichloromethane-methanol (1:1 v/v) extract from the stem bark of Polyscias fulva Hiern (Araliaceae)', *BMC Complementary and Alternative Medicine*, 13, p. 95. doi: 10.1186/1472-6882-13-95

Park, M.-J. et al. (2007) 'Antifungal activities of the essential oils in Syzygium aromaticum (L.) Merr. Et Perry and Leptospermum petersonii Bailey and their constituents against various dermatophytes', *Journal of Microbiology*, 45(5), pp. 460–465.

Pasero, D. et al. (2020) 'A challenging complication following SARS-CoV-2 infection: a case of pulmonary mucormycosis', *Infection*, pp. 1–6. doi: 10.1007/s15010-020-01561-x

Patel, A. K. et al. (2017) 'Mucormycosis at a tertiary care centre in Gujarat, India', *Mycoses*, 60(6), pp. 407–411. doi: https://doi.org/10.1111/myc.12610

Peeri, N. C. et al. (2020) 'The SARS, MERS and novel coronavirus (COVID-19) epidemics, the newest and biggest global health threats: what lessons have we learned?', *International Journal of Epidemiology*, 49(3), pp. 717–726. doi: 10.1093/ ije/dyaa033

Pinto, E. et al. (2013) 'Antifungal activity of Ferulago capillaris essential oil against Candida, Cryptococcus, Aspergillus and dermatophyte species', *European Journal of Clinical Microbiology & Infectious Diseases*, 32(10), pp. 1311–1320.

Potenza, L. et al. (2016) 'Mucorales-specific T cells in patients with hematologic malignancies', *PloS one*, 11(2), p. e0149108. doi: 10.1371/journal.pone.0149108

Prakash, H. and Chakrabarti, A. (2019) 'Global epidemiology of mucormycosis', *Journal of Fungi (Basel, Switzerland)*, 5(1). doi: 10.3390/jof5010026.

Prasad, C. S. et al. (2010) 'In vitro and in vivo antifungal activity of essential oils of Cymbopogon martini and Chenopodium ambrosioides and their synergism against dermatophytes', *Mycoses*, 53(2), pp. 123–129. doi: 10.1111/j.1439-0507.2008.01676.x

Pyun, M.-S. and Shin, S. (2006) 'Antifungal effects of the volatile oils from Allium plants against Trichophyton species and synergism of the oils with ketoconazole', *Phytomedicine: International Journal of Phytotherapy and Phytopharmacology*, 13(6), pp. 394–400. doi: 10.1016/j.phymed.2005.03.011

Roden, M. M. et al. (2005) 'Epidemiology and outcome of zygomycosis: A review of 929 reported cases', *Clinical Infectious Diseases*, 41(5), pp. 634–653. doi: 10.1086/432579

Santos, D. A. and Hamdan, J. S. (2007) 'In vitro activities of four antifungal drugs against Trichophyton rubrum isolates exhibiting resistance to fluconazole', *Mycoses*, 50(4), pp. 286–289. doi: https://doi.org/10.1111/j.1439-0507.2007.01325.x

Sen, M. et al. (2021) 'Epidemiology, clinical profile, management, and outcome of COVID-19-associated rhino-orbital-cerebral mucormycosis in 2826 patients in India – Collaborative OPAI-IJO study on mucormycosis in COVID-19 (COSMIC), Report 1', *Indian Journal of Ophthalmology*, 69(7), pp. 1670–1692. doi: 10.4103/ijo.IJO_1565_21

Shafique, S. and Shafique, S. (2008) 'Antifungal activity of n-hexane extracts of Datura metel against Ascochyta rabiei', *Mycopath*, 6(1&2), pp. 31–35.

Sharma, Y. et al. (2021) 'In-vitro and in-silico evaluation of the anti-chikungunya potential of Psidium guajava leaf extract and their synthesized silver nanoparticles', *VirusDisease*. doi: 10.1007/s13337-021-00685-4

Singh, D. N. et al. (2006) 'Antifungal anthraquinones from Saprosma fragrans', *Bioorganic & Medicinal Chemistry Letters*, 16(17), pp. 4512–4514. doi: 10.1016/j.bmcl.2006.06.027

Skiada, A., Pavleas, I. and Drogari-Apiranthitou, M. (2020) 'Epidemiology and diagnosis of mucormycosis: An update', *Journal of Fungi* . doi: 10.3390/jof6040265

Skiada, A. and Petrikkos, G. (2009) 'Cutaneous zygomycosis', *Clinical Microbiology and Infection: The Official Publication of the European Society of Clinical Microbiology and Infectious Diseases*, 15 Suppl 5, pp. 41–45. doi: 10.1111/j.1469-0691.2009.02979.x

Soković, M. D. et al. (2006) 'Antifungal activity of the essential oil of Mentha. x piperita', *Pharmaceutical Biology*, 44(7), pp. 511–515.

Soliman, M. et al. (2019) 'Disseminated mucormycosis with extensive cardiac involvement', *Cureus*, p. e4760. doi: 10.7759/ cureus.4760

Spellberg, B., Edwards, J. and Ibrahim, A. (2005) 'Novel perspectives on mucormycosis: Pathophysiology, presentation, and management', *Clinical Microbiology Reviews*, 18(3), pp. 556–569. doi: 10.1128/CMR.18.3.556-569.2005

Tepe, B. et al. (2004) 'Antimicrobial and antioxidative activities of the essential oils and methanol extracts of Salvia cryptantha (Montbret et Aucher ex Benth.) and Salvia multicaulis (Vahl)', *Food chemistry*, 84(4), pp. 519–525.

Vachirayonstien, T. et al. (2010) 'Molecular evaluation of extracellular activity of medicinal herb Clinacanthus nutans against herpes simplex virus type-2', *Natural Product Research*, 24(3), pp. 236–245. doi: 10.1080/14786410802393548

Wali, U., Balkhair, A. and Al-Mujaini, A. (2012) 'Cerebro-rhino orbital mucormycosis: An update', *Journal of Infection and Public Health*, 5(2), pp. 116–126. doi: 10.1016/j.jiph.2012.01.003

Worldometer (2022) 'COVID live update: 160,416,106 cases and 3,333,785 deaths from the Coronavirus', *Worldometer*. Available at: www.worldometers.info/coronavirus/#countries.

Xu, K. et al. (2020) '[Management of corona virus disease-19 (COVID-19): the Zhejiang experience]', *Zhejiang da xue xue bao. Yi xue ban = Journal of Zhejiang University. Medical Sciences*, 49(1), pp. 147–157. doi: 10.3785/j.issn.1008-9292.2020.02.02

Yin, C. H. et al. (2004) 'Clinical analysis of 146 patients with critical severe acute respiratory syndrome in Beijing areas', *Clinical Journal of Emergency Medicine*, 1(13), pp. 12–14.

Zhang, Y. et al. (2003) 'Hospital acquired pneumonia occurring after acute stage of the serious SARS and its treating strategies', *Chinese Journal of Nosocomiology*, 11(13), pp. 1081–1087.

15 Mucormycosis
An Emerging Opportunistic Infection in the Post-COVID Era

Ashok Gupta,[1] Anuragini Gupta,[2] and Shruti Baruah[3]
[1]Director and Head, Department of Otorhinolaryngology Head and Neck Surgery, Fortis Hospital, Mohali, India
[2]Attending Consultant, Department of Otorhinolaryngology Head and Neck Surgery, Fortis Hospital, Mohali, India
[3]Senior Resident, Department of Otorhinolaryngology Head and Neck Surgery, Fortis Hospital, Mohali, India

CONTENTS

15.1 INTRODUCTION

In 1876, German scientist Fürbinger first described pulmonary mucormycosis as a lethal fungal infection during an autopsy [1]. Thereafter, the first case of disseminated mucormycosis with central nervous system involvement was described by Paltauf in a cancer patient [2]. In May 2021, rhino-orbital-cerebral mucormycosis (ROCM), the most rampant form of mucormycosis, was declared an epidemic in a large number of Indian states after the second wave of SARS CoV-2 infection. ROCM was first described by Gregory et al. in a series of three patients who had the classical syndrome of uncompensated diabetes mellitus, unilateral orbital cellulitis along with total ophthalmoplegia, and blackish discoloration of skin and mucosa in 1943 [2].

DOI: 10.1201/9781003358909-15

Mucormycosis refers to infection by fungi of the order *Mucorales*, class *Zygomycetes*. These fungi are ubiquitous saprophytes that usually grow in dead decaying vegetable matter or soil [3]. In humans, Mucor have a predilection for the nose, paranasal sinuses, and lungs. The fungal spores enter the human body by inhalation, ingestion, or direct inoculation [7]. Classically, mucormycosis has been reported in uncontrolled diabetic patients with ketoacidosis and in patients with chronic neutropenia, post-organ transplantation, and in those undergoing steroid treatments. ROCM is associated with a high morbidity and mortality due to the angioinvasive property of the fungus, thereby causing microthromboemboli, which consequently result in extensive tissue necrosis [5]. The overall mortality of the types of mucormycosis is pulmonary mucormycosis: 50–70%; rhinoorbital: 30–70%; CNS involvement: >80%; and disseminated: >90% [6]. However, with the introduction of lipid and liposomal formulations of amphotericin B since the 1990s, and the increased awareness of clinical progression of the disease, the overall mortality rate has declined to 27–30% [2].

15.2 INCIDENCE

Prior to the Covid-era, Indian literature reported the prevalence rate of mucormycosis as 0.14 cases per 1000 population [9]. A 10-year study by Manesh et al. reported an annual incidence of 18.4 cases in Southern India [8]. Although the exact incidence is not known due to a lack of adequate population-based studies, even prior to the Covid-2019 pandemic, global data showed that India had a 70 times higher prevalence of mucormycosis than the rest of the world [9]. The second wave of Covid-19 in 2021 brought a new devastating spurt in the incidence of mucormycosis, more commonly ROCM, with a total of 28,200 cases by mid-2021. Out of these, 86% of affected cases had a reported history of recent or concurrent Covid infection [10]. In a study conducted by the senior author and colleagues, in a tertiary healthcare centre, a total of 20 patients of Covid-associated mucormycosis were diagnosed and managed within a span of 60 days after the second wave of the Covid-19 pandemic.

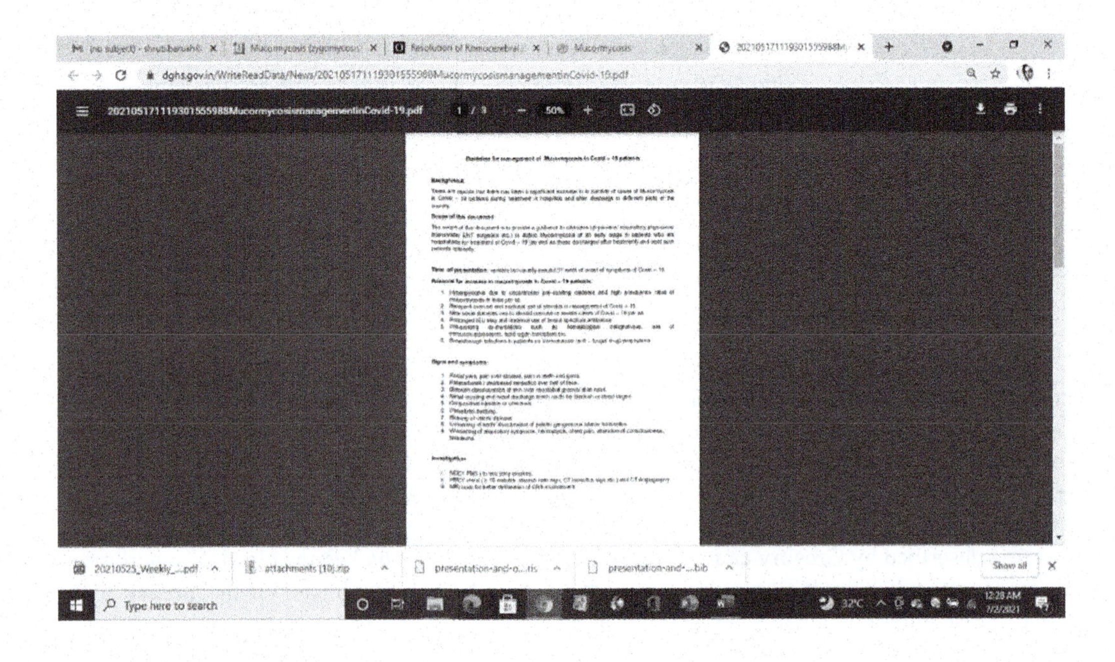

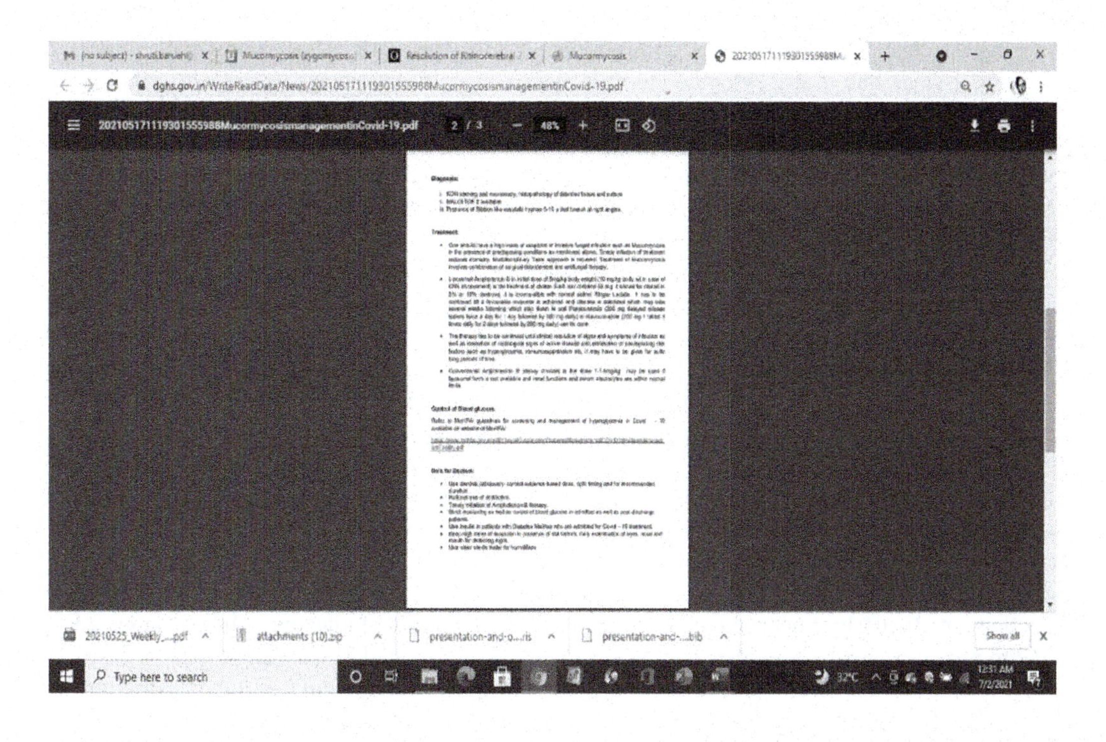

15.3 ETIOLOGICAL MYCOLOGY

There are two orders of zygomycetes that cause disease in humans: mucorales and the entomophthorales. The entomophthorales are true pathogens, affecting mainly immunocompetent hosts. While entomophthorales generally do not invade blood vessels and rarely disseminate, mucorales are infamous for their angioinvasive properties, which often lead to thrombosis, resulting in infarction of involved tissues, along with tissue destruction mediated by a number of fungal proteases, lipases, and mycotoxins such as rhizoxin by *Rhizopus spp.* [6·7].

Three genera of mucoraceae are most commonly responsible for mucormycosis, Rhizopus, Mucor, and Absidia, while less common causative agents include those under the families of *Mortierellaceae, Cunninghamellaceae, Saksenaeaceae,* and *Syncephalastraceae* [4]. Rhizopus oryzae is the most common organism isolated from patients with mucormycosis (70% of all cases of mucormycosis) and Cunninghamella bertholettiae carries the highest mortality rates among reported cases of infection with mucorales [6]. The infection is considered to be caused by asexual spore formation [2].

15.4 RISK FACTORS AND UNDERLYING DISEASES

I. Diabetes mellitus: In both the pre- and post-Covid eras, diabetes mellitus has played a key role in the causation of mucormycosis. Before the widespread use of steroids for the management of SARS Covid viral pneumonia, most world literatures have held uncontrolled diabetes, frequently associated with ketoacidosis, accountable for more than 60% of mucormycosis cases [9]. In a study of 20 cases of Covid-associated mucormycosis conducted by the senior author and colleagues, a 100% association with diabetes mellitus with or without acidosis was seen. Diabetes mellitus, especially with acidosis, appears to weaken the initial containment of the infection by polymorphonuclear leukocytes by means of phagocytic activity. Polymorphonuclear neurtrophils (PMNs) are less effective in removing hyphae in a glucose-rich, acidic milieu, which also facilitates fungal growth, and the infection thus becomes more established. There has also been reported evidence that suggests that SARS-CoV virus induces damage to pancreatic islets, along with increased insulin resistance due to the cytokine storm it manifests in its later stage, eventually resulting in acute insulin-dependent diabetes mellitus and diabeticketocidosis (DKA) [12]. The pathophysiology involved in such acidemic and hyperglycemic states appears to be the induction of the endothelial receptor glucose-regulated protein (GRP 78) and the mucorales adhesin spore coat protein homologs (CotH), creating a "perfect storm" for increased adhesion and penetration of mucorales to the endothelium [11].

II. Prolonged steroid therapy: The exact mechanism by which corticosteroids impair phagocyte activity has yet to be determined but the immune suppression caused by corticosteroids has resulted in an increase in the cases of opportunistic fungal infections [9]. During the crisis of the second wave of Covid in India, the inadequate knowledge and over-the-counter availability of steroids gave rise to a grave situation, wherein irrational and erratic use of systemic steroids paved the way for notorious opportunistic infections such as mucormycosis. Additionally, mucorales might possess unique virulence factors that enable them to exploit this state of immunosuppression and physiologic impairment [6]. In a study conducted by the senior author and colleagues, a history of steroid therapy either oral or intravenous was present in 100% of patients.

III. Post-covid immune suppression (lymphopenia): Persistent lymphopenia in COVID-19 patients is now being suspected as a crucial factor in the etiogenesis of opportunistic fungal infections, particularly mucormycosis [9]. However, not enough studies have been conducted in order to provide conclusive evidence for lymphopenia as an isolated causal factor.

IV. Renal disease: Both post-renal transplant and CKD patients have a higher vulnerability to mucormycosis. Patients receiving dialysis are treated with the iron chelator deferoxamine, which alone makes them susceptible to infection by mucorales. In post-renal transplant patients, the immunosuppressive agents used to prevent rejection weaken the innate response to mucorales. It has been suggested that cyclosporin has less effect on PMNs than prednisone or azathioprine [2].

V. Prolonged mechanical ventilation, ICU stay: While originally it was a community acquired infection, iatrogenic or nosocomial mucormycosis has notoriously been on the rise in recent times. Contaminated endotracheal tube connecters, surgical dressings, nasogastric tubes, oxygen prongs and masks, and nonsterile adhesive tapes in ICU setups, especially during the Covid pandemic, are some of the incriminated sources of nosocomial mucormycosis [6].

VI. Iron overload: Studies have shown an alteration in iron metabolism in severe COVID-19 disease. The pathophysiology suspected is that cytokines, especially IL-6, due to severe infection and DKA, stimulate ferritin synthesis. This further downregulates iron export, which results in intracellular iron overload, thus exacerbating the disease process. Mucorales have the unique ability to acquire iron from the host and iron is essential for their cell growth and development. For example, in patients undergoing deferoxamine therapy, iron removed by it is captured by siderophores on *Rhizopus* species and this iron helps in the growth of the fungi. Acidotic conditions (pH 7.3–6.6) further decrease the iron-binding capacity of normal serum, and this alteration abolishes an important host defense mechanism which permits growth of mucorales like *R. oryzae* [12].

VII. Leukemia and other cancers: Classically, rhinoorbital mucormycosis is not a common entity in leukemia. When it occurs, mucormycosis is usually in its disseminated or pulmonary forms. In its disseminated state in leukemia, it usually has poor prognosis. The pathogenesis is attributable to immature white blood cells in sera of leukemic patients and the inability to fight off any infections in the neutro/leukopenic state [2]. Again, the virulence factor of mucorales to take advantage of hosts' immunosuppressive state might play a role here. Polymorphonuclear neutrophil chemotactic deficit and phagocytosis deficit are the key defective defense mechanisms in such cases [14]. In the study by the senior author and colleagues only one patient of a concurrent cancer (high-grade ovarian cancer) developed ROCM post SARS-COV2 infection.

Myriad other theories are also being suggested such as prolonged use of unhygienic masks, use of unhygienic tap water with humidifiers, injudicious use of industrial grade oxygen, overcrowding, repeated traumatic sampling, etc.; however, further research on these is required before their etiological roles are established definitely.

VIII. Virulence factors of mucorales: Aside from the multiple favorable host factors, a large part of the etiological process is dependent on the multiple virulence factors mucorales possess such as their unique ability to acquire iron from the host, the secretion of siderophores like *Rhizoferrin,* the secretion of lytic enzymes such a*spartic proteinases* that play a major role in their angioinvasive and necrotizing properties, and the ability to produce mycotoxins by certain species of *Rhizopus.*

15.5 ETIOPATHOGENESIS

In the immunologically competent host, the fungal spores that have entered the host's body are contained by a phagocytic response of the host's innate immunity.

The mechanism by which innate immunity helps in the defense against mucorales has been reported in studies as the upregulation in toll-like receptor 2 expression on exposure of neutrophils to *R. oryzae hyphae*. This in turn results in a proinflammatory gene expression with rapid induction of NF-ꝁB pathway-related genes. Thus, both mononuclear and polymorphonuclear phagocytes of immunocompetent hosts can kill mucorales by means of oxidative metabolites and cationic peptides such as defensins [6].

In the immunosuppressed host, a robust response does not occur, thus germination ensues and hyphae develop. Several pathophysiological mechanisms have been proposed to explain the loco-regional growth and spread of mucormycosis. Expression of both GRP-78 (glucose receptor protein-78) of endothelial cells as well as the fungal ligand CotH (spore-coating homologue protein) and their mutual interaction is the most important and established molecular mechanism, which enhances and establishes the angioinvasive and necrotic properties of mucorales. This cellular process flourishes in a host environment of hyperglycemia, low pH, high iron, acidosis, and decreased phagocytic activity of leucocytes [13].

Another important defense mechanism of the host is the skin and mucosal barrier. The pathogens of mucormycosis are typically incapable of penetrating intact mucosa and skin. Thus, any breach or trauma to the mucosa plays a considerable role in establishment of the fungal spores in the host's body. This has been the cause of the recent shift of mucormycosis becoming a nosocomial infection rather than a community-acquired one. Contaminated surgical dressings, nasogastric tubes, oxygen prongs and masks, and nonsterile adhesive tapes are some of the incriminated sources of nosocomial mucormycosis [6].

Mucormycosis is characterized by extensive angioinvasion, which results in microvascular thrombosis subsequently causing tissue necrosis, which gives the classical blackening of affected tissues.

This angioinvasion is possibly the means by which the organism disseminates hematogenously to other target organs. Consequently, there must be damage of and penetration through endothelial cells or the extracellular matrix proteins lining blood vessels. This is likely to be the critical step in the pathogenetic and dissemination process of mucormycosis [6]. A study by Bouchara et al. used immunofluorescence to reveal that mucor spores interacted exclusively with laminin and type IV collagen thus establishing the theory of fungal adherence to epithelial basement membranes [15].

15.6 CLINICAL PRESENTATION

The clinical presentation varies, depending on the source of infection. The disease manifestations reflect the mode of transmission and based on these, there are five variants of mucormycosis: namely, ROCM, pulmonary mucormycosis, cutaneous mucormycosis, gastrointestinal mucormycosis, and

TABLE 15.1
Manifestations of the Various Types of Mucormycosis

Type	Organs Affected	Symptoms	Warning Signs
RHINO-ORBITO-CEREBRAL MUCORMYCOSIS (ROCM)	NOSE, PARANASAL SINUSES, NASOLACRIMAL SAC AND DUCT, ORBIT, BRAIN	Unilateral facial swelling/pain/numbness Headache Nasal or sinus congestion/heaviness Black lesions on nasal bridge/eye/upper inside of mouth Unilateral swelling of eye with ptosis/proptosis	Pain and redness around the eye/nose Double vision/loss of vision Restricted eye movements Loosening of teeth Altered mental state/loss of consciousness Facial Palsy
PULMONARY MUCORMYCOSIS	LUNGS	Refractory fever on broad-spectrum antibiotics, dry cough, progressive dyspnea, pleuritic chest pain	Severe breathlessness, bloody sputum
CUTANEOUS MUCORMYCOSIS	SKIN AND SOFT TISSUE	Redness, induration, then black eschar at trauma/puncture site of skin, muscle pain with deeper involvement	Black eschar formation
GASTROINTESTINAL MUCORMYCOSIS	STOMACH, SMALL AND LARGE INTESTINE	Fever, abdominal cramps, bleed per rectum	Bleed per rectum, sudden collapse due to intestinal perforation
DISSEMINATED MUCORMYCOSIS	MULTIPLE ORGANS	As per organs involved	

disseminated mucormycosis (Table 15.1). Among these, rhinocerebral and pulmonary are the most common manifestations, as mucorales generally have a predilection for the nose, paranasal sinuses, and lungs.

In ROCM, the invasive fungal rhinosinusitis usually starts in the middle turbinate and involves the maxillary and ethmoid sinuses in early stages. It becomes rapidly progressive, extending into neighboring tissues, the periorbital region of the face, and ultimately into the orbit and brain [7].

Infections due to bandages, adhesives, or contaminated wound dressings are mostly cutaneous. Percutaneous exposure in immunocompromised patients has led to disseminated disease. Ingestion of contaminated tablets or food, and the use of tongue depressors, are responsible for gastrointestinal mucormycosis [16].

Mucormycosis is a disease with a high rate of morbidity and mortality. The reported overall mortality rates are pulmonary mucormycosis: 50–70%, rhinocerebral: 30–70%; CNS involvement: >80%; and disseminated: >90%.

15.7 RHINO-ORBITO-CEREBRAL MUCORMYCOSIS

ROCM mucormycosis can have a wide spectrum of clinical presentations varying from sinonasal, ophthalmic to neurological. These are usually of various degrees and are accompanied by constitutional symptoms of fever and deteriorated general health status due the general poor immunologic status of these patients [14]. The initial presenting symptoms may be non-specific and require a high

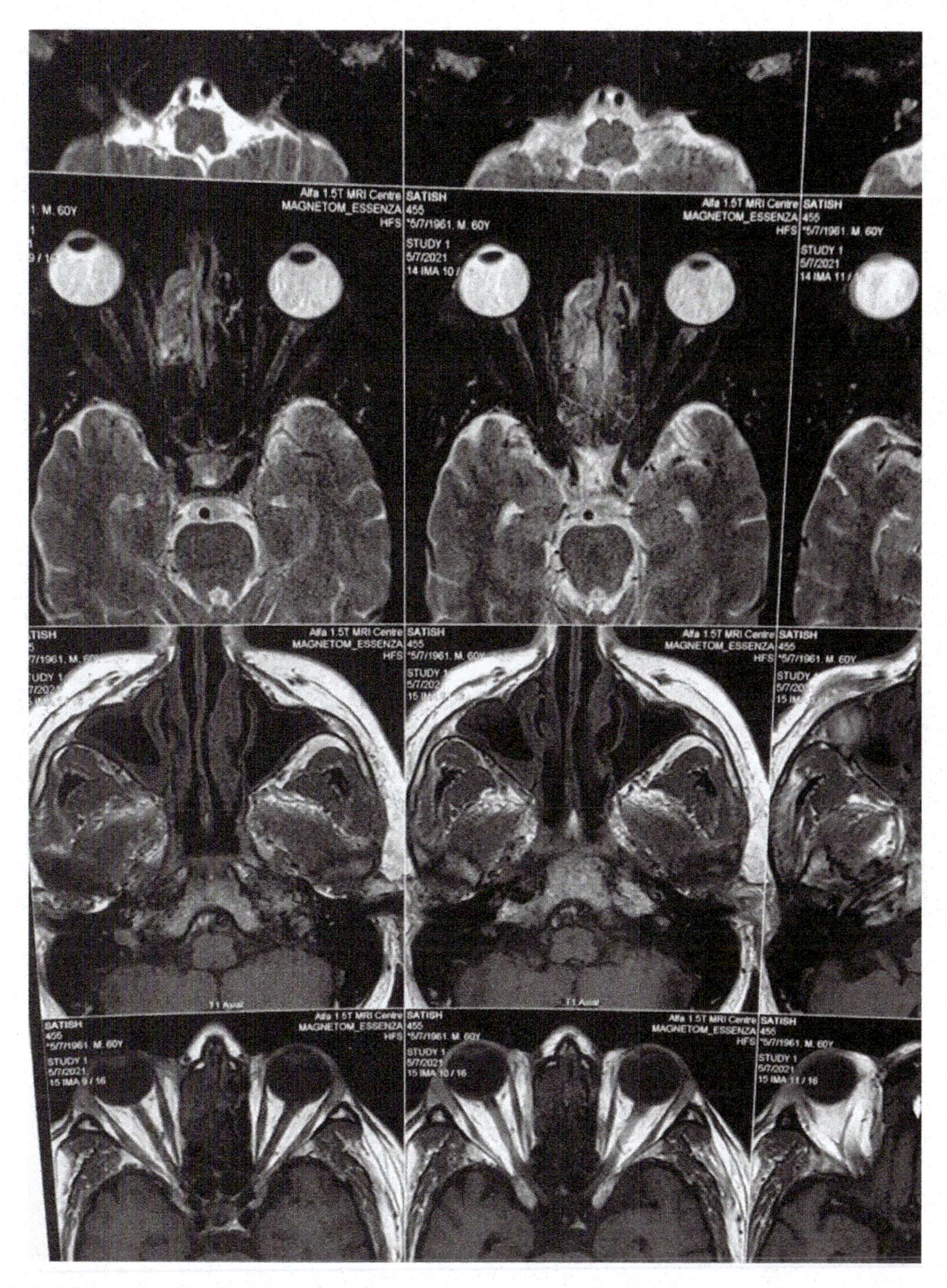

FIGURE 15.1A MR T1 weighted axial image showing disease in right ethmoidal gallery with mucosal thickening in right maxillary sinus.

degree of suspicion along with good history taking for early diagnosis. Leon et al. in their study of mucormycosis in diabetic patients described the following as the "red flags/warning signs" of ROCM mucormycosis: cranial nerve palsies, vision loss, diplopia, dental pain and numbness, proptosis of eye, periorbital swelling, orbital apex syndrome, or a palatine ulcer (Figures 15.1A–1C).

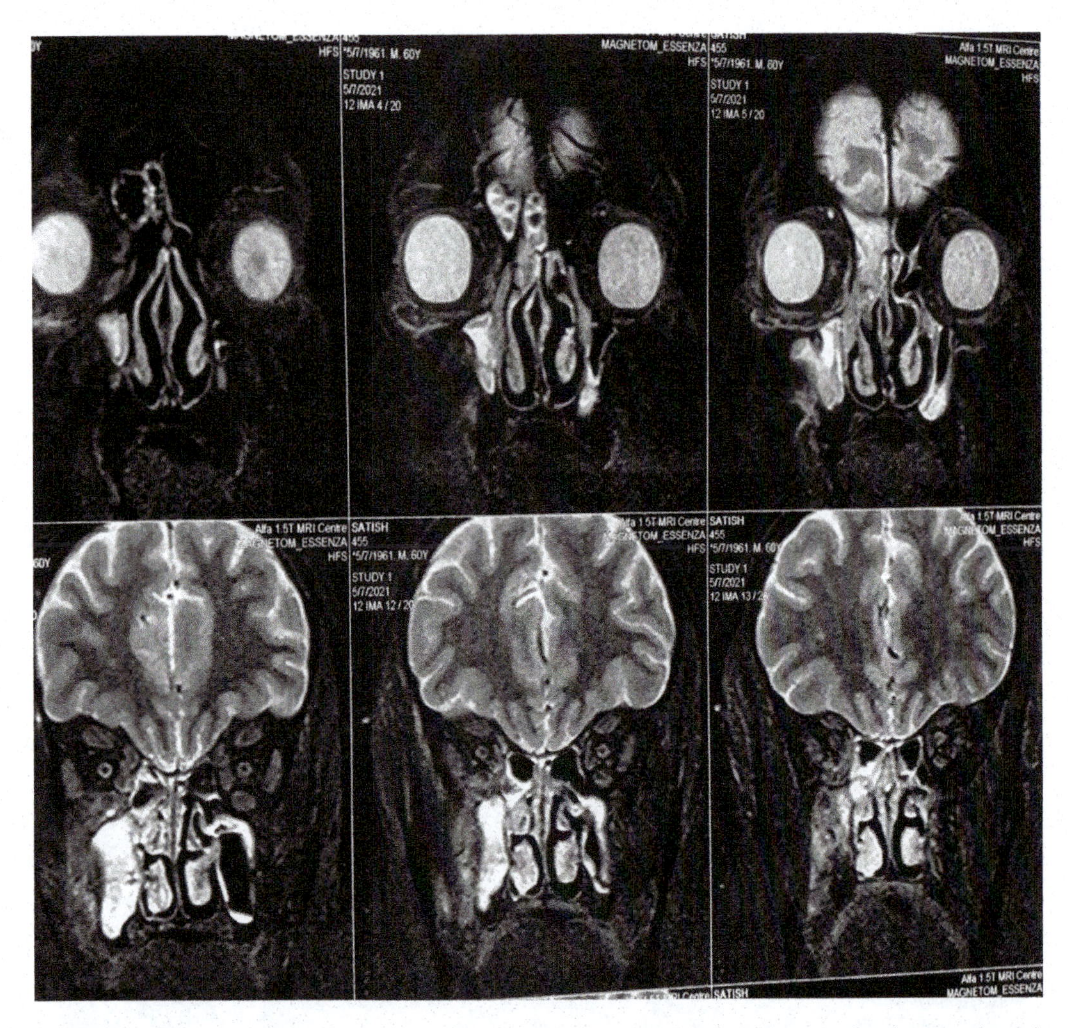

FIGURE 15.1B T2 weighted MR image showing hyperintense right heterogenous involvement in right maxillary and ethmoid sinus.

The presence of any of these clinical findings is an indicator of rapidly progressing angioinvasion and spread of disease [17]. A timely diagnosis and prompt intervention is key to reducing the high morbidity and mortality of ROCM mucormycosis. A study by the senior author and colleagues showed headache as the most consistent and early symptom reported in 86.67%; however, since all patients in the study were with concurrent or past history of SARS-Covid the cause of headache could not be singled out as mucormycosis. Periorbital swelling followed by facial/dental pain and numbness were the next most commonly presented symptoms in this study. In a large-scale study on diabetic patients by Bhansali et al., ophthalmoplegia (89%) was the most frequent clinical presentation followed by proptosis (83%) [5]. The first presentation of mucormycosis with intracranial complication is rare but not unheard of. In the senior author's study only 6.25% intracranial complications were seen at presentation as opposed to 20% in the study by Bhansali et al. [5]. Another rare but dreaded complication is an internal carotid artery thrombosis, which manifests as a massive cerebral infarction on imaging.

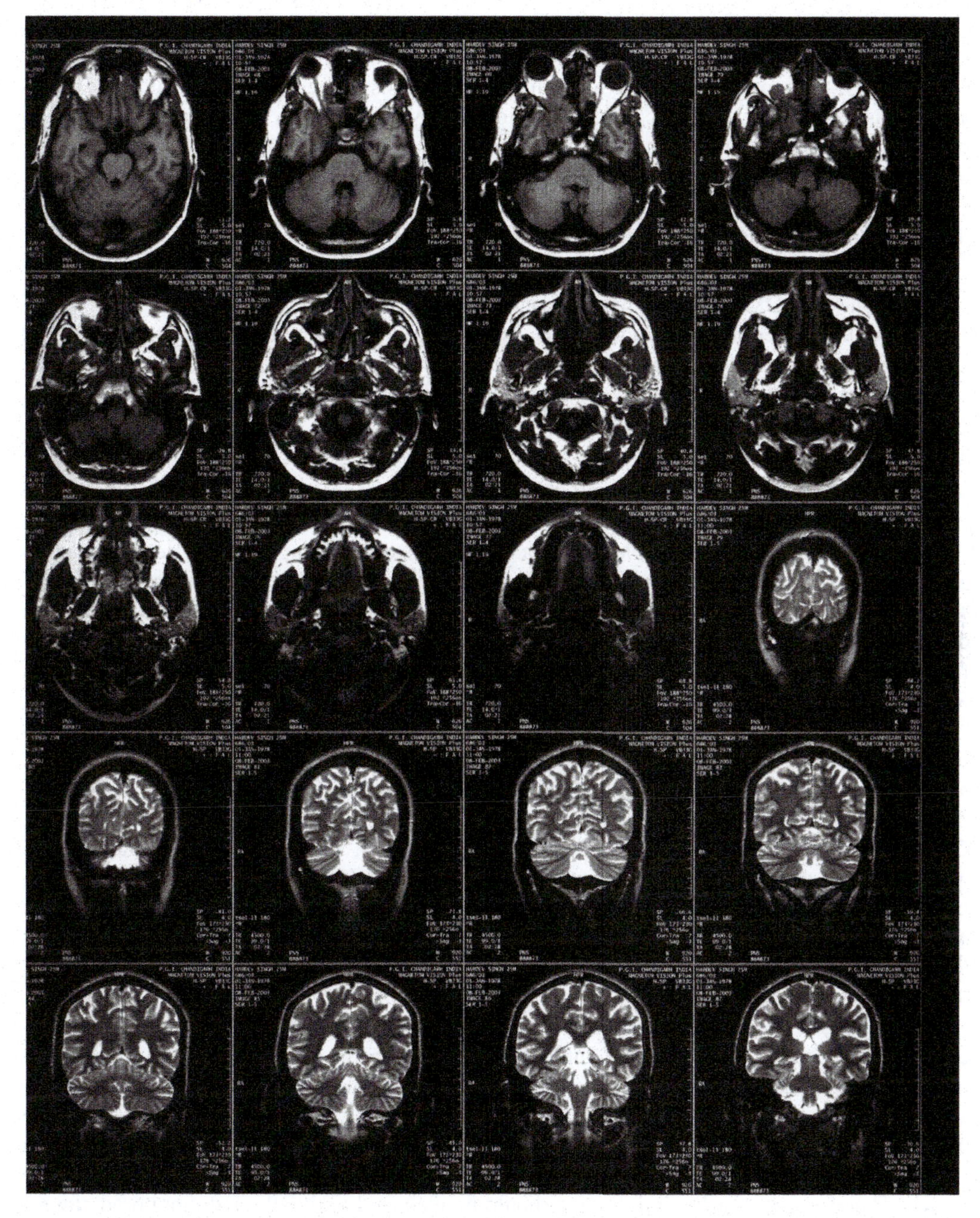

FIGURE 15.1C NCCT PNS showing involvement of right orbit.

15.8 CLINICAL DIAGNOSIS

Timely diagnosis of mucormycosis has shown to increase survival as well as reduce the need for or extent of surgical resection, facial and orbital disfigurement, and overall morbidty [16]. In the study by senior author and colleagues a high rate of suspicion in the post-Covid period and early screening of high-risk groups during the pandemic helped in avoiding orbital exenteration in 100% of cases.

Bhansali et al. described the factors associated with poor survival in ROCM as delay in diagnosis and treatment, hemiparesis, bilateral sinus involvement, and facial necrosis [5].

Optimally, diagnosis of ROCM mucormycosis consists of recognition of high-risk individuals, assessment of the clinical manifestations, early screening by imaging modalities, and prompt initiation of diagnostic methods based on endoscopy, tissue microscopy, histopathology, or cultures.

Proposed "high-risk" category of individuals in the study on Covid-associated ROCM by senior the author and colleagues included those with any one or more of the following:

1. Uncontrolled DM with or without ketoacidosis.
2. Any other underline comorbidities such as post-organ transplant, cancers, and other chronic illnesses with a history of prolonged systemic steroids.
3. Prolonged ventilator support with severe Covid illness (grossly elevated D-Dimer, CTSS score > 15, elevated ferritin and CRP levels).

The study revealed that all the patients (100%) in the study group fell in either one or more of these criteria.

A thorough clinical assessment including a detailed history, a comprehensive general examination with special attention to loosening of dentition, proptosis/ptosis of eyes, cranial nerve palsies, necrotic areas in the face, and/or palate. A diagnostic nasal endoscopy plays a vital role in the diagnosis of mucormycosis. Findings of discoloration, ulceration, extensive crusting, and eschar formation mainly involving middle and inferior turbinates is the classical nasal endoscpic picture of ROCM mucormycosis.

A fungal tissue wet mount in potassium hydroxide–calcofluor (KOH) would reveal broad ribbon-like hyphae, 10 to 20 μ across, right-angled branching, with absence or paucity of hyphal septation. With a sensitivity of 90% a presumptive diagnosis of mucormycosis based on a high clinical suspicion and a KOH-microscopy positivity for mucor alone can be made to initiate early treatment. Certain fungal stains, such as periodic acid-Schiff and Gridley, stain mucor poorly and therefore have no role in diagnosis [18]. Fungal culture in Sabourard's Dextrose Agar media has a sensitivity of only 50% and is relevant only in knowing the species of mucorale causing the disease. In newer diagnostic tests, matrix-assisted laser desorption ionization-time of flight mass spectrometry (MALDI-TOF) identification of cultured mucorales is a promising method for those laboratories that are accordingly equipped [16].

Imaging has a definite role particularly in the assessment of extent of the disease. A non-contrast enhanced CT scan with fine-cut (2–3 mm) slices in the axial and coronal plane is the initial investigation of choice. CT scans are helpful in defining individual variations in sinus architecture, evaluation of bony defects, and possible periorbital spread. An initial CT picture of minimal thickening in the maxillary and/or ethmoid gallery could rapidly progress to an extensive disease with widespread destruction of bony walls of sinuses and lamina papyraceae breach. Therefore, even a limited disease should be managed vigilantly and monitored closely for spread. The role of MRI is superior once disease has spread intraorbital or intracranially [18] (Figures 15.2A–B).

15.9 TREATMENT

The mainstay of treatment for mucormycosis remains early initiation of systemic antifungal therapy in the form of injection amphotericin B and optimum surgical debridement, along with correction of underlying predisposing risk factors, if possible [6].

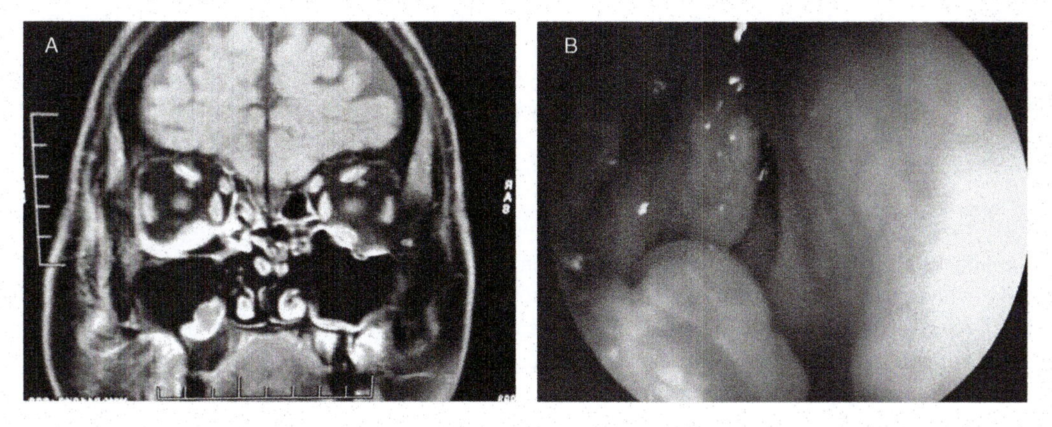

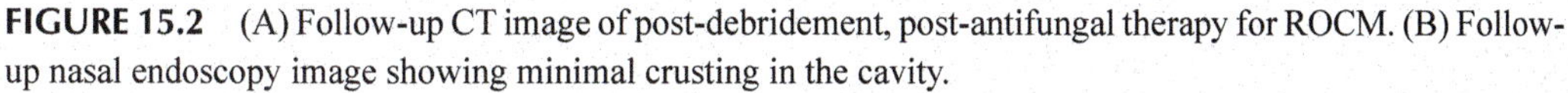

FIGURE 15.2 (A) Follow-up CT image of post-debridement, post-antifungal therapy for ROCM. (B) Follow-up nasal endoscopy image showing minimal crusting in the cavity.

15.9.1 Antifungals

A timely initiation of antifungal, especially those that are effective against zygomycetes (e.g., inj amphotericin B), remains the treatment of choice. Chamilos et al. demonstrated that a delayed amphotericin B-therapy (i.e., initiating treatment even >/=6 days after diagnosis) resulted in a two-fold increase in mortality rate [19]. The dilemma with the use of the highly nephrotoxic amphotericin B is in patients with chronic kidney disease, post-renal transplant, and those with diabetic nephropathy. The conventional formulation of amphotericin B(AmB) is given as a 1 mg test dose in 30 ml of 5% dextrose intravenously over several hours. If this is well tolerated, the therapeutic dose of 0.5–1 mg/kg body weight is given up to a total dose of 2–4 g attained over 6 weeks. The nephrotoxicity of conventional formulations of amphotericin is reversible and thus requires a regular monitoring of renal function and dose adjustment [18].

15.9.1.1 Lipid and Liposomal Amphotericin B

With the introduction of the newer and safer lipid and liposomal formulations of amphotericin B there has been a significant rise in the survival rates of the disease [20]. The mechanism of action of amphotericin B is based on the binding of the AmB molecule to the fungal cell membrane ergosterol. This produces an aggregate that creates a transmembrane channel, allowing the cytoplasmic contents to leak out, leading to apoptosis. Inj. AmB probably produces renal injury by a variety of mechanisms. Administration of amphotericin B encapsulated in liposomes (L-AMP-B) significantly decreases this toxicity, while retaining and perhaps increasing efficacy. Liposomal encapsulation of AMP-B also appears to enhance its delivery to fungi, infected organs, and to phagocytes, thus accounting for the increased efficacy.

15.9.1.2 Adjunctive Antifungal Agents

Despite several studies on the effect of various antifungals, the role of any other systemic antifungals such as azoles remains debatable. Newer antimycotics like fluconazole, voriconazole, and itraconazole have not been found to have reliable activity against mucormycosis. However, recent studies have shown posaconazole as a reliable drug for step-down and maintenance therapy. Posaconazole is a broad-spectrum azole available in both parenteral and oral formulations. Dosage is 200 mg four times per day or alternatively, posaconazole delayed-release tablets (300 mg every 12 hours on first day, then 300 mg once daily) taken with fatty food [22].

15.9.1.3 Combination Drug Therapy

To date there exists no convincing data or study to support any form of combination antifungal therapy, and on the contrary, combination therapy is not recommended in the established treatment guidelines for mucormycosis [22].

15.9.2 Surgical Debridment

The aim of surgical debridement in the management of mucormycosis is to remove all necrotic tissue when possible and open all sinuses to encourage adequate ventilation and drug delivery of the affected areas. Ideally, debridement should be done until normal healthy mucosa with bleeding is encountered. In the study by the senior author and colleagues, endoscopic sinonasal debridement using microdebrider was performed on all patients who had microscopy-proven disease or any patient suspected of having fungal invasion.

Endoscopic debridement is more advantageous than open surgery as it gives better and magnified visualization of the nasal cavity and sinuses, with equally good surgical control of disease and much less morbidity and disfigurement to patient. Orbital exenteration may also be avoided by timely endoscopic debridement [18]. In the study by the senior author and colleagues orbit-sparing early debridements were done and patients kept on close monitoring until completely disease free, thus avoiding untoward cosmetic disfigurements.

15.9.3 Adjuvant Treatment

Addressing the underlying risk factor is of key importance in the management of mucormycosis. Managing blood glucose levels, iron levels as well as cessation of immune modulating drugs are essential in controlling the disease progression [6]. GM-CSF, granulocyte–macrophage colony-stimulating factor, has shown promising results in patients of ROCM zygomycosis with uncontrolled diabetes when combined with surgical debridement and amphotericin B.

Decongestant nasal drops such as xylometazoline and oxymetazoline also have a role in augmenting the disease process and hence must be timely discontinued.

15.9.4 Supportive Treatment

Adequate saline irrigation of the sinonasal tracts by means of alkaline saline douching, saline nasal drops, or even by the yogic jal neti technique using saline water ensures a healthy and functional sinonasal mucosa.

15.10 PROGNOSIS AND FOLLOW-UP

Mucormycosis even today remains an aggressive fungal infection with guarded prognosis. An epidemic of such a dreaded disease poses a situation of much dilemma for medical science. Thus, a robust reversal of immune compromise status or other underlying high-risk factors apart from the standard treatment protocol can drastically reduce morbidity and improve the mortality rates in mucormycosis. Over and above surgical debridement and amphotericin B therapy, regular follow-up with nasal endoscopy and suction clearance of crusts under cover of a step-down antifungal, until complete clearance of disease is achieved, which can be documented on a CT paranasal sinus, is essential to increasing survival rates.

REFERENCES

1. Furbringer P. Beobachtungen uber lungenmycose beim menschen. *Arch Pathol Anat Physiol Klin Med.* 1876;66330–66365.

2. Yohai RA, Bullock JD, Aziz AA, Markert RJ. Survival factors in rhino-orbital-cerebral mucormycosis. *Surv Ophthalmol*. 1994;39(1):3–22. doi:10.1016/s0039-6257(05)80041-4
3. Bank H, Shibolet S, Gilat T, Altmann G, Heller H. Mucormycosis of head and neck structures. A Case with Survival. *Br Med J*. 1962;1(5280):766–768. doi:10.1136/bmj.1.5280.766
4. Spellberg B, Edwards J, Jr, Ibrahim A. Novel perspectives on mucormycosis: Pathophysiology, presentation, and management. *Clin Microbiol Rev*. 2005;18:556–569.
5. Bhansali A, Bhadada S, Sharma A, Suresh V, Gupta A, Singh P, Chakarbarti A, Dash RJ. Presentation and outcome of rhino-orbital-cerebral mucormycosis in patients with diabetes. *Postgrad Med J*. 2004;80:670–674.
6. Ibrahim AS, Spellberg B, Walsh TJ, Kontoyiannis DP. Pathogenesis of mucormycosis.
7. Ribes JA, Vanover-Sams CL, Baker DJ. Zygomycetes in human disease. *Clin Microbiol Rev*. 2000 Apr;13(2):236–301. doi: 10.1128/CMR.13.2.236. PMID: 10756000; PMCID: PMC100153.
8. Manesh A, Rupali P, Sullivan MO, Mohanraj P, Rupa V, George B, Michael JS. Mucormycosis – A clinicoepidemiological review of cases over 10 years. *Mycoses*. 2019;62:391–398. doi: 10.1111/myc.12897
9. Yadav S, Rawal G. Mucormycosis in Covid-19 – A burgeoning epidemic in the ongoing pandemic
10. Prakash H, Chakrabarti A, Epidemiology of Mucormycosis in India. Microorganisms. 2021;9:523.
11. John, Teny M et al. When uncontrolled diabetes mellitus and severe COVID-19 converge: The perfect storm for mucormycosis. *J Fungi (Basel)*. 15 Apr 2021;7(4):298. doi: 10.3390/jof7040298
12. Artis WM, Fountain JA, Delcher HK, Jones HE. A mechanism of susceptibility to mucormycosis in diabetic ketoacidosis: Transferrin and iron availability. *Diabetes*. 1982 Dec;31(12):1109–1114. doi: 10.2337/diacare.31.12.1109. PMID: 6816646.
13. Radhika S, Swain S, Ray A, Wig, N. COVID-19 associated mucormycosis: An epidemic within a pandemic. *QJM: An International Journal of Medicine*. 2021. doi:10.1093/qjmed/hcab165/6295692.
14. Kermani W, et al. ENT mucormycosis. Report of 4 cases. *European Annals of Otorhinolaryngology, Head and Neck diseases*. 2015. http://dx.doi.org/10.1016/j.anorl.2015.08.027
15. Bouchara JP, Oumeziane NA, Lissitzky JC, Larcher G, Tronchin G, Chabasse D. Attachment of spores of the human pathogenic fungus Rhizopus oryzae to extracellular matrix components. *Eur J Cell Biol*. 1996;70(1):76–83.
16. Skiada A, Pavleas I, Drogari-Apiranthitou M. Epidemiology and diagnosis of mucormycosis: An update. *J Fungi (Basel)*. 2 Nov 2020;6(4):265. doi: 10.3390/jof6040265
17. Corzo-León DE, Chora-Hernández LD, Rodríguez-Zulueta AP, Walsh TJ. Diabetes mellitus as the major risk factor for mucormycosis in Mexico: Epidemiology, diagnosis, and outcomes of reported cases. *Med Mycol*. 2018;56(1):29–43. doi: 10.1093/mmy/myx017
18. Singh I, Gupta V, Gupta SK, Sunil G, Kumar M, Singh A. Our experience in endoscopic management of mucormycosis: A case series and review of literature. *International Journal of Otorhinolaryngology and Head and Neck Surgery*. 2017;3:465. 10.18203/issn.2454-5929.ijohns20171217.
19. Chamilos G, Lewis RE, Kontoyiannis DP. Delaying amphotericin B-based frontline therapy significantly increases mortality among patients with hematologic malignancy who have zygomycosis. *Clin Infect Dis*. 2008;47(4):503–509. doi: 10.1086/590004
20. Spellberg B, Ibrahim AS. Recent advances in the treatment of mucormycosis. *Curr Infect Dis Rep*. 2010;12(6):423–429.
21. Laniado-Laborín R, Cabrales-Vargas MN. Amphotericin B: Side effects and toxicity. *Rev Iberoam Micol*. 2009;26(4):223–227. doi: 10.1016/j.riam.2009.06.003
22. Cornely OA al: ESCMID and ECMM joint clinical guidelines for the diagnosis and management of mucormycosis 2013. *Clin Microbiol Infect*. 2014; 20(S3):5.
23. Julia B. Garcia-Diaz, Leonardo Palau, George A. Pankey, Resolution of rhinocerebral zygomycosis associated with adjuvant administration of granulocyte-macrophage colony-stimulating factor, *Clin Infect Dis.*, 15 June 2001;32(12):e166–e170. https://doi.org/10.1086/320767

16 Role of DRDO in Covid-19 Containment

K. P. Mishra and A. K. Singh
Defence Research and Development Organization (DRDO)-HQ, New Delhi, India

CONTENTS

DOI: 10.1201/9781003358909-16

16.1 INTRODUCTION

The onset of COVID-19 due to SARS-CoV-2 encouraged the development of innovative products and rapid technological advancements. The Defence Research and Development Organization (DRDO), India in preparedness to fight against the coronavirus disease (COVID-19) came up with a variety of products such as sanitizers and disinfectant devices, personal protective gears, COVID testing facilities, medical oxygen plants ventilators, shelters, makeshift COVID hospitals, and mobile apps to help overcome the ongoing pandemic. This chapter describes critical equipment and technologies developed by DRDO for management of COVID-19 pandemic.

16.2 SARS-COV-2 AND COVID-19

Human beings suffer from two types of diseases: i) non-communicable, which are not transferred from one person to another such as heart disease, diabetes, and neurological disorders; and ii) communicable, which can spread from one person to another. Communicable diseases are also known as infectious diseases caused by pathogenic microorganisms. The disease-causing pathogens are broadly classified into five main types: viruses, bacteria, fungi, protozoa, and worms. The whole world is presently suffering with the human coronavirus called Severe Acute Respiratory Syndrome Coronavirus-2 (SARS-CoV-2) causing a disease known as COVID-19.

What exactly are viruses? Viruses are completely different from bacteria. A bacterium is a living thing; most of them have all of the components required for their own survival and reproduction. Viruses are not really living organisms as they can not carry out any of the functions on their own that we consider to be connected with life. They do not have the ability to replicate themselves without being inside of a living cell. The virus contains instructions that tells a cell to make more copy of the virus itself (Andersen et al., 2020).

Coronaviruses are a family of viruses that can infect both humans and animals. In humans, so far seven coronaviruses are known to cause respiratory infections ranging from the common cold to more severe diseases such as Severe Acute Respiratory Syndrome (SARS), which circulated in China in 2003, and Middle East Respiratory Syndrome (MERS), which circulated in the Middle East (Saudi Arabia) in 2012. The most recently discovered coronavirus causes COVID-19, which was declared a pandemic by the WHO on March 11, 2020.

Bats are considered natural hosts of these viruses, although several other species of animals are also known to act as sources. For instance, MERS-CoV is transmitted to humans from camels, and SARS-CoV is transmitted to humans from civet cats. The closest relative of SARS-CoV-2 among human coronaviruses is SARS-CoV, with 79% genetic similarity. However, among all known coronavirus sequences, SARS-CoV-2 is most similar to bat coronavirus RaTG13, with ~96% similarity, and coronavirus sequences in the pangolin (a scaly anteater mammal) with ~91.02 % similarity (Andersen et al., 2020).

SARS-CoV-2 is a single-stranded RNA virus that has about 29903 RNA letters (nucleotide sequences), which make about 14 genes that make ~27 proteins. It includes non-structural proteins such as RNA-dependent RNA polymerase or RdRp that make more RNA copy of the virus – a target of remdesivir, the most promising drug for COVID-19. Structural proteins include spike (S), envelop (E), membrane protein (M), and nucleocapsid (N) proteins. The S proteins make crown-like spikes and give coronaviruses their name. Others are accessory proteins (Gallaher, 2020; Yan et al., 2020).

The virus spreads from person to person (human-to-human transmission), and seems to be transmitted mainly via small respiratory droplets through sneezing, coughing, or when people interact with each other for some time in close proximity which is usually less than a metre. When these droplets are inhaled, or they land on the surfaces that other people may come in contact with, the people can then get infected when they touch their nose, mouth, or eyes. The virus can survive on different surfaces for several hours (copper, cardboard) up to a few days (plastic and stainless steel). However, the amount of viable virus diminishes over time and may not always be in enough numbers to induce infection. The incubation period for COVID-19 (i.e., the time between exposure to the virus and onset of symptoms) is between 1–14 days.

Transmission of the virus can occur from symptomatic and presymptomatic people. Later, it was found that asymptomatic people also spread the virus by shedding it through sneezing, coughing, spitting, etc. Public toilets could also be a main source of virus transmission as many studies have reported the presence of coronavirus in stool and saliva.

SARS-CoV-2 infects upper and lower respiratory tract cells, where it binds to the host cell receptor known as ACE2 (Lukassen et al., 2020).

These receptors are mainly expressed on type 2 pneumocytes in lungs, intestine, and other organs. The moment the virus gets entry into the host cell, it starts replicating using host cell machinary and increases its numbers. When the host cell identifies the entry of intruders by recognizing its ssRNA, it triggers Type I interferon production to protect from viral infection. In a healthier immune surveillance mechanism, after recognition of viral particles, immediately Type I interferon is secreted and virus replication inhibited. However, delayed IFN-I production may lead to further recruitment of inflammatory cells such as monocytes, macrophages, and neutrophils. These cells secrete a huge number of proinflammatory cytokines known as a cytokine storm that damages the lung alveoli causing SARS. The cytokine storm is also responsible for sepsis, which can damage multiple organ systems (Lokugamage et al., 2020).

The immune system of young individuals has an excellent surveillance mechanism and as soon it recognizes COVID-19 virus entry inside the cell the IFN-I signaling cascade activates and neutralizes the virus before it evades the immune system. However, in older individuals due to immune senescence and comorbid conditions the innate immune response is not as good at recognizing the virus and inducing the interferon response. The delay in interferon response leads to the recruitment of inflammatory cells that occurs in the lung that secretes cytokines like TNF-alpha, IL-1 beta, IL-6, and IL-8, which further damages the tissues of the lungs (Pedersen and Ho, 2020; Hirano and Murakami, 2020). The exaggerated activation of the immune system in elderly individuals is one of the causes of high mortality. Some clinical trials are being done with anti-sepsis drugs in COVID-19 patients.

When the first line of host defence is defeated with the virus, simultaneously the second line of defence also known as adaptive immune response comes into action. It involves T and B lymphocytes. B cells make IgM and IgG antibodies specifically to neutralize the virus. This reaction takes a little time to develop. In the case of COVID-19, antibody responses to SARS-CoV-2 are generated within 19 days after symptom onset. Seroconversion for IgM and IgG occurs simultaneously or sequentially. A low titer of antibodies is also responsible for high mortality rate among elderly individuals.

Various types of vaccines were developed to fight with COVID-19. These are the i) Nucleic acid vaccine (mRNA/DNA)-based; ii) Viral vectors-based; iii) Protein-based; and iv) Attenuated whole virus vaccine. The potential vaccines being used for immunizations are the BioNTech-Pfizer (produced by German biotech firm BioNTech and US pharmaceutical firm Pfizer), Moderna (American biotech company Moderna), Oxford-AstraZeneca (Swedish-British drugmaker in partnership with the University of Oxford) Sputnik V Vaccine (Russia), COVAXIN (Bharat Biotech,

India) etc. The diet also plays an important role in enhancing immunity since a balanced diet with high fiber content strengthens gut microbes that strengthen immunity. Sleep is also a proven immune booster. Yoga and moderate exercise are also suggested for good immunity. A strong immune system can defeat the coronavirus very easily.

16.3 ROLE OF DRDO IN CONTAINING THE SPREAD OF COVID-19

The DRDO has been tracking the spread of Coronavirus (COVID-19) since the WHO announced it as a public health emergency. The DRDO started the first week of March 2020 to enhance efforts to create countermeasures to stop the spread of the disease in India. By that time, there were more than 30 affected individuals in India. It also started focusing on creating mass supply solutions for critical medical needs. As a result of this focused approach, the DRDO developed several items that are deployed in the 'war against corona', as follows.

16.4 HAND AND SURFACE SANITIZER

To address the need for WHO-compliant and -certified hand sanitization solution for personal and surface decontamination, DRDO laboratories prepared hand sanitizer compliant with WHO guidelines for local production. More than 150,000 bottles of sanitizer-based on isopropyl alcohol/ethanol have been produced in-house and supplied to various government departments (Figure 16.1).

16.5 BODY SUITS AND PERSONAL PROTECTIVE EQUIPMENT (PPE)

The Personal Protective Equipment (PPE) having a specific type of fabric with coating was developed by the DRDO scientists to keep medical, paramedical, and other personnel engaged in combating COVID-19 safe from the deadly virus. The Defence Research and Development Establishment (DRDE), Gwalior developed Bio Suits, which were produced by three industry partners, namely M/s Shiva Texyarn, Coimbatore; M/s Arvind Mills, Ahmedabad and M/s Aeronav, Noida. Biosuits

FIGURE 16.1 Hand sanitizer developed by DRDO.

FIGURE 16.2 Personal protective equipment (PPE) MK-III.

were also developed by Aerial Delivery Research and Development Establishment (ADRDE), Agra, and Institute of Nuclear Medicine and Allied Sciences (INMAS), Delhi to protect medical professionals handling COVID-19 patients. The suits were subjected to rigorous testing for textile parameters and protection against synthetic blood. Efforts were made to ramp up production to 15–20 thousand PPEs per day during COVID pandemic. A special sealant was prepared by DRDO as an alternative to seam sealing tape based on the sealant that is used in submarine applications. Bio-suits prepared using this glue for seam sealing by an industry partner cleared testing at Southern India Textile Research Association (SITRA) Coimbatore. This may be a game changer for the textile industry and this glue can be produced in bulk through industry to support the seam sealing activity by suit manufacturers (Figure 16.2).

16.5.1 Face Masks

The DRDE, Gwalior developed five layered N-99 masks using a nanoweb filter layer. The masks were manufactured with an aim to produce two lakh N-99 masks per week through the industry during COVID pandemic (Figure 16.3). The INMAS, Delhi also designed a three-ply surgical mask and supplied it to various police departments for their protection from infection.

Various types of masks are being used including three-ply surgical masks, N-95, and N-99. According to the Center for Disease Control (CDC), USA, coronaviruses can be transmitted from asymptomatic people (who lack symptoms) as well as pre-symptomatic people (who eventually develop symptoms) in addition to symptomatic people. This means people working in close proximity and carrying the coronavirus but not exhibiting symptoms can spread it through speaking, coughing, or sneezing. Thus, the CDC recommends wearing cloth face coverings in public settings where other social distancing measures are difficult to maintain, especially in areas with significant community-based transmission. The cloth face coverings recommended for public use, however, are surgical masks or N-95; N99 respirators are recommended for healthcare workers and other medical first responders. Face cloth coverings should be routinely washed depending on the frequency of use. One can use a washing machine for properly washing a face covering. While removing the face covering, the nose, eyes, and mouth should not be touched and hands should be washed immediately.

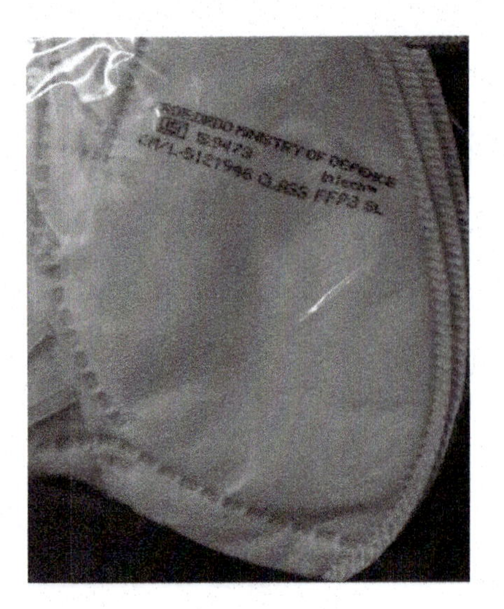

FIGURE 16.3 N-99 respirator.

16.5.2 N95 and N99 Masks

Healthcare workers taking care of coronavirus patients must wear N95 or N99 masks. These masks were designed to fit tightly around the nose and mouth and if worn properly block out at least 95% and 99% of small airborne particles (0.3 micron), respectively. Although N95 and N99 masks are available for the public to purchase, there is no recommendation from health agencies for the general public to wear them.

16.5.3 How to Wear and Remove N95 or N99 Mask (Donning & Doffing)

There is a proper way to wear masks. First, hold the mask in your palm and place it on your face covering nose and mouth. Pull the bottom strap and put it behind your head below the ears; then take the upper strap and put it behind towards the crown of your head. To obtain a tight seal, the nose piece of the respirator is molded over the nose.

While donning N95 or N99 masks make sure that you are not touching the front of the respirator. To remove your mask, tilt your head forward and use two hands to grab the bottom strap, pull to the sides, then over your head. Next, use both hands to grab the upper strap, pull to the sides, then over your head. Keep tension on the upper strap as you remove it, which will let the mask fall forward. Dispose of the mask.

16.5.4 Surgical Masks

Three-ply surgical masks are made up of a meltblown polymer, polypropylene, that is placed between the outer and inner layer of nonwoven fabric – a material made from short staple fibre and long fibres that are bonded together by mechanical, chemical, heat, or solvent treatment. The middle melt-blown polymer acts as the filter to prevent pathogens from entering or exiting the mask. Pleats are used to expand the mask so that it covers the area from the nose to the chin. Masks are fastened to the head using elastic straps or ear loops. These are loose-fitting disposable masks and help stop large droplets from being spread by the person who is wearing it. Wearing surgical masks may provide some degree of protection. These masks should not be used for a longer duration. The inside layer will only stay dry for a short time as the inner lining is eventually going to absorb water, sweat, and spit and becomes ineffective once damp. Therefore, they are not to be used for long periods.

16.5.5 Protective Face Shields

The Research Centre Imarat (RCI), Hyderabad and Terminal Ballistics Research Laboratory (TBRL), Chandigarh developed a face protection mask for healthcare professionals handling COVID-19 patients. Its lightweight construction makes it convenient for comfortable wear for long time periods. For face protection, this design makes use of readily available Over-Head Projection (OHP) film in A4 size.

16.5.6 Quality of Face Masks

For face masks the following standards may be applicable.

16.5.6.1 ASTM 2100

This specification covers the classifications, performance requirements, and test methods for the materials used in the construction of medical face masks that are used in healthcare services such as surgery and patient care.

16.5.6.2 EN 14683

This European Standard specifies construction and performance requirements and test methods for surgical masks intended to limit the transmission of infective agents from staff to patients.

16.5.6.3 NIOSH

Surgical N95 – A NIOSH-approved N95 respirator – has also been cleared by the Food and Drug Administration (FDA) as a surgical mask. N95 masks filter at least 95% of airborne particles.

16.5.6.4 ISO 10993

This standard certifies the test methods of biocompatibility for face masks. Classification of medical devices depends on the nature and duration of contact with the patient.

16.5.7 Masks as Fashion During COVID-19

Masks seem to have become a fashion statement around the world, and are now one of the new essentials of everyday life. Several online boutiques now sell quality fabric masks by famous designers. People are also looking for more aesthetic versions of masks for daily wear. Masks may remain the newest fashion accessory in the post-COVID era.

16.5.8 Disposal

During the COVID-19 crisis, discarded masks have been found on streets and other public places and have become a big environmental hazard adding a further source of virus transmission. People should be aware of the impact of improperly disposing of masks or any biowaste (e.g., gloves, PPEs) to prevent further transmission of the coronavirus and other infectious diseases.

People should avoid throwing used masks and gloves on the streets and should not mix disposed masks and gloves with dry and wet waste. If they are thrown on the roads, they may serve as the potential source of virus transmission. Keep discarded masks separately in a biohazard labelled bin and dispose of them in a safe manner. For proper disposal, used masks should be put in separate containers and transported to incineration plants.

16.5.9 Testing and Certification of Samples of PPE

The INMAS, Delhi has been authorized to perform laboratory testing of PPE submitted by prospective manufacturers in India. A laboratory test called the Synthetic Blood Penetration Resistance Test is conducted and a Test Report is issued for the same by INMAS.

16.6 UV-BASED SANITIZATION EQUIPMENT

16.6.1 Ultra Swachh for Disinfection of PPE and Other Materials

The INMAS-DRDO developed a disinfection unit called Ultra Swachh to disinfect a wide range of materials, PPEs, electronics items, fabrics, etc. The system uses an advanced oxidative process comprising a multiple barrier disruption approach using ozonated space technology for disinfection. The system is double layered with specialized ozone sealant technology assuring trapping of ozone for complete disinfection. It also has catalytic converter to ensure environmentally friendly exhaust (i.e., only oxygen and water). The system is in compliance with international standards for industrial, occupational, personal, and environmental safety. The Ultra Swachh comes in two variants: Ozonated Space and Trinetra Technology. Trinetra technology, comprised of ozonated space and radical dispenser. Treatment is optimized with automation for quick disinfection cycle. The system operates on 15 ampere, 220 volts, 50 hertz power supply. The system provides various safety features such as emergency shutdown, door interlocks, dual door, delay cycle, and leak monitors, etc., to ensure safe operations for long durations. The dimensions of the industrial cabinet are 7'x4'x3.25' to disinfect large quantities at a time. Cabinets of different sizes re available for industry (Figure 16.4).

FIGURE 16.4 Ultra Swachh for disinfection of PPEs and other materials.

FIGURE 16.5 Automated systems for decontamination of N95 face masks.

16.6.2 Automated System for Decontamination of N95 Face Masks

DEBEL, Bengaluru developed an automated system for decontamination of face masks. This system works on the principle of 'ultraviolet (UV-C) germicidal irradiation for killing the bacteria' and virus. UV-C radiation is one of the technologies recommended by Centers for Disease Control and Prevention vide a report on 'Decontamination and Reuse of Filtering Face piece Respirators' (Figure 16.5).

16.6.3 UV Sanitization Box and Handheld UV Device

The DIPAS and INMAS, both DRDO laboratories in Delhi, designed and developed an ultraviolet C light-based sanitization box (Figure 16.6) and a handheld UV-C (ultraviolet light with wavelength 254 nanometres) device (Figure 16.7). The UV-C has a shorter and more energetic wavelength of light.

The UV-C is particularly good at destroying genetic material of SARS-CoV-2. The handheld device disinfects office and household objects like chairs, files, mail, and food packages. These measures can reduce the transmission of the coronavirus.

16.7 OTHER SANITIZATION EQUIPMENT AND SYSTEMS

16.7.1 Germiklean-Dry Heat Sanitization System

In the Covid-19 pandemic, prevention through sanitization is the most effective way to remain safe. With this in mind, Germiklean developed a product to sanitize uniforms, canes, polycarbonate shields, files, papers, metals, ceramic items, etc. Germiklean is a dry heat-based system that uses hot air at 70°C for 10 minutes to sanitize items placed inside.

The systems consist of a mild steel powder coated box with a working chamber made up of stainless steel (Grade 304) and a heating chamber (Figure 16.8).

16.7.2 Evaporating Fogger For Sanitization

An evaporator type fogging device was developed by the DIPAS for sanitization of workspaces and associated accessories. The fogger can be used with any type of chemical suitable for particular disinfection purposes (Figure 16.9).

FIGURE 16.6 UV-based sanitization box.

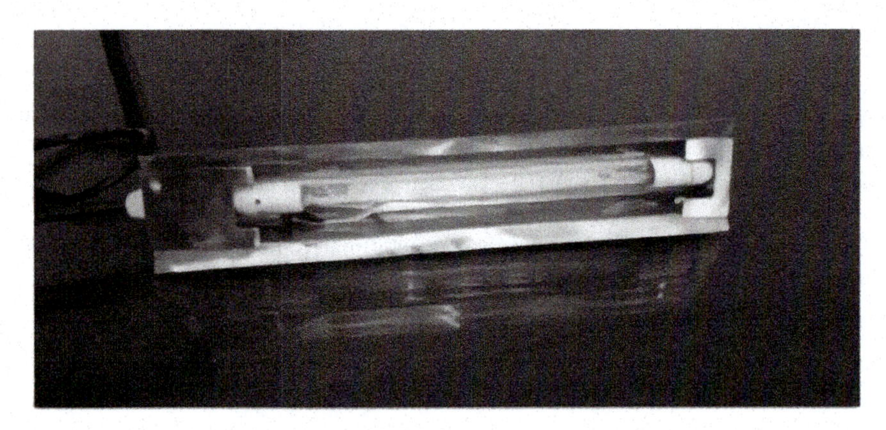

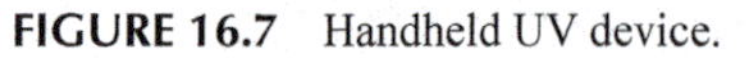

FIGURE 16.7 Handheld UV device.

16.7.3 Vehicle Sanitization Enclosure

The Vehicles Research Development Establishment (VRDE)-DRDO developed a vehicle sanitization enclosure conceptualized using locally available material based on a four-man tent. Since it is a very lightweight system with a portable canopy, it can be made operational in less than 3 hours. The tent canopy through which the vehicles are passed is filled with a disinfecting mist created by an electrically operated positive displacement pump. A separate tank with 500-liter capacity for storage of disinfectant is used that requires refilling after 200 vehicles are disinfected. The system is noise free and requires a 10 minute break after every 4 hours of operation. The system can be utilized at any location including entry location for sanitization of vehicles. Hospitals and administrative offices having high ingress and egress can deploy this system (Figure 16.10).

FIGURE 16.8 Germiclean.

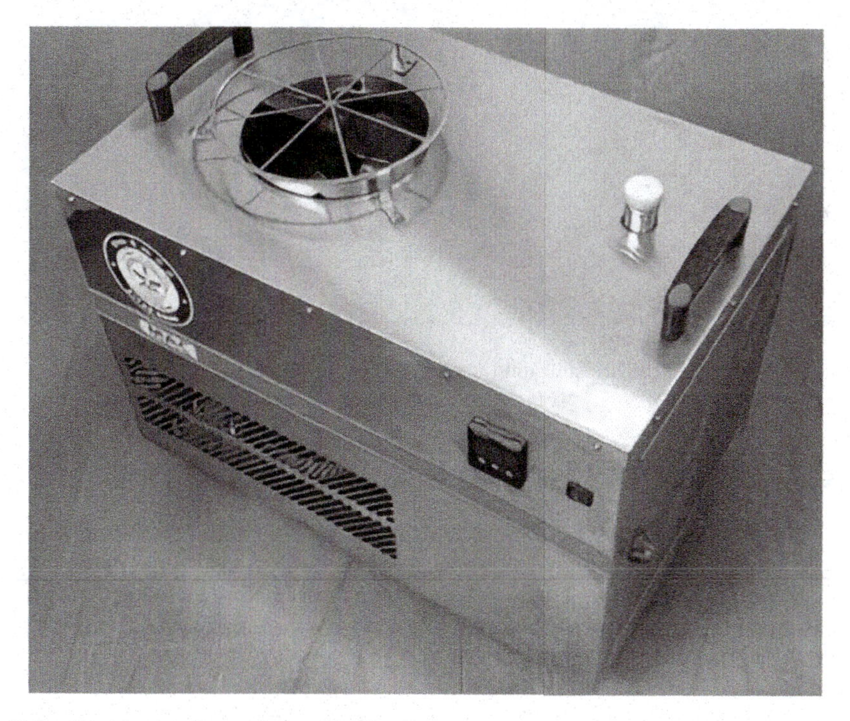

FIGURE 16.9 Evaporating fogger for sanitization.

FIGURE 16.10 Vehicle sanitization enclosure.

FIGURE 16.11 Fog-based vehicle sanitization system.

16.7.4 Vehicle Sanitization System (Fog Based)

Another vehicle sanitization enclosure has been developed comprising a fog generator and a flexible hose with adapter. The fog generator generates fog of 1 to 5 μm aerosols. The fog generator is operated for 30 s to 60 s depending on the volume inside the vehicle. The fog generator is operated using a foot switch. It takes 3–4 seconds for the fog to start accumulating in the vehicle and within 30 s the vehicle is filled with the fog (Figure 16.11).

16.7.5 Shoe and Driveway Sanitizer

Shoes have high potential for spreading the virus from one place to another. Similarly, car tires are also potentially highly infective. The INMAS, Delhi innovated a solution to prevent the spread of the coronavirus through shoes and car tires. This solution is based on a PVC mat with threading to retain moisture. Artificial grass is the second choice. However, coir and rubber mats are not suitable. A sodium hypochlorite gel of 150 ppm is used as the disinfectant as it has advantages of enhanced stability, better moisture retention, and non-chlorine composition. This leaves a minimum footprint and can be wiped off easily. 200 ppm solution can be used for cars.

16.7.6 Mobile Area Sanitization System

Based on the experience of dust suppression systems for use in deserts, the Defence Laboratory, Jodhpur (DLJ-DRDO) joined the fight against Covid-19 by developing a 'Mobile Area Sanitization System' that uses sodium hypochlorite solution to sanitize large areas. Two variants have been developed: one is for outdoor use that is mounted on a 'B' class vehicle and the other one is mounted on a battery operated cart for indoor use. The former can spray to a distance of 6–7 m and the latter to 2–3 m. This will ensure safety and hygiene of large areas.

16.7.7 Personnel Sanitization Enclosure (PSE)

The Vehicle Research Development Establishment (VRDE), Ahmednagar, a DRDO laboratory, designed a full-body disinfection chamber called a PSE. One at a time personnel decontamination is intended for this walkthrough enclosure. It is a portable system with a soap and hand sanitizer dispenser. At the entrance, a foot pedal is used to initiate the decontamination. An electrically powered pump produces a hypo sodium chloride disinfection mist as soon as the person enters the chamber. The mist spray is calibrated for a 25-second operation and shuts off automatically when that time has passed. According to protocol, everybody being disinfected within the chamber must keep their eyes closed.

16.7.8 Automatic Mist-Based Sanitizer Dispensing Unit

The Centre for Fire Explosive & Environment Safety (CFEES), Delhi along with HPO1, using its expertise in mist technology for fire suppression, developed an automatic mist-based sanitizer dispensing unit. The device is activated by an ultrasonic sensor and operates without making physical contact. A single fluid nozzle with low flow rate is used to generate aerated mist to dispense the hand sanitizer. This sanitizes the hands with minimum wastage. Using an atomizer, only 5–6 ml sanitizer is released for 12 seconds in one operation that gives a full cone spray over both palms so that disinfection operation of hands is complete (Figure 16.12).

16.8 COVID-19 SAMPLE TESTING

At the DRDO, six out of eight life science cluster laboratories are involved in COVID-19 testing using real-time PCR machine in addition to developing other anti-COVID-19 measures. DRDE, Gwalior has been testing samples since March 2020. The Delhi-based laboratories DIPAS and INMAS have testing facilities and clinical samples are being tested in the newly created BSL-2 facility. DFRL, Mysuru deployed its design patented Mobile Containment (BSL-3) Laboratory, PARAKH, to Mysore Medical College and Research Institute. It is also helped in the fabrication of three more units of PARAKH at IIT Delhi, Guwahati and Chennai in collaboration with DBT-BIRAC. In the northeast region, DRL, Tezpur established a COVID-19 testing facility. Apart from that DIBER, Haldwani joined hands with Haldwani Medical College and ran a COVID-19 testing facilty by providing human resources and RT-PCR machines. In Leh (Ladakh) DIHAR established a COVID-19 testing facilty to augment testing capacity of the region.

FIGURE 16.12 Automatic mist-based sanitizer dispensing unit.

16.9 VENTILATORS

Patients with Covid-19 may require ventilators because they tend to develop acute respiratory distress syndrome (ARDS). The Society for Biomedical Technology (SBMT), a DRDO funded and managed initiative, and Defence Bioengineering and Electromedical Laboratory (DEBEL), Bengaluru, developed a ventilator by using existing technologies like breath regulators, pressure/flow sensors, etc. DEBEL developed the critical components of the ventilator, which are produced with the help of local industry. Defence PSU, M/s BEL joined the efforts for large-scale production of ventilators. Production can reach a capacity of 10,000 ventilators per month (Figure 16.13). The DRDO also developed the Multi-Patient Ventilation (MPV) Kit, which helps convert a single ventilator to allow providing treatment to multiple patients in an emergency. The MPV Kit has been tested and works satisfactorily.

16.10 MEDICAL OXYGEN PLANT (MOP)

The Medical Oxygen Plant (MOP) is a technology that is an offshoot of the On-Board Oxygen Generation System (OBOGS) project for medical-grade oxygen generation onboard Tejas fighter aircraft. It utilizes the pressure swing adsorption (PSA) technique and molecular sieve technology to generate oxygen directly from atmospheric air. The oxygen generator components were developed by the DEBEL and the technology has been transferred to industry (Figure 16.14). This plant can provide oxygen supply to hospitals in urban and rural areas. Installation of MOP helps to avoid hospitals' dependency on scarce oxygen cylinders, especially in high altitude and inaccessible remote areas. Its benefits include reduced logistics of transporting cylinders to these areas, low cost, and continuous and reliable oxygen supply available around the clock. The facility can be used for filling cylinders in addition to direct installations at hospitals.

16.11 COVID SAMPLE COLLECTION KIOSK (COVSACK)

The DRDL, Hyderabad developed a COVID Sample Collection Kiosk (COVSACK) in consultation with the doctors of Employees' State Insurance Corporation (ESIC), Hyderabad. Healthcare

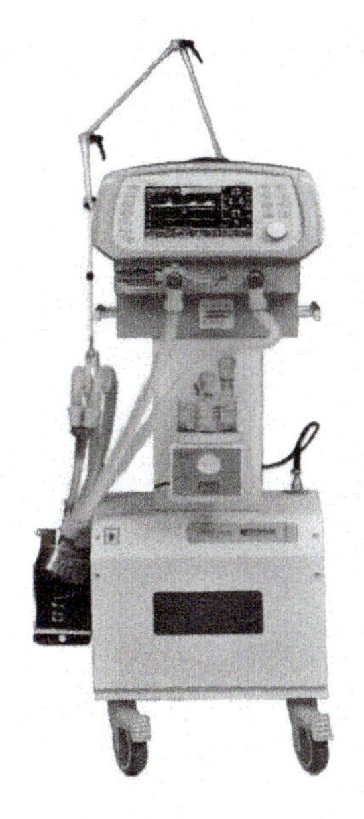

FIGURE 16.13 Ventilator.

FIGURE 16.14 Medical oxygen plant.

professionals can use the COVSACK, a kiosk, to collect COVID-19 samples from patients who may be infected. The patient being tested walks into the kiosk and a nasal or oral swab is taken by a healthcare professional from the outside through the built-in gloves (Figure 16.15).

Healthcare professionals can use the COVSACK, a kiosk, to collect COVID-19 samples from patients who may be infected. The shielding screen of the kiosk cabin protects the healthcare worker from aerosol/droplet transmission while taking the sample. This reduces the requirements of PPE change by healthcare workers.

16.12 MOBILE BSL3 VIRAL RESEARCH AND DIAGNOSTIC LABORATORY (VRDL) LAB

The Mobile BSL3 VRDL lab was developed by the DRDO, Ministry of Defence, Govt. of India in collaboration with ESIC Medical College & Hospital, Hyderabad with due permission of Indian Council of Medical Research (ICMR) and Govt. of Telangana. This bio-safety level-2 and level-3 lab was set up in a record time of 15 days, compared to the usual six months needed. This testing

FIGURE 16.15 COVID-19 sample collection kiosk (COVSACK).

facility can process more than 1,000 samples in a day and enhances the country's capabilities to fight COVID-19.

16.13 DIFFERENT TYPES OF INFLATABLE SHELTERS

16.13.1 Negative Pressure Inflatable Isolation Shelter for Ten Occupants

A negative pressure shelter intended for isolating and treating patients without the risk of spreading the contamination to others was developed by DEBEL-DRDO. This system is suitable for isolating COVID-19 patients since the system is based on negative pressure and the materials used have passed the synthetic blood penetration test. The system consists of negative pressure-based air handling unit and inflatable multiple chambers. The system has five air sterilizer units and ducts are uniformly distributed to provide filtered air (Figure 16.16). It covers a total area of approximately 1,000 sq ft and is manufactured out of two-layer water- and airproof fabric. The structure has separate rooms for decontamination and medical waste and is equipped with modular restrooms and is illuminated with sufficient light throughout.

16.13.2 Anywhere Erectable Isolation Shelters

The Research & Development Establishment (Engrs.) (R&DE) (E), Pune developed various shelters, green power sources, chemical toilets, and quick erectable medical complexes under various projects related to field defence and CBRN protection. These products with minor modifications/customization can be utilized for activities pertaining to containment of COVID-19. These products are useful especially in remote locations, where there is no medical/electrical facility available (Figure 16.17).

FIGURE 16.16 Negative pressure inflatable isolation shelter for ten occupants.

FIGURE 16.17 Anywhere erectable isolation shelters.

16.14 METRICS

The Institute for System Studies and Analysis, Delhi developed METRICS (Mathematical Estimation for Tracking Infections of COVID-19 Spread in India) and generates a daily estimation report based on data available.

16.15 CONCLUSION

The efforts made by the DRDO in bringing these technologies together to fight against COVID-19 has helped limit the spread of the coronavirus and significantly helped to reduce the death toll. These products were developed in a very short time span and used technologies from various industries. Many industries that develop components and sub-components for these primary industries also take part. About 200 industries play in a role in producing the equipment and products discussed here combating COVID-19.

REFERENCES

Andersen KG, Rambaut A, Lipkin WI, Holmes EC and Garry RF (2020). The proximal origin of SARS-CoV-2. *Nat Med* 26: 450–452.

Gallaher WR (2020). A palindromic RNA sequence as a common breakpoint contributor to copy-choice recombination in SARS-COV-2. *Arch Virol* 165: 2341–2348.
Hirano T and Murakami M (2020). COVID-19: A new virus, but a familiar receptor and cytokine release syndrome. *Immunity* 52: 731–733.
Lokugamage KG, Adam H, Craig S, Ricardo R and Menachery VD (2020). SARS-CoV-2 is sensitive to type I interferon pretreatment. *J Virol* 94: e01410–01420.
Lukassen S, Chua RL, Trefzer T, Kahn NC, Schneider MA, Muley T, Winter H, Meister M, Veith C, Boots AW, Hennig BP, Kreuter M, Conrad C and Eils R (2020). SARS-CoV-2 receptor ACE2 and TMPRSS2 are primarily expressed in bronchial transient secretory cells. *EMBO J* 39: e105114.
Pedersen SF and Ho YC (2020). SARS-CoV-2: a storm is raging. *J Clin Invest* 130: 2202–2205.
Wu F et al. (2020). A new coronavirus associated with human respiratory disease in China. *Nature* 579: 265–269.
Yan R, Zhang Y, Li Y, Xia L, Guo Y and Zhou Q (2020). Structural basis for the recognition of SARS-CoV-2 by full-length human ACE2. *Science* 367: 1444–1448.

17 Food and Nutrition Security as Preventive and Therapeutic Health Care

Now and Beyond the Covid-19 Pandemic

Geeta Mehra[1] *and Rajneesh Mehra*[2]

[1]Department of Food Science, Mehr Chand Mahajan DAV College for Women, Chandigarh, India

[2]Amity Global Business School, Mohali, Chandigarh, India

CONTENTS

17.1 SARS-COV-2

Coronavirus disease 2019 (COVID-19) is a new disease caused by a novel coronavirus that has not previously been seen in humans. Coronaviruses are a group of related RNA viruses that cause diseases in mammals and birds. In humans and birds, they cause respiratory tract infections that can range from mild to lethal. The virus is thought to be natural and of animal origin, through spill-over infection.

Amongst various measures suggested by experts for prevention against the SARS-VoV-2 infection is maintenance of a good diet and exercise routine in order to build immunity. Food and nutrition play a major role in providing the capability for resistance against many diseases, including SARS-CoV-2. Ensuring food safety and food security can help fight against acquiring infections.

DOI: 10.1201/9781003358909-17

17.2 FOOD AS SUSTAINABLE PREVENTIVE HEALTHCARE THERAPY

Food is one of the basic human needs. It fulfils social, psychological, and physiological requirements. It plays nourishment, physiological, psychological, social, and therapeutic roles. Good nutrition supports immune systems to fight pathogens and helps to avoid chronic inflammation following an infection.

Food supplements are used for treatment and prevention of cardiovascular diseases, obesity, diabetes, gastrointestinal diseases, cancer, diarrhoea, ulcers, allergy, osteoporosis, night blindness, beriberi, goitre, anaemia, rickets, and scurvy to name a few.

17.2.1 Learning from the COVID-19 Pandemic

Supplements can help increase immunity, which in turn can help to fight SARS-CoV-2 viral infection as well. For instance:

- Vitamin D offers protection against viral respiratory infections.
- Vitamin C has important anti-inflammatory, immunomodulating, antioxidant, antithrombotic, and antiviral properties.
- Vitamin E has antioxidant properties, protecting the body from oxidative damage. It also has the potential to influence both innate and adaptive immunity.
- Zinc boosts antiviral immunity and curbs inflammation.
- Herbal tea/decoction (kadha) made from basil, cinnamon, black pepper, dry ginger, and raisins has been suggested in order to build immunity using herbs available in every Indian kitchen.

Management of COVID-19 is mainly supportive as to date there is no effective antiviral treatment. Vitamin C levels in serum and leukocytes are depleted during the acute stage of infection owing to increased metabolic demands. High-dose vitamin C supplement helps to normalise both serum and leukocytes vitamin C levels (Abobaker et al., 2020).

Further, the Ministry of Ayush, Govt. of India released immunity boosting, self-care measures based on Indian traditional wisdom and Ayurvedic system of treatment as preventive therapy. The recommendations are summarized in Table 17.1.

17.2.2 Food Supplements as Cure

SARS-CoV-2 is widely transmitted through direct or indirect contact. At present, there is no specific treatment against this virus. Only prevention and supportive strategy can mitigate the ill effects. Extensive research on SARS-CoV-2 has focused on transmission, symptoms, structure, and its structural proteins as that can facilitate identification of potential inhibitors. The quick identification of potential inhibitors and immune-boosting functional food ingredients is crucial to combat this pandemic disease **Invalid source specified**. A well-balanced diet keeps the body healthy, strengthens the immune system, and reduces the risk of chronic disease. It also helps in managing SARS-CoV-2 infection **Invalid source specified**. Studies have emphasized the fact that a balanced diet that contains sufficient amounts of macronutrients and diverse micronutrients is necessary for the immune system to fight against any infection, including COVID-19 **Invalid source specified.**

WHO guidelines suggest that in order to maintain a strong immune system, one should eat fresh and unprocessed foods, stay hydrated, and avoid foods that can cause obesity, diabetes, cardiovascular diseases, and cancer **Invalid source specified.** A proper diet can ensure that the body is in a proper state to defeat the virus. However, along with dietary management guidelines food safety management and good food practices are compulsory **Invalid source specified**.

TABLE 17.1
Immunity Boosting, Self-Care Measures

COVID-19 Crisis
Summary of Ministry of AYUSH Recommendations on Immunity Boosting Measures for Self-Care

Measures for Enhancing Immunity	Ayurvedic Immunity Enhancing Tips	Simple Ayurvedic Procedures	Actions During Dry Cough/ Sore Throat
✓ Drink warm water throughout the day ✓ Daily practice of yogasana, pranayama, and meditation for at least 30 minutes ✓ Spices like haldi (turmeric), jeera (cumin), dhaniya (coriander) and lahsun (garlic) recommended in cooking	✓ Take Chyavanprash 10 mg (1tsf) in the morning. Diabetics should take sugar-free Chyavanprash ✓ Drink herbal tea/decoction (kadha) made from tulsi (basil), dalchini (cinnamon), kalimirch (black pepper), shunthi (dry ginger), and munakka (raisin) – once or twice a day/add jaggery (natural sugar) and/or fresh lemon juice to your taste, if needed. ✓ Golden milk – half teaspoon haldi (turmeric) powder in 150 ml hot milk – once or twice a day.	✓ **Nasal application** – Apply sesame oil/coconut oil or ghee in both the nostrils (PratimarshNasya) in morning and evening. ✓ **Oil pulling therapy** – Take 1 tablespoon Sesame or coconut oil in mouth. Do not drink, swish in the mouth for 2 to 3 minutes and spit it off followed by warm water rinse. This can be done once or twice a day.	✓ Steam inhalation with fresh pudina (mint) leaves or ajwain (caraway seeds) can be practices once in a day. ✓ Lavang (clove) powder mixed with natural sugar/honey can be taken 2–3 times a day in case of cough or throat irritation. ✓ These measures generally treat normal dry cough and sore throat. However, it is best to consult doctors if these symptoms persist.

Functional foods provide their benefits in the diet during effective treatment of diseases by enhancing the physiological response and/or reducing the risk of disease **Invalid source specified**. Nutrition is vital for maintaining skeletal muscle and avoiding metabolic disturbances – when patients spend around two weeks in the ICU, these can become very serious issues. Respiratory difficulties add an extra layer of complexity, preventing patients from eating effectively **Invalid source specified.**

While there is no evidence to suggest that food or dietary supplements can help in prevention of diseases like COVID-19, supplements certainly have a role to play in improving immunity in general **Invalid source specified.** Research has shown that supplementing with certain vitamins, minerals, herbs, and other substances can help improve immune response and potentially protect against illness **Invalid source specified.**

17.3 SUSTAINABLE DEVELOPMENT GOALS

The 17 Sustainable Development Goals (SDGs) are an urgent call for action by all countries in a global partnership. These were agreed upon by the member countries of the United Nations in 2015 as a part of the 2030 Sustainable Development Agenda **Invalid source specified.** The COVID-19 pandemic that has spread rapidly and extensively around the world since late 2019 has had profound implications for food security and nutrition. It has resulted in undermining the right to food and stalling efforts to meet Sustainable Development Goal (SDG) 2: "Zero Hunger," SDG 3: "Good Health and Well-being," and SDG 12: "Responsible Consumption and Production." **Invalid source specified.**

17.4 CHALLENGES

The current pandemic, as mentioned, has impacted achievement of the second Sustainable Development Goal of "Zero Hunger." The stated aim of this goal is to "end hunger, achieve food security and improved nutrition and promote sustainable agriculture." The pandemic is being seen as an additional threat to food systems apart from conflict, climate shocks, and locust crisis. It is also increasing food insecurity. A United Nations report on SDGs suggested that food insecurity has worsened with children and small-scale food producers being hit the hardest **Invalid source specified**.

17.4.1 Hunger

Hunger is defined as a condition in which a person does not have sufficient food to meet basic nutritional needs for a sustained period. In the field of hunger relief, the term hunger is used in a way that goes beyond the common desire for food that all humans experience **Invalid source specified**.

17.4.2 Hidden Hunger

Hidden hunger refers to a lack of vitamins and minerals. It is state that occurs when the quality of food people eat does not meet their nutrient requirements. This occurs when food is deficient in micronutrients like vitamins and minerals that are necessary for growth and development. It is estimated that approximately 2 billion people suffer from vitamin and mineral deficiencies across the world, thus making them vulnerable and potential targets for such pandemics like the COVID-19 (WHO, 2021).

Approximately 2 billion people suffer from vitamin and mineral deficiencies across the world, thus making them vulnerable and potential targets for such pandemics like COVID-19 **Invalid source specified**.

17.4.3 Food Production

A food system involves many activities and processes that get the food from farm to fork. These activities comprise production, processing and packaging, distribution, and retailing, and consuming and disposing of waste. Despite modernization and usage of technology at each stage, there is still a lot of wastage of food. Weather patterns (like unseasonal rains or a lack thereof) and environmental and ecological disasters (floods, droughts, forest fires, landslides, pests) can lead to loss of crops during the production stage of food. There are pilferages and leakages in the processing and packaging stage of food systems. Accidents, poor transportation facilities, and inadequate and unhygienic storage conditions can lead to loss of processed and packaged food. Finally, there is a lot of wastage of food at the consumption stage of any food system either due to poor planning or due to even lack of awareness.

17.4.4 Demand and Supply Gap

Agriculture, with its allied sectors, is the largest source of employment in India. About 70% of India's rural households still depend primarily on agriculture for their livelihood, with 82% of farmers being small and marginal. While achieving food sufficiency in production, India still has 25% of the world's hungry people. The incidence of poverty is now pegged at nearly 30%. As per the Global Nutrition Report (2020), India ranks 114th out of 132 countries on under-5 stunting, 120th out of 130 countries on under-5 wasting, and 170th out of 185 countries on prevalence of

TABLE 17.2
Demand and Supply Gap of Food in India

Year	Population of India	Food Grain Production in India (kgs)	Per Capita per Annum Requirement of Food (kgs) (Assuming RDA 2200 Kcal/day ~ 1.317 kg Food)	Per Capita per Annum Availability of Food Grains in India (kgs)	Per Capita per Annum Deficit (-)/ Surplus(+) (Availability – Requirement) kgs
2020	1,38,00,04,385	292,000,000,000	480.705	211.59	-269.11
2019	1,36,64,17,754	285,200,000,000	480.705	208.72	-271.98
2018	1,35,26,42,280	285,000,000,000	480.705	210.70	-270.01
2017	1,33,86,76,785	275,100,000,000	480.705	205.50	-275.20
2016	1,32,45,17,249	251,600,000,000	480.705	189.96	-290.75
2015	1,31,01,52,403	252,000,000,000	480.705	192.34	-288.36

Source: Compiled from Govt. of India's published data and calculated by the authors.

anaemia. Anaemia continues to affect 50% of women including pregnant women and 60% of children in the country **Invalid source specified.** Table 17.2 provides an estimate of the food deficit in India, which is one of the major causes of food and nutrition insecurity.

17.4.5 Food Insecurity

India's performance on food insecurity is not something to be proud of. A recent report from the Food and Agriculture Organization (FAO) of the United Nations titled "State of Food Security and Nutrition in the World" shows that it has the largest population of food insecure people. In fact, the prevalence of food insecurity increased by 3.8% in 2019 as compared to 2014. Estimates extrapolated from the report show that while 27.8% of India's population suffered from moderate or severe food insecurity in 2014–2016, the proportion rose to 31.6% in 2017–2019. The number of food insecure people grew from 42.65 crore in 2014–1016 to 48.86 crore in 2017–2019. India accounted for 22% of the global burden of food insecurity, the highest for any country, in 2017–2019. **Invalid source specified.**

17.4.6 Factors for Food Shortage

India has a paradox in terms of food production and food shortage-linked food insecurity. Various studies have pointed out that there are numerous factors that contribute to hunger and malnutrition. These include inadequate food distribution, social and economic policies, inequality, and poverty. This is a case of "scarcity" amid abundance. The country exports food even as millions remain hungry. In India, the soils have become deficient in minerals, thus impacting the quality of food being consumed.

Another paradox results from the fact that people in India seem to be consuming more calories but the intake of micronutrients has fallen, pointing to the existence of malnutrition and various deficiency-related ailments prevalent in the country. Another factor that may be the cause of shortage of food is the cropping systems that have been promoted by the Green Revolution since the 1970s **Invalid source specified.**

Invalid source specified. identified three categories of the factors that caused food shortages – physical and biological factors, sociocultural influences, and political and economic forces. Climactic conditions, soil types and quality, pests and the resultant diseases, weeds and animals, and the quality of seeds are the physical and biological factors that can cause global food shortages. Sociocultural influences such as land holdings, access to irrigation systems, farm labour availability, and overall dietary trends in the population of the country can change cropping patterns, thus impacting the availability of food. Similarly, lack of farm incentives, domestic and international trade regulations of agricultural commodities, etc., can affect access to food.

17.4.7 Wastage Due to Farm-to-Fork Losses

Adequate food production may not be sufficient to ensure food security as not all food produced is consumed. A large quantity of food is lost or wasted. One estimate suggests that a reduction of 25% in lost or wasted food globally can feed around 87 crore hungry people. A huge majority of such people are in India (around 20 crore)! Maximum food loss happens during transit from farm to fork, especially to urban markets **Invalid source specified.**

Authors have attempted to differentiate between food loss and food wastage. Food loss has been defined as taking place from the farm up to the retail level, whereas food wastage takes place at subsequent stages of the food value chain – namely, retail, food service, and household levels. It has been observed that the issue of food wastage has not been rigorously studied in India. Two studies conducted by the Central Institute of Post-Harvest Engineering and Technology, Ludhiana (CIPHET) in 2005 and 2015 measured the losses incurred at various stages of production and transportation in cereals, pulses, oilseeds, plantation crops, spices, vegetables, fruits, milk, fisheries, poultry, and meat. The studies found that overall losses were much lower than the general perception. Perishable crops had higher losses at the farm operations stage than at storage stage. For instance, it was reported that farm operations losses in the case of rice and wheat were 4.67% and 4.07%, respectively, and the storage losses for both crops were only 0.86%. It was concluded that "reduction of losses during harvesting and threshing as well as sorting and grading of produce is as important as setting up modern storage/cold chain facilities at various levelsof marketing after the produce has been sold by the farmers" **Invalid source specified.**

17.5 IMPACT OF FOOD AND NUTRITION INSECURITY

Population studies of India indicate that the demographic transition is in the final stage. Demographic transition is a phenomenon "which refers to the historical shift from high birth rates and high infant death rates in countries with minimal technology, education (especially of women) and economic development, to low birth rates and low death rates in societies with advanced technology, education and economic development" **Invalid source specified**. The stages of the demographic transition are:

- Stage 1 characterized by high birth and death rates.
- Stage 2 where birth rates remain high, but death rates fall rapidly.
- Stage 3 when birth rates start falling and death rates decline slows down.
- Stage 4 where both birth rate and death rate are low.
- Stage 5 when birth rates start rising again and death rates remain low.

17.5.1 Learning from the Covid-19 Pandemic

It is estimated that "the ultimate size of India's population when population stabilization is achieved will be about 1.72 billion around the year 2060." In general, food access, food availability, and food adequacy at the individual level are influenced by several factors, the most important of which is poverty. Other factors include the national and international economic environment, population growth, infrastructure, the climate, the level of donor aid commitment and intervention, access to appropriate training and job skills, asset base, conflict and access to pasture, and the quality of diet, health, and sanitation **Invalid source specified.**

The current pandemic is bound to have an impact on not only the existing demographic profile but also on future generations as the infection and its treatment may have a long-term physiological influence that can affect fertility and death rates. This has been observed in earlier epidemics like the Zika virus in South America.

17.6 SOLUTIONS: FOOD AND NUTRITION SCARCITY TO SECURITY

It is essential that food and nutrition are given the importance that is due in order to not only facilitate convalescence but also to make the country future ready against such viral spreads that can quickly turn fatal if people in general do not gather immunity. In the present circumstances, food has to act as medicine and vice versa. Some of the steps suggested by various authors, including the present ones, are:

- Increase food production.
- Food processing.
- Food storage facilities.
- Govt. initiatives.
- Reduce food wastage.
- Improve accessibility to food.
- Improve delivery of nutrients.
- Maintain healthy lifestyles by eating healthy, nutrient-dense foods.
- Make fitness training as part of the healthy lifestyle in order to reduce the burden of the lifestyle diseases that have been identified as co-morbidities in people infected with COVID-19.
- Go back to roots in terms of harnessing traditional wisdom for food preparation in order to derive therapeutic value from the ingredients.

The journey might seem to be challenging and overwhelming, but it is worth the cause of sustaining humanity and the planet.

BIBLIOGRAPHY

Abobaker, Anis, Aboubaker Alzwi, and Alsalheen Hamed A. Alraied. 2020. "Overview of the possible role of vitamin C in management of COVID 19." Pharmacological Reports 1517–1528. doi:https://doi.org/10.1007/s43440-020-00176-1.

Aman, Faseeha, and Sadia Masood. 2020. "How nutrition can help to fight against COVID-19 Pandemic." Pakistan Journal of Medical Sciences. 36 (COVID19-S4):COVID19-S121-S123. doi: https://doi.org/10.12669/pjms.36.COVID19-S4.2776.

Bansal, Vaishali. 2020. "More evidence of India's food insecurity." The Hindu, 24 August. www.thehindu.com/opinion/lead/more-evidence-of-indias-food-insecurity/article32424037.ece.

Coelho-Ravagnani, Christianne de Faria, Flavia Campos Corgosinho, Fabiane La Flor Ziegler Sanches, Carla Marques Maia Prado, Alessandro Laviano, and Jo~aoFelipMota. 2020. "Dietary recommendations during the COVID-19 pandemic." Nutrition Reviews (Oxford University Press) 0 (0): 1–14. doi:10.1093/nutrit/nuaa067.

Davis, Charles Patrick. 2020. What's a Virus? Viral Infection Types, Symptoms, Treatment. 10 June. www.onhealth.com/content/1/viral_infections.

Department of Economic and Social Affairs. 2021. Sustainable Development: Goal 2. sdgs.un.org/goals/goal2.

———. n.d. THE 17 GOALS. Accessed 2021. https://sdgs.un.org/goals.

DeRose, Laurie, Ellen Messer, and Sara Millman. 1998. Who's hungry? And how do we know? Food shortage, poverty, and deprivation. New York: United Nations University Press. http://collections.unu.edu/eserv/UNU:2380/nLib9280809857.pdf.

EMRO, WHO. 2021. Nutrition Advice for Adults During the COVID-19 Outbreak. www.emro.who.int/nutrition/news/nutrition-advice-for-adults-during-the-covid-19-outbreak.html#:~:text=Eat%20fresh%20and%20unprocessed%20foods%20every%20day&text=Daily%2C%20eat%3A%202%20cups%20of,%E2%88%923%20times%20per%20week.

Food and Nutrition Security as Preventive and Therapeutic Healthcare: Now & Beyond COVID-19 Pandemic.

Henderson, Emily. 2020. "The Role of Nutrition in Recovery from COVID-19." News-Medical. 24 June. Accessed May 20, 2021. www.news-medical.net/news/20200624/The-Role-of-Nutrition-in-Recovery-from-COVID-19.aspx.

HLPE. 2020. Impacts of COVID-19 on food security and nutrition: Developing effective policy responses to address the hunger and malnutrition pandemic. Rome: FAO. doi:https://doi.org/ 10.4060/cb1000en.

Hussain, Siraj. 2021. "How Much of India's Agricultural Produce Is Wasted Annually? "The Wire. 01 January. https://thewire.in/agriculture/ india-agricultural-produce-wasted.

Khadka, Shyam. 2017. Reducing food waste vital for India's food security. March. www.downtoearth.org.in/blog/food/reducing-food-waste-vital-for-india-s-food-security-57345.

Kubala, Jillian. 2021. The 15 Best Supplements to Boost Your Immune System Right Now. www.healthline.com/nutrition/immune-boosting-supplements.

Lange, Klaus W. 2021. "Food Science and COVID-19." Food Science and Human Wellness 10: 1–5. doi:https://doi.org/10.1016/j.fshw.2020.08.005.

Min. of Food Processing Industry, Govt. of India. 2020. Estimates of Food Production – All India. March. https://mofpi.nic.in/documents/statistics.

Min. of Health and Family Welfare, Govt. of India. 2020. Ayurveda's Immunity Boosting Measures for Self-Care during COVID 19 Crisis. April. www.mohfw.gov.in/pdf/ImmunityBoostingAYUSHAdvisory.pdf

Nuwer, Rachel. 2020. Why the World Needs Viruses to Function. 18 June. www.bbc.com/future/article/20200617-what-if-all-viruses-disappeared.

Press Trust of India. 2020. India Ranks 94 in Global Hunger Index 2020, Placed in "Serious" Category. 17 October. https://theprint.in/india/india-ranks-94-in-global-hunger-index-2020-placed-in-serious-category/525558/.

Rai, Rajesh Kumar, Sandhya Kumar, Madhushree Sekher, Bill Pritchard, and Anu Rammohan. 2014. "A Life-Cycle Approach to Food and Nutrition Security in India." Public Health Nutrition 944–949. doi:10.1017/S1368980014001037.

Singh, Pushpendra, Manish Kumar Tripathi, Mohammad Yasir, Ruchi Khare, Manoj Kumar Tripathi, and Rahul Shrivastava. 2020. "Potential Inhibitors for SARS-CoV-2 and Functional Food Components as Nutritional Supplement for COVID-19: A Review." Plant Foods for Human Nutrition (Springer Nature) 75: 458–466. doi:https://doi.org/10.1007/s11130-020-00861-9.

Todhunter, Colin. 2019. "The Paradox of Food Scarcity Amid Abundance in India and What Needs to Be Done." Outlook India, 28 November. https://poshan.outlookindia.com/story/ poshan-news-the-paradox-of-food-scarcity-amid-abundance-in-india-and-what-needs-to-be-done/343202.

WHO. 2021. Nutrition. www.who.int/nutrition/topics/WHO_ FAO_ICN2_videos_hiddenhunger/en/#:~:text=Hidden%20hunger%20is%20a%20lack,for%20their%20growth%20and%20development.

Wikipedia contributors. 2021. "COVID-19." Wikipedia, The Free Encyclopedia. 18 May. https://en.wikipedia.org/w/index.php?title=COVID-19&oldid=1023813407.

———. 2021. "Demographic Transitions." Wikipedia, The Free Encyclopedia. 7 May. https://en.wikipedia.org/w/index.php?title=Demographic_transition&oldid=1021927126.

——— 2021. "Disease." Wikipedia, The Free Encyclopedia. 12 May. https://en.wikipedia.org/w/index.php?title=Disease&oldid=1022864776.

——— 2021. "Hunger." Wikipedia, The Free Encyclopedia. https://en.wikipedia.org/w/index.php?title=Hunger&oldid=1023796HAPTER

18 Potential Role of Natural Essential Oils in the Management of Respiratory Distress in Patients Infected with SARS-CoV-2

An Overview

Sukender Kumar,[1] Harpal S. Buttar,[2] Samander Kaushik,[3] and Munish Garg[1]*

[1]Department of Pharmaceutical Sciences, Maharshi Dayanand University, Rohtak, Haryana, India

[2]Department of Pathology and Laboratory Medicine, University of Ottawa, Faculty of Medicine, Ottawa, Ontario, Canada

[3]Centre for Biotechnology, Maharshi Dayanand University, Rohtak, Haryana, India

*Corresponding author: Munish Garg, Email

CONTENTS

18.1 INTRODUCTION

As per scientific reports, the novel coronavirus pandemic started with an outbreak of pneumonia of unknown epidemiology from Wuhan, China in December 2019 and rapidly spread worldwide using the modern world's travel modes. The virus is now called severe acute respiratory syndrome

DOI: 10.1201/9781003358909-18

coronavirus-2 (SARS-CoV-2) because of genetic similarities to SARS-CoV 2003, and the respiratory sickness is called Covid-19 (Wu et al., 2020).

To date, the pandemic has infected more than 630 million people around the world, and around 6.5 million have lost their lives as per official data (WHO, 2022); however, this figure can be much more as some are either asymptomatic or have not been tested. It is said that this is the deadliest pandemic in the last 100 years. Scientists all over the world are trying hard to develop an effective preventive or therapeutic regimen, but so far nothing has officially been declared as an approved remedy. Thus, physicians are treating patients on symptomatic observations (Silva et al., 2020; Kaushik et al., 2020).

Medicinal plants play a significant role in controlling various viral infections such as Herpes Simplex Viruses, Dengue, Chikungunya, and Encephalitis (Mohan et al., 2020; Kaushik et al., 2021; Saleh and Kamisah, 2020; Kaushik et al., 2018; Sharma et al., 2019; Kaushik et al., 2020; Bhimaneni et al., 2020). Essential oils of plant origin are complex mixtures of volatile compounds, containing terpenes as the largest group of chemical constituents that may have pharmacological activity (Bhimaneni and Kumar, 2020; Benyoucef et al., 2020). They also contain acids, oxides, phenolic ethers and aromatic alcohol, ester, and aldehydes (Can Baser and Buchbauer, 2010), which are extracted by distillation, mechanical methods, enfleurage, and solvents, from special cells, ducts or cavities, or glandular hairs (Garozzo et al., 2009). A large number of publications have confirmed the antiviral activity of essential oils against respiratory viruses both in the liquid and gaseous phase (Usachev et al., 2013; Horváth and Acs, 2015; Catella et al., 2021; Choi Hwa-Jung, 2018). In the case of SARS CoV-2 infections, the most affected epithelial cells of nasal passage and lungs (Zeigler et al., 2020) could be directly targeted using inhalation of essential oils containing antiviral and/or virucidal constituents, which may act synergistically to provide relief from infection by targeting viral fusion and entry to host cells, helicases, proteases, replication, and translation (Catella et al., 2021; Loizzo et al., 2008; Nadjib, 2020). Further, as inflammation and stress contribute to disease severity, the anti-inflammatory and stress-managing properties of oils may also potentially assist in managing this viral infection (Kunnumakkara et al., 2021; Rehman et al., 2021; Shahid et al., 2020).

At present, the whole world is facing the SARS-CoV-2 pandemic without any treatment and it could take some considerable time to find any cure. In this situation, a short-term solution to treat a large number of seriously ill patients is highly required. Hence, repurposing natural essential oils directly for the most affected cells is perhaps an effective short-term solution. Here we give an overview of current knowledge of the structure, infection strategy, molecular biology, and drug targets of SARS-CoV-2, and the potential of natural essential oil inhalation therapy in the management of this deadly infection.

18.2 DECIPHERING MOLECULAR BIOLOGY AND POTENTIAL STRATEGIES FOR DRUG THERAPY AGAINST SARS-COV-2

The SARS-CoV-2 is a positive-sense, single-stranded RNA virus of the *Coronaviridae* family. The genome of SARS-CoV-2 encodes 16 non-structural proteins (nsp1-16), four structural proteins (spike, membrane, envelope, and nucleocapsid), and eight accessory proteins (3a, 3b, 6, 7a, 7b, 8b, 9b, and 14) (Wu et al., 2020). The spike glycoproteins that cover the virus facilitate viral entry by binding to the cell receptor angiotensin-converting enzyme (ACE)-2, followed by protein priming by host proteases, mainly transmembrane serine protease (TMPRSS)-2, furin, and may be another lung proteases, which activate the fusion of virus and human cell (Hoffmann et al., 2020; Walls et al., 2020).After entering the cell, the virus synthesizes viral RNA and polyprotein, successively assembles them, and forms the new virus particles. Researchers have further identified the cells that co-express the genes ACE2 and TMPRSS-2 by filtering through the data and shortlisted the cell types in the: 1) Nasal passage (mucus-producing goblet secretory cells); 2) Lungs (type II pneumocytes lining the alveoli); and 3) Intestine (ileal absorptive enterocytes). Strikingly, the genes' ACE2 use airway epithelial cells that could be a potential drug target (Zeigler et al., 2020).

Like influenza, SARS-CoV-2 is an enveloped virus and must be uncoated to enter the cytoplasm, but often results in "cytokine storms" followed by inflammation, lung injury, pneumonia, and sometimes death. The term "cytokine storm" used here describes the unchecked systemic overproduction of cytokines that contribute to disease severity by uncontrolled inflammation and many other types of pathologies including vascular leakage, coagulopathy, organ dysfunction, and transaminitis (Hoffmann et al., 2020; Turnquist et al., 2020). The inflammatory cytokines levels, mainly IL-6 and TNF-α, predict disease severity and have been suggested to be considered for management and treatment of Covid-19 (Del Valle et al., 2020; Ye et al., 2020). As these inflammatory responses make infection severe, hence controlling the inflammatory cytokines, managing infection at the mostly infected epithelial cells of the airway directly, inhibition of virus entry and replication, could be the potential strategies for rapid mitigation, management, and treatment of this deadly infection.

18.3 APPROACHES TO USING ESSENTIAL OILS IN THE MANAGEMENT OF SARS-COV-2 INFECTION

18.3.1 Antiviral Effects of Essential Oils in Managing SARS-CoV-2 Infection

The SARS-CoV-2 spike glycoproteins bind to host cell receptors, just like SARS-CoV, and then enter the host cell, much like influenza (Hoffmann et al., 2020). Hence, essential oils with antiviral activity against SARS-CoV and influenza have relevance to SARS-CoV-2. Many studies have shown the antiviral effects of essential oils against the influenza virus (Usachev et al., 2013; Horváth and Acs, 2015; Catella et al., 2021; Choi Hwa-Jung, 2018). In one study, 11 essential oils were investigated as antivirals against the influenza virus in which three oils were found most effective in Madin-Darby canine kidney (MDCK) cells without any cytotoxicity at a concentration of 100 μg/mL. These three oils mainly contained 1,8-cineole, linalool, β–pinene, linalayl acetate, α-terpineol, trans-anethole, and estragole. This study suggested linalool as the main constituent responsible for antiviral activity as it was present in all most effective oils in a significant quantity (Choi Hwa-Jung, 2018). Tea tree and eucalyptus oils showed strong anti-influenza activity both in aerosol and vapour form (Usachev et al., 2013). The vapour of the essential oils obtained from the bergamot, eucalyptus, and their pure constituents (i.e., citronellol and eugenol) showed strong anti-influenza activity at only 10 minutes of exposure while the vapours from the geranium, lemongrass, and cinnamon leaves oil took 30 minutes for the same activity. Although the liquid form of the essential oils from *Cinnamomum zeylanicum, Citrus bergamia, Cymbopogon fexuosus,* and *Thymus vulgaris* exhibited complete inhibitory activity against influenza virus (i.e., 100% inhibition), and the vapour form was found safer against the epithelial cell monolayers. This study concluded that the vapour form of these essential oils could be potentially useful in influenza treatment (Vimalanathan and Hudson, 2014). Anise oil can also be useful in the treatment of respiratory complaints associated with the cold and as an antiviral in influenza virus infection (Choi Hwa-Jung, 2018; ESCOP Monographs, 2003). This oil can be used 50–200 μL in a single dose, three times daily, but not for more than two weeks (ESCOP Monographs, 2003). In a study, 221 phytochemicals and essential oil constituents were screened *in-vitro* for antiviral effect against severe acute respiratory syndrome-associated coronavirus (SARS-CoV). Ten diterpenoids, two sesquiterpenoids, and two triterpenoids were potent anti-SARS-CoV inhibitors at concentrations between 3.3 and 10 μM (Wen et al., 2007).

The closest to an applicable study, conducted by Loizzo et al. (2008) reported the virucidal effect of distilled oil extracted from *Laurusnobilis* berries against SARS-CoV (the 2002/2003 outbreak virus). Interestingly, a docking study indicated that garlic essential oil prevents the entry of SARS-CoV-2 into the human body by strongly interacting with the ACE2 protein and other main proteases of the virus. Allyl disulfide and allyl trisulfide exerted the strongest anti-coronavirus activity as compared to other constituents of the oil (Thuy et al., 2020). Further, Da Silva and colleagues (2020) screened the activity of 171 essential oils constituents in treatment of SARS-CoV-2 infection using molecular docking techniques. They screened compounds against different viral proteins including main viral proteases (Mpro), ADP-ribose-1-phosphatase (SARS-CoV-2 ADRP), spike proteins (SARS-CoV-2 rS), endoribonuclease (SARS-CoV-2 Nsp15/NendoU), RNA-dependent RNA polymerase (SARS-CoV-2 RdRp), and human angiotensin-converting enzyme (hACE2) protein. Three compounds, (*E,E*)-α-farnesene, (*E,E*)-farnesol, and (*E*)-nerolidol, were found to have the potency of inhibiting viral replication as indicated by their binding with SARS-CoV-2 Mpro. (*E,E*)-farnesol showed the best docking scores against SARS-CoV-2 RdRp. These studies strongly address the antiviral activity of essential oil components against corona viruses *in-vitro*, but human data is not present and the mechanism of action is also not significantly explained (Table 18.1).

18.3.2 Anti-Inflammatory and Immunomodulatory Effects of Essential Oils in Managing SARS-CoV-2 Infection

Studies have shown that the hyper-inflammation and coagulopathy in airways make SARS-CoV-2 infection severe and sometimes result in the death of patients. This overactive inflammatory response to SARS-CoV-2 contributes to the production of cytokines, tumour necrosis factor (TNF), profound lymphopenia, and substantial mononuclear cell infiltration in the lung and other vital organs of the body (Merad et al., 2020). Hence, managing these inflammatory reactions could improve the recovery rate of patients. The authors summarized evidence that established the key role of essential oils in targeting inflammation and hyper-reactivity of the airways, mucus secretion, and cough (Horváth and Acs, 2015; Banner et al., 2011). Linalool, the main constituent of many essential oils, exhibited protective effects on inflammation of lung cells in an *in-vivo* lung injury model (Huo et al., 2013). Clinical studies were conducted to study the anti-inflammatory effects of eucalyptus oil and its active compound 1,8-cineole (extracted from eucalyptus) in treatments of inflammatory airway diseases (Juergens et al., 2020; Sadlon and Lamson, 2010). The inhalation therapy of cineole (plasma concentrations of 1.5 μg/ml) exerted strong anti-inflammatory effects by blocking pro-inflammatory cytokines release in lung inflammation and infections. Cineol also has mucolytic and bronchodilatory properties (Juergens et al., 2020). Furthermore, eucalyptus oil induces activation of human monocyte-derived macrophages (MDMs) and dramatically increases their phagocytic ability both *in vitro* and *in vivo* (Serafino et al., 2008). Diallylsulphide, a major active constituent of garlic oil, has been found to bring Nrf2 activation in lung MRC-5 cells and has been suggested to be useful in oxidative stress-induced lung injury (Ho et al., 2012). This Nrf2 activation may potentially regulate the cytokine storm in COVID-19 patients (McCord et al., 2020). These findings give clues about the possible beneficial role of anti-inflammatory and immunomodulatory effects of essential oils in managing SARS-CoV-2, but more detailed scientific studies are required to establish their efficacy.

18.3.3 Antianxiety and Stress-Reducing Effects of Essential Oils in the Management of SARS-CoV-2 Infection

Studies have found significant psychological distress (stress, anxiety, and depression) among the general public, health professionals, and Covid-19 patients worldwide during this pandemic, which is a major obstacle in the recovery of patients (Rehman et al., 2021; Shahid et al., 2020). Literature carries strong evidence in support of using essential oils for managing mental distress to achieve psychological well-being. Plenty of studies have proven the anxiolytic effects of the essential oils of lavender (*Lavandula angustifolia*) in all three main application ways (Zhang and Yao, 2019; Moss et al., 2003; Kheirkhah et al., 2014). It can be used with citrus oils, often sweet orange (*Citrus sinensis*) or lemon (*Citrus limon*), for reducing stress and anxiety (Perry et al., 2006; Lehrner et al., 2005; Goes et al., 2012). Inhalation of the bitter orange (*Citrus aurantium*) and sweet orange (*Citrus sinensis*) oils reduced anxiety and stress in female patients and dental patients, respectively (Chaves et al., 2017). Oral administration (50–150 mg/kg for 30 days) of lemon (*Citrus limon*) oil to male Swiss mice exhibited anxiolytic effect (Lopes et al., 2011). The essential oils of rose (*Rosa

TABLE 18.1
In Vitro Antiviral Activity Studies Done with Essential Oils against Influenza Virus, SARS-CoV-1, and SARS-CoV-2

Essential Oils	Bioactive Compounds	Active Phase and Concentration	Experimental Model	References
Eucalyptus (*Eucalyptus polybractea*)	Cineole (eucalyptol)	Aerosol and vapour	H11N9 virus infection in MDCK cells	Usachev et al., 2013
Eucalyptus (*Eucalyptus globulus*)	1,8-Cineole, α-pinene, camphor, limonene	Vapour and liquid (100% inhibition at 50 μL/mL)	H1N1 virus infection in MDCK and human lung epithelial cells	De Groot and Schmidt, 2016; Vimalanathan and Hudson, 2014
Tea Tree (*Melaleuca alternifolia*)	Terpinen-4-ol γ-terpinene 1,8-cineole, α-terpinene, α-terpineol, p-cymene	Aerosol and vapour	H11N9 virus infection in MDCK cells	De Groot and Schmidt, 2016; Usachev et al., 2013; Tisserand and Young, 2014
		Liquid (0.2% v/v))	H11N9 virus infection in MDCK cells	Pyankov et al., 2012
Cinnamon leaf (*Cinnamomum zeylanicum*)	Eugenol, benzyl benzoate, (E)-cinnamyl acetate, (E)-cinnamaldehyde, linalool	Liquid (100% inhibition at 3.1 μL/mL) and vapour	H1N1 virus infection in MDCK and human lung epithelial cells	De Groot and Schmidt, 2016; Vimalanathan and Hudson, 2014
Lemongrass (*Cymbopogon flexuosus*)	Geranial Neral	Liquid (100% inhibition at 3.1 μL/mL)	H1N1 virus infection in MDCK and human lung epithelial cells	Vimalanathan and Hudson, 2014
Geranium leaf (*Pelargonium graveolens*)	Citronellol Geraniol	Liquid (100% inhibition at 3.1 μL/mL)	H1N1 virus infection in MDCK and human lung epithelial cells	Vimalanathan and Hudson, 2014
Thyme (*Thymus vulgaris*)	Thymol, carvacrol, geraniol, linalool 1,8-cineole, terpenyl acetate, borneol, α-terpineol, p-cymene	Liquid (100% inhibition at 3.1 μL/mL)	H1N1 virus infection in MDCK and humanlung epithelial cells	De Groot and Schmidt, 2016; Vimalanathan and Hudson, 2014
Bergamot (*Citrus bergamia*)	Linalool, limonene, γ-Terpinene, α-pinene α-terpinene	Liquid (100% inhibition at 3.1 μL/mL) and vapour	H1N1 virus infection in MDCK and human lung epithelial cells	Vimalanathan and Hudson, 2014
Anise (*Pimpinella anisum*)	α–Pinene, limonene, linalool, *trans*-Anetholecharvicol	Liquid (52% inhibition at 100 μg/mL)	Influenza A/WS/33 virus infection in MDCK cells	Horváth and Ács, 2015
Berries of *Laurus nobilis* L.	β-Ocimene, 1,8-cineole, α-pinene, and β-pinene	Liquid (50% inhibition at 130 μg/mL)	SARS-CoV-1 (isolate FFM-1 from a patient) infection in Vero cells	Loizzo et al., 2008
Berries of *Juniperus oxycedrus* L. ssp. *Oxycedrus*	α-Pinene, β-myrcene, α-phellandrene, limonene	Liquid (50% inhibition at 170 μg/mL)	SARS-CoV-1 (isolate FFM-1 from a patient) infection in Vero cells	Loizzo et al., 2008
Garlic (*Allium sativum* L.)	Allyl disulfide and allyl trisulfide	Liquid	Molecular docking modeling (SARS-CoV-2)	Thuby et al., 2020

TABLE 18.2
Anxiolytic Clinical Trials Done with Essential Oils in Men and Women

Essential Oils	Bioactive Compounds	Treatment Mode	Subjects Studied	Reference
Lavender (*Lavandula angustifolia*)	Not reported	Inhaled oil diffused through a diffuser pad (5 drops per 5 minutes)	Healthy adults (57 males and 39 females)	Moss et al., 2003
	Not reported	Oral (80mg/day for 10 weeks)	Anxiety patients (159 women and 53 men)	Kheirkhah et al., 2014
Sweet orange (*Citrus sinensis*)	Limonene and myrcene	Inhalation, diffused through an electrical dispenser	Anxiety patients, aged 18–77 years old (100 women, 100 men)	Lehmer et al., 2005
Bitter orange (*Citrus aurantium*)	Limonene	Inhalation (5%, v/v solution, diffused by nebulization)	patients experiencing crack withdrawal (N = 51)	Chaves et al., 2017
Rose (*Rosa damascena*)	Not reported	Inhalation of air impregnated with oil for 90 second (8.3 ppm, v/v)	20 female university students	Igarashi et al., 2014
Sandalwood (*Santalum album*)	Not reported	Skin application; (massage with 1% oil for 20 minutes, once a week for 4 weeks)	patients (N = 750)	Kyle, 2006
Bergamot oil (*Citrus bergamia*)	Limonene, linalyl acetate, γ-terpene, α and β-pinene, linalool	Inhalation; essential oils were evaporated with water (0.1%, v/v, 15 minutes)	42 Healthy female students	Watanabe et al., 2015

damascena) or orange (*Citrus sinensis*) induced physiological and psychological relaxation by olfactory stimulation (Igarashi et al., 2014). Many clinical trials showed that the inhalation therapy of essential oils could act on the nervous system using the neurological pathway and/or pharmacological pathway to combat stress and anxiety (Table 18.2).

18.4 INSIGHTS FOR USING ESSENTIAL OILS THERAPY AGAINST SARS-COV-2 INFECTION

Although studies have reported the antiviral activity of essential oils against respiratory viruses in the liquid phase (Usachev et al., 2013; Horváth and Acs, 2015; 15. Garozzo et al., 2009; Catella et al., 2021; Choi Hwa-Jung, 2018), inhalation therapy with essential oils is becoming more frequent in the case of respiratory tract infections. The multi-faceted action of essential oils and their volatile constituents can directly target infected cells of the respiratory tract (Horváth and Acs, 2015). Also, several studies have reported significant antiviral activity of the vapours or "gaseous" phase of essential oils, which sometimes appear superior to the liquid phase of the oil (Usachev et al., 2013; Vimalanathan and Hudson, 2014; Inouye et al., 2001; Tyagi and Malik, 2011; Hudson et al., 2011). Inouye et al. (2001) found essential oils and their constituents active in the vapour state. The aerosols of tea tree and eucalyptus oil were found to be active against the influenza A virus. In another study, the anti-influenza activity of five essential oils was investigated in the vapour phase, and the vapour phase was found to be more active than the liquid phase (Vimalanathan and Hudson, 2014). The current literature provides strong evidence in support of inhalation therapy with essential oils as potentially effective antivirals against SARS-CoV-2 infection. Inhaled essential oil constituents mediate effects via two pathways: the neurological pathway (via olfactory nerves) and the pharmacological pathway (through the bloodstream) (Perry et al., 2006; Lehrner et al., 2005; Goes et al., 2012). For future research, the authors propose three approaches for the application of essential oils against Covid-19: 1) use as an air disinfectant to stop aerosol transmission; 2) use as antivirals by defusing to patients rooms and houses; and 3) nebulised to the patient for respiratory and psychological support.

Essential oils can be added to a diffuser or vaporiser to decontaminate the air in patients' rooms and at houses to prevent further aerosol transmission of the virus. A patent by Vail and Vail (2006) describes the use of essential oils of *Eucalyptus globulus, E. citriodora, E. radiatto,* and *Melaleuca alternifolia* in the treatment of severe acute respiratory syndrome and recommend the inhalation of essential oils to reduce the risk of infection in public places. Tea tree and eucalyptus oils were found highly efficient in inactivating more than 95% of airborne influenza viruses within 5–15 min of exposure suggesting highly effective natural antiviral agents for disinfecting applications (Pyankov et al., 2012).

According to a case report (Kamyar, 2009), the room of a patient infected with the respiratory syncytial virus was nebulized every six hours with a mixture of essential oils containing *Lavandulalatifolia, Thymus mastichina, Balsam abies,* and *Menthapiperita,* and the patient passively inhaled the vapours. As a result, the oxygen requirement of the patient was reduced to 1.5 litres per minute within 12 hours. Therefore, inhalation therapy of essential oils can provide immediate respiratory support to patients, which is very important in the case of severe Covid-19 patients with breath shortness. The anti-inflammatory, antioxidant, immunomodulatory, and antiviral properties of essential oils may further contribute to their activity against Covid-19 (Asif et al., 2020; Becker, 2020). The European Pharmacopoeia (2004) also officially described the use of more than 25 essential oils for the treatment of respiratory tract diseases. Among them, the use of the essential oil of eucalyptus, peppermint, tea tree, anise, and thyme were repeatedly reported.

18.5 DISCUSSION

Based on the evidence obtained from the published literature and the current pandemic, the application of inhalation therapy with natural essential oils would be highly useful in the management of SARS-CoV-2 infection. However, existing evidence on essential oils is based on the claims of computer-aided docking (Silva et al., 2020; Nadjib, 2020) and a few in vitro studies (Wen et al., 2007; Choi Hwa-Jung, 2018; Vimalanathan and Hudson, 2014; Loizzo et al., 2008). No *in-vivo* or well-controlled clinical studies are available to establish the safety and efficacy against SARC-CoV-2 infection. In most situations, the influenza virus has been used as a proxy to get some insights about Covid-19 because of some of the similarities/resemblances with coronaviruses like viral uncoating during cell entry, cytokine storm, and increase in the levels of pro-inflammatory molecules (Merad et al., 2020; Asif et al., 2020; Becker, 2020). Hence, antiviral mechanisms of essential oils against influenza may hypothetically translate to SARS-CoV-2 infection, but do not provide strong support toward the concentrations or dosages, safety, effectiveness, inhalation delivery systems, etc. In addition, essential oils are extremely potent, and they must be diluted to bearable concentrations in inhalation preparations and for suitable blends. Therefore, caution should be exercised in calibrating the effective therapeutic concentration of essential oils to stop SARS-CoV-2 infection. The duration and frequency of inhalation and risks associated must also be precisely determined prior to the initiation of essential oil therapy. It remains to be ascertained if combination therapy with a mixed blend of oils would be more effective than a single oil therapy.

18.6 CONCLUSIONS

The studies described in this chapter show that essential oils may play some supportive role in the management of Covid-19 due to their antiviral, anti-inflammatory, and antianxiety effects. Patients may benefit from essential oil inhalation therapy that can directly target infected epithelial cells of the naso-respiratory system. The present chapter provided some in vitro and in vivo study support for the potential role of inhalation therapy with essential oils in the management of SARS-CoV-2 infection at this critical moment of our time. Future well-designed clinical trials are needed before definitive and conclusive claims can be made about the therapeutic effect of plant-derived essential oils in reducing distress parameters in SARS-CoV-2 infected patients. However, essential oil therapy should not be used in patients allergic to perfumes and essential oils as well as by asthmatic patients.

REFERENCES

Asif M, Saleem M, Saadullah M, Yaseen HS, Al Zarzour R (2020). COVID-19 and therapy with essential oils having antiviral, anti-inflammatory, and immunomodulatory properties. *Inflammo pharmacology* 28(5):1153–1161. doi: 10.1007/s1078 7-020-00744-0

Banner KH, Igney F, Poll C (2011). TRP channels: emerging targets for respiratory disease. *Pharmacol Ther* 130(3):371–384. doi: 10.1016/j.pharmthera.2011.03.005

Becker S (2020). Essential Oils and Coronaviruses. https://tisserandinstitute.org/ essential-oils-coronavirus/

Benyoucef F, Dib ME A, Tabti B, Zoheir A, Costa J, Muselli A (2020). Synergistic effects of essential oils of ammoides verticillata and satureja candidissima against many pathogenic microorganisms. *Anti-Infective Agents* 18(1):72–78. https://doi.org/10.2174/221135 2517666190227161811

Bhimaneni SP, Kumar A (2020). Abscisic acid, a plant hormone, could be a promising candidate as an anti-japanese encephalitis virus (JEV) agent. *Anti-Infective Agents* 18:326–331. https://doi.org/10. 2174/ 2211352518666200108092127

Can Baser KH, Buchbauer G (2010). Handbook of Essential Oils: Science, Technology, and Application, 2nd edn. New York: Taylor & Francis Group: CRC Press.

Catella C et al. (2021). Virucidal and antiviral effects of Thymus vulgaris essential oil on feline coronavirus. *Res Vet Sci* 137:44–47. doi: 10.1016/j.rvsc.2021.04.024.

Chaves Neto G et al.(2017). Anxiolytic effect of citrus aurantium L. in crack users. *Evid Based Complement Alternat Med* 2017:7217619. doi: 10.1155/2017/7217619

Choi HJ (2018). Chemical Constituents of Essential Oils Possessing Anti-Influenza A/WS/33 Virus Activity. *Osong Public Health Res Perspect* 9(6):348–353. doi: 10.24171/j.phrp.2018.9.6.09

De Groot AC, Schmidt E (2016). Essential oils, Part III: chemical composition. *Dermatitis* 27(4):161–169. doi: 10.1097/DER.000 0000000000193

Del Valle DM et al. (2020.) An inflammatory cytokine signature predicts COVID-19 severity and survival. *Nat Med* 26(10):1636–1643. doi: 10.1038/s41591-020-1051-9

ESCOP Monographs (2003). The Scientific Foundation for Herbal Medicinal Products, 2nd edn. Thieme, Stuttgart: New York.

European Pharmacopoeia (2004). Directorate for the Quality of Medicines of the Council of Europe, 5th edn. Vol 2. Strasbourg, France, 1004, 1108, 1570, 2206, 2534, 2569

Garozzo A, Timpanaro R, Bisignano B, Furneri PM, Bisignano G, Castro A (2009). In vitro antiviral activity of Melaleuca alternifolia essential oil. *Lett Appl Microbiol* 49(6):806–808. doi: 10.1111/ j.1472-765X.2009.02740.x

Goes TC, Antunes FD, Alves PB, Teixeira-Silva F (2012). Effect of sweet orange aroma on experimental anxiety in humans. *J Altern Complement Med* 18(8):798–804. doi: 10.1089/ acm.2011.0551

Ho CY, Cheng YT, Chau CF, Yen GC (2012). Effect of diallyl sulfide on in vitro and in vivo Nrf2-mediated pulmonic antioxidant enzyme expression via activation ERK/p38 signaling pathway. *J Agric Food Chem* 60(1):100–7. doi: 10.1021/jf203800d

Hoffmann M et al. (2020). SARS-CoV-2 cell entry depends on ACE2 and TMPRSS2 and is blocked by a clinically proven protease inhibitor. *Cell* 181(2):271–280.e8. doi: 10.1016/j.cell.2020. 02.052

Horváth G, Ács K (2015). Essential oils in the treatment of respiratory tract diseases highlighting their role in bacterial infections and their anti-inflammatory action: a review. *Flavour Fragr J* 30(5):331–341. doi: 10.1002/ffj.3252

Hudson J, Kuo M, Vimalanathan S (2011). The antimicrobial properties of cedar leaf (Thuja plicata) oil; a safe and efficient decontamination agent for buildings. *Int J Environ Res Public Health* 8(12):4477–4487. doi: 10.3390/ijerph8124477

Huo M et al. (2013). Anti-inflammatory effects of linalool in RAW 264.7 macrophages and lipopolysaccharide-induced lung injury model. *J Surg Res* 180(1):e47–54. doi: 10.1016/j.jss. 2012.10.050

Igarashi M, Ikei H, Song C, Miyazaki Y (2014). Effects of olfactory stimulation with rose and orange oil on prefrontal cortex activity. *Complement Ther Med* 22(6):1027–1031. doi: 10.1016/j.ctim. 2014.09.003

Inouye S, Yamaguchi H, Takizawa T (2001). Screening of the antibacterial effects of a variety of essential oils on respiratory tract pathogens, using a modified dilution assay method. *J Infect Chemother* 7(4):251–254. doi: 10.1007/s101560170022

Juergens LJ, Worth H, Juergens UR (2020). New perspectives for mucolytic, anti-inflammatory and adjunctive therapy with 1,8-cineole in COPD and asthma: review on the new therapeutic approach. *Adv Ther* 37(5):1737–1753. doi: 10.1007/s12325-020-01279-0

Kamyar MH (2009). United States Patent Application Publication: Essential Oil Diffusion, Patent US20090169487 A1, 02 July 2009.

Kaushik S, Dar L, Kaushik S, Yadav JP (2021). Identification and characterization of new potent inhibitors of dengue virus NS5 proteinase from Andrographis paniculata supercritical extracts on in animal cell culture and in silico approaches. *J Ethnopharmacol* 267:113541. doi: 10.1016/j.jep.2020.113541

Kaushik S, Jangra G, Kundu V, Yadav JP, Kaushik S (2020). Anti-viral activity of Zingiber officinale (Ginger) ingredients against the Chikungunya virus. *Virus Disease* 31(3):1–7. doi: 10.1007/s13337-020-00584-0

Kaushik S, Kaushik S, Sharma V, Yadav JP (2018). Antiviral and therapeutic uses of medicinal plants and their derivatives against dengue viruses. *Pharmacog Rev* 12:177–185. doi: 10.4103/phrev.phrev_2_18

Kaushik S, Kaushik S, Sharma Y, Kumar R, Yadav JP (2020). The Indian perspective of COVID-19 outbreak. *Virus Disease* 31(2):1–8. https://doi.org/10.1007/s13337-020-00587-x

Kheirkhah M, Vali Pour NS, Nisani L, Haghani H (2014). Comparing the effects of aromatherapy with rose oils and warm foot bath on anxiety in the first stage of labor in nulliparous women. *Iran Red Crescent Med J* 16(9):e14455. doi: 10.5812/ircmj.14455

Kunnumakkara AB et al. (2021). COVID-19, cytokines, inflammation, and spices: how are they related? *Life Sci* 119201. doi: 10.1016/j.lfs.2021.119201. Epub ahead of print.

Kyle G (2006). Evaluating the effectiveness of aromatherapy in reducing levels of anxiety in palliative care patients: results of a pilot study. *Complement Ther Clin Pract* 12(2):148–155. doi: 10.1016/j.ctcp.2005.11.003

Lehrner J, Marwinski G, Lehr S, Johren P, Deecke L (2005). Ambient odors of orange and lavender reduce anxiety and improve mood in a dental office. *Physiol Behav* 15;86(1–2):92–95. doi: 10.1016/j.physbeh.2005.06.031

Loizzo MR et al. (2008). Phytochemical analysis and in vitro antiviral activities of the essential oils of seven Lebanon species. *Chem Biodivers* 5(3):461–470. doi: 10.1002/cbdv.200890045

Lopes CLM et al. (2011). Sedative, anxiolytic and antidepressant activities of Citrus limon (Burn) essential oil in mice. *Pharmazie* 66(8):623–627.

McCord JM, Hybertson BM, Cota-Gomez A, Gao B (2020). Nrf2 activator PB125® as a potential therapeutic agent against COVID-19. bioRxiv:2020.2005.2016.099788. https://doi.org/10.1101/ 2020.05.16.09978 8

Merad M, Martin JC (2020). Pathological inflammation in patients with COVID-19: a key role for monocytes and macrophages. *Nat Rev Immunol* 20(6):355–362. doi: 10.1038/s41577-020-0331-4

Mohan S et al. (2020). Bioactive natural antivirals: an updated review of the available plants and isolated molecules. *Molecules* 25(21):4878. doi: 10.3390/molecules25214878

Moss M, Cook J, Wesnes K, Duckett P (2003). Aromas of rosemary and lavender essential oils differentially affect cognition and mood in healthy adults. *Int J Neurosci* 113(1):15–38. doi: 10.1080/00207450390161903

Nadjib BM (2020). Effective antiviral activity of essential oils and their characteristic terpenes against coronaviruses: an update. *J Pharmacol Clin Toxicol* 8:1138.

Perry N, Perry E (2006). Aromatherapy in the management of psychiatric disorders: clinical and neuropharmacological perspectives. *CNS Drugs* 20(4):257–280. doi: 10.2165/00023210-200620040-00001

Pyankov OV, Usachev EV, Pyankova O, Agranovski IE (2012). Inactivation of airborne influenza virus by tea tree and eucalyptus oils. *Aerosol Sci Tech* 46(12):1295–1302. https://doi.org/10. 1080/02786826.2012.708948

Rehman U et al. (2021). Depression, anxiety and stress among Indians in times of Covid-19 lockdown. *Community Ment Health J* 57(1):42–48. doi: 10.1007/s10597-020-00664-x

Sadlon AE, Lamson DW (2010). Immune-modifying and antimicrobial effects of Eucalyptus oil and simple inhalation devices. *Altern Med Rev* 15(1):33–47. PMID: 20359267.

Saleh MSM, Kamisah Y (2020). Potential medicinal plants for the treatment of dengue fever and severe acute respiratory syndrome-coronavirus. *Biomolecules* 11(1):42. doi: 10.3390/biom 11010042

Serafino A et al. (2008). Stimulatory effect of Eucalyptus essential oil on innate cell-mediated immune response. *BMC Immunol* 9:17. doi: 10.1186/1471-2172-9-17

Shahid R et al. (2020). Assessment of depression, anxiety and stress among covid-19 patients by using dass 21 scales, *Journal of Medical Case Reports and Reviews* 3(06). https://doi.org/10.15520/jmcrr.v3i06.189

Sharma V, Kaushik S, Pandit P, Dhull D, Yadav JP, Kaushik S (2019). Green synthesis of silver nanoparticles from medicinal plants and evaluation of their antiviral potential against chikungunya virus. *Appl Microbiol Biotechnol* 103:881–891. https://doi.org/10.1007/ s00253-018-9488-1

Silva JKRD, Figueiredo PLB, Byler KG, Setzer WN (2020). Essential oils as antiviral agents. Potential of essential oils to treat SARS-CoV-2 infection: An in-silico investigation. *Int J Mol Sci* 21(10):3426. doi: 10.3390/ijms21103426

Thuy BTP et al. (2020). Investigation into SARS-CoV-2 resistance of compounds in garlic essential oil. *ACS Omega* 5(14):8312–8320. doi: 10.1021/acsomega.0c00772

Tisserand R, Young R (2014). Essential Oils Safety, 2nd edn. Churchill Livingstone Elsevier: London.

Turnquist C, Ryan BM, Horikawa I, Harris BT, Harris CC (2020). Cytokine storms in cancer and COVID-19. *Cancer Cell* 38(5):598–601. doi: 10.1016/j.ccell.2020.09.019

Tyagi A, Malik A (2011). Antimicrobial potential and chemical composition of Eucalyptus globulus oil in liquid and vapour phase against food spoilage microorganisms. *Food Chemistry* 126:228–35. doi: 10.1016/ j.foodchem.2010.11.002

Usachev EV, Pyankov OV, Usacheva OV, Agranovski IE (2013). Antiviral activity of tea tree and eucalyptus oil aerosol and vapour. *J Aerosol Sci* 59:22. https://doi.org/10.1016/j.jaerosci. 2013.01.004

Vail WB, Vail ML (2006). United States patent: methods and apparatus to prevent, treat and cure infections of the human respiratory system by pathogens causing severe acute respiratory syndrome (SARS), US 20067048953.

Vimalanathan S, Hudson J (2014). Anti-influenza virus activity of essential oils and vapors. *American Journal of Essential Oils and Natural Products* 2:47–53.

Walls AC, Park YJ, Tortorici MA, Wall A, McGuire AT, Veesler D (2020). Structure, function, and antigenicity of the SARS-COV-2 spike glycoprotein. *Cell* 181(2):281–292.e6. doi: 10.1016/j. cell.2020.02.058

Watanabe E, Kuchta K, Kimura M, Rauwald HW, Kamei T, Imanishi J (2015). Effects of bergamot (Citrus bergamia (Risso) Wright & Arn.) essential oil aromatherapy on mood states, parasympathetic nervous system activity, and salivary cortisol levels in 41 healthy females. *Forsch Komplement med* 22(1):43–49. doi:https://doi.org/ 10.1159/000380989

Wen CC et al. (2007). Specific plant terpenoids and lignoids possess potent antiviral activities against severe acute respiratory syndrome coronavirus. *J Med Chem* 50(17):4087–4095. doi: 10.1021/ jm070295s.

World Health Organization (WHO) (2022). WHO Coronavirus Disease (COVID-19) Dashboard. https://covi d19.who.int/

Wu A et al. (2020). Genome composition and divergence of the novel coronavirus (2019-nCoV) originating in China. *Cell Host Microbe* 27(3):325–328. doi: 10.1016/j.chom.2020.02.001

Ye Q, Wang B, Mao J (2020). The pathogenesis and treatment of the "Cytokine Storm" in COVID-19. *J Infect* 80(6):607–13. doi: 10.1016/j.jinf.2020.03.037

Zhang N, Yao L (2019). Anxiolytic effect of essential oils and their constituents: a review. *J Agric Food Chem* 67(50):13790–13808. doi: 10.1021/acs.jafc.9b00433

Ziegler CGK et al. (2020). SARS-CoV-2 receptor ACE2 is an interferon-stimulated gene in human airway epithelial cells and is detected in specific cell subsets across tissues. *Cell* 181(5):1016–1035.e19. doi: 10.1016/j.cell.2020.04.03

19 Traditional Herbal Medicines for Accelrating Research in and Biomedical Sciences in the Era of Covid-19

M. C. Sidhu[1] *and A. S. Ahluwalia*[2]
[1]Department of Botany, Panjab University, Chandigarh, India
Email: mcsidhu@gmail.com
[2]Eternal University, Baru Sahib, District Sirmaur, Himachal Pardesh, India

CONTENTS

19.1 INTRODUCTION

Plants are an integral component of different lifeforms on planet Earth. These are continuously facilitating human life with a variety of resources including as medicines. Plant-based traditional medicines have been part and parcel of human life since the beginning of existence. Plant-based medicines are also referred to as herbal medicines in the pharmaceutical sectors and are being used throughout the world in different proportions and formulations (mixtures of different plant extracts). All groups of plants including herbs, shrubs, and trees contribute to human health. According to the available literature, common as well as some serious health troubles can be treated using herbal medicines. The whole plant as such or its different parts including root, stem, leaves, bark, seeds, flowers, and fruits are used depending upon their medicinal significance. The medicinal bioactivity of plants is due to the presence of secondary metabolites, which are said to be produced by the plant as a defense response. These secondary metabolites are also called as phytochemicals or phytoconstituents. The medicinal value of a particular species depends upon the number and quantity of phytoconstituents it carries. Due to their affordability, cost effectiveness, availability, and few or no side effects, herbal medicines are used by 80% of the human population across the world (Ekor, 2014).

Interest in traditional medicines around the globe is continuously increasing and efforts are also underway for their regularization. The traditional Indian medicine system (ayurveda) is still alive. The emergence of resistance in certain microbes, high cost, and side effects of modern medicines have revived interest in alternative and complementary medicines. Based on the scientific facts some therapies have successfully been promoted but yet need extensive evidence to support these

DOI: 10.1201/9781003358909-19

claims. The expansion of these medicines requires time, study, and financial resources (Pandey et al., 2013).

While advancements in the modern world have increased the life expectancy of humans, microorganisms causing diseases keep on changing their races and sometimes become resistant to antibiotics. COVID-1, as a key example, has threatened human life throughout the world. The ministry of AYUSH, Government of India, has recommended 'kadha' to boost immunity and provide protection against the virus. The Prime Minister of India, Hon'ble Sh. Narendra Modi Ji has also advised the nation to follow the guidelines issued by the ministry of 'AYUSH.' The number of plant species that are currently used in herbal preparations is limited. Studies conducted by various research groups in different parts of the world have gathered additional information on plant species that can be used for the preparation of new, alternative, and more effective medicines. Today the whole world is in search of ways to treat Covid-19, and unexplored plant species that are used by different sections of society for healthcare issues may be one such avenue.

19.2 QUANTIFICATION AND ASSORTMENT OF PHYTOCHEMICALS

Plant species that have yet to be explored for phytoconstituents need detailed analysis to determine their phytochemicals. It is important to determine the specific parts of the plant with the highest quantity of the phytochemicals of medicinal interest. Adopting this approach, only specific parts of the plant can be collected instead of wasting the entire plant. This will prevent the unnecessary uprooting of the whole plant. Subsequently it will help in conservation of plant genetic resources. Some medicinal plants show seasonal variation in quantity of phytoconstituents in question. Thus, it is important to evaluate the species in different seasons. The members of some plant species differ in their morphology due to availability of nutrients or different ecological conditions. These morphotypes of plant species can be assessed for variation in the phytochemicals. Sharma and Sidhu (2017) reported three morphotypes (prostrate, semi-erect, and erect) of *Eclipta alba* having a constant chromosome number 2n=22 in all forms. Different phytoconstituents like alkaloids, amino acid, carbohydrates, flavonoids, glycosides, phenolics, tannins, etc., were reported using two solvents in these three forms. Prostrate form was better than the other forms in terms of phytoconstituents.

Morphologically different plants of the same species may or may not possess the same chromosome number. Plants revealing alteration in chromosome numbers are referred to as cytotypes. The cytotypes of the same plants of a particular species are likely to differ in type and amount of the phytoconstituents of medicinal value and thus require critical analysis. These plants fall into the category of aneuploids having additional or fewer chromosomes than standard genetic organization. In other cases, members of the same plant species may have different ploidy levels (diploid, triploid, tetraploid, pentaploid, hexaploidy, etc.). The plants with different sets of chromosomes are supposed to be valuable raw materials for the pharma companies. Autotetraploid have four copies of the gene as compared to diploids with two copies. The additional copies of the genes may produce the required protein in double the amount. Therefore, species have to be worked out for cytological parameters to find the best one. Sidhu and Sharma (2016) compared the morphological, cytological, and phytochemical properties of three forms (diploid, tetraploid, and hexaploid) of *Solanum nigrum L.* Fruit color and chromosome numbers (diploid: 2n=24, tetraploid: 2n=48 and hexaploid: 2n=72) were the major distinguishing features. This plant species is widely used in different parts of the world in traditional medicines, but few people are aware of the cytological differences between the three forms. Tetraploid and hexaploid contains four and six copies of a particular gene, respectively. The increase in copy number of the gene may result in increased amount of the phytochemicals of medicinal interest.

The study of seasonal variation and the analysis of morphotypes and cytotypes for good quality, quantity, and variety of phytochemicals can provide additional materials to the medicine world, and reduce the burden on the species that are already in use and facing threat of extinction. This approach will be of great significance towards the conservation of biodiversity in general and plant diversity in particular. This is an important aspect but can only be achieved if research groups include scientists of different related disciplines. The role of the plant taxonomist is of the utmost value as the correct identification of the species is the primary requirement. Only a trained taxonomist can determine the morphotypes and cytotypes. Similarly, it is easy for a pharmacist to classify and quantify phytoconstituents. This mix of disciplines will encourage collaborative research in basic sciences and biomedical sciences within and between research institutes. The available literature has revealed the significance of medicinal plants and their parts for the management of various human health-related issues, but more work is to be done.

19.3 DIABETES

The number of diabetic people around the globe was 425 million people in 2017, and is likely to reach 629 million by 2045 due to unhealthy lifestyle, physical inactivity, and obesity (Zulcafli, et al., 2020).

Information related to the antidiabetic activity of plant species has been compiled by researchers. In one study, 88 plant species from five families were reported. In another study, a total of 81 plant species from 12 families were documented. Amongst the various plant parts of these species, leaves have been utilized in almost 50% of the total medical preparations (Sidhu and Sharma, 2013a,b). This information is useful for herbal practitioners, biomedical researchers, and pharmaceutical industries.

The leaf extract of *Ficus krishnae* was studied for its antidiabetic activity in alloxan-induced diabetic rats. Leaf extract was able to reduce the blood glucose level comparable to the standard drug, glibenclamide, in diabetic rats. Fourier Transform Infrared analysis of glibenclamide and leaf powder showed common absorption spectra. This revealed that leaf extract contains a molecule that is close to glibenclamide. Raman spectra of both glibenclamide and leaf powder also showed some similarities. Further studies are required for the isolation of the compound imparting antidiabetic activity and its clinical validation (Sidhu and Sharma, 2014).

Similar information was also documented from the district Mandi of Himachal Pradesh. A total of 25 plant species belonging to 17 families were used to treat diabetes. The majority of plants (28%) were trees followed by herbs and shrubs (24% each), climber (20%), and liana (4%). Fourteen species were wild and leaves were the preferred plant part (40%) to be used in herbal preparations. The medicinal preparations included juice, decoction, cooked, powder, or paste form (Sidhu and Thakur, 2015). A survey was conducted to document antidiabetic plants. A total of 500 respondents were contacted and interviewed to gather information related to the anti-diabetic plant species from district Una, Himachal Pradesh. A single plant or its parts or a combination of different plant species in polyherbal forms are used by the natives to manage diabetes. People were using 84 plant species in different preparations to take care of diabetes (Gupta et al., 2017).

19.4 CANCER

Cancer is a serious disease and causes a significant number of deaths throughout the world. Presently cancer is treated by different means like surgery, medicine, and radiation. All of these treatments have side effects. Thus, the search for herbal medicines likely to minimize these ill effects of other medicines is ongoing. Kaur (2012) compiled information related to plant species with anticancer properties. There were a total of 1287 plant species belonging to 178 families. The majority of (98) of plant species were from the family *Fabaceae*. Most of these plant species have yet to be explored at the scientific level for their bioactivity. Thus, a detailed study of phytoconstituents of these species and clinical trials are needed.

19.5 COVID-19 AND HERBAL MEDICINES

In the history of modern medicines and after the outbreak of COVID-19, the Drug Controller General of India has allowed Sun Pharmaceutical along with International Centre for Genetic Engineering and Biotechnology (ICGEB), New Delhi, Indian Institute of Integrative Medicine (IIIM), Jammu, Council of Scientific and Industrial Research (CSIR), New Delhi and Department of Biotechnology (DBT), New Delhi to conduct trials. We are going to undertake trials of the plant-based drug (phytopharmaceutical) to combat COVID-19. This is a good opportunity for researchers to provide scientific bases to traditional knowledge systems.

A single plant may possess a number of medicinally important phytochemicals, thus exploration of plant species for this purpose is required. Recently even the Ministry of AYUSH recommended the use of herbal formulations of *Tinospora cordifolia, Andrographis paniculata, Zizyphusjujube*, etc., to combat COVID-19 (Vellingiri et al., 2020).

The World Health Organization (WHO, 2020) has considered *Artemisia annua* a possible candidate to treat COVID-19, but first it has to be tested for efficiency and undesirable side effects. Any medicine to be tested around the world generally adopts the standards procedures. The WHO is also collaborating with the institutes of traditional medicines for safety and clinical efficacy. This organization continuously supports the exploration of traditional healthcare practices. Many plant-based substances have been suggested to treat COVID-19. The use of any such proposed substance without robust investigations related to quality, safety, and efficiency may land users in serious trouble.

Ang et al. (2020) reviewed the pattern of the use of herbal medicines to treat COVID-19. In this review, some medicinal plant (or parts) preparations individually or in different combinations were used to treat vomiting, cough, cold, as anti-inflammatory, antiasthmatic, antiepileptic, antiviral, etc. All these preparations were effective at boosting immunity in general and at reducing or minimizing the effects of COVID-19, but not recommended directly for its treatment. Thus, clinical trials are required to evaluate the safety and efficacy of these medicines before their use as an alternative treatment for COVID-19.

According to Ang et al. (2020b) a combination of herbal medicine and Western medicine has the potential to treat COVID-19. However, additional and exclusive Randomized Controlled Trials (RCT) are needed to assess the effectiveness and adverse effects of herbal medicines used to treat COVID-19.

Allium cepa (onion), *Allium sativum* (garlic), and *Zingiber officinale* (ginger) are major staples of kitchens in different parts of the world. They are antibacterial, antiviral, anti-inflammatory, antioxidant, and antitumor in nature and also help the immune system deal with various pathogens. Garlic and onions possess 70 phytochemicals that reduce the risk of hypertension. Ginger improves blood circulation and digestion. Antioxidant, anti-inflammatory, and immune-modulation potential of garlic, ginger, and onion are beneficial in COVID-19 patients during the early stages (Deng et al., 2020).

Out of 100 screened patients, 42 were included in the trial: 28 in the Chinese Herbal Medicine (CHM) plus standard care group and 14 in the standard care alone group. All 42 patients were women and 83% of them were more than 65 years of age. Out of 42, two patients died during the trials. The death rate was lower in the patients of CHM plus group than the standard care alone group. The effects of CHM in the treatment of COVID-19 have been reported from around 200 countries of the world, but require thorough clinical trials (Ye and G-Champs, 2020).

The COVID-19 virus is phylogenetically related to the SARS coronaviruses. Both of these viruses use a similar receptor called angiotensin-converting enzyme 2 (ACE2) to enter the host cell. Heidary et al. (2020) discussed the potentiality of Persian herbal medicines to treat COVID-19. Ziai and Heidari (2015) reported the plant extracts of *Cerasus avium* (L.) Moench, *Alcea digitata* (Boiss.) Alef, and *Rubia tinctorum* L had 100% inhibitory effect on ACE. Based on this, Heidary et al. (2020) suggested the use of these herbal products or any other with similar mechanisms to treat

COVID-19. The worldwide outbreak of COVID-19 may be due to the mutative nature of the virus. Therefore, further detailed studies are necessary to validate this possibility.

Tillu et al. (2020) reported *Withaniasomnifera* (Ashwagandha), *Tinospora cordifolia* (Guduchi), *Asparagus racemosus* (Shatavari), *Phyllanthus embelica* (Amalaki), and *Glycyrrhiza glabra* (Yashtimadhu) as potential immune modulators. These can additionally be considered for the treatment of COVID-19. During clinical study, Ashwagandha roots showed immune homeostasis effect. Thus, these roots can be used to improve the immunity of the host of COVID-19. There are still several gaps in the management of COVID-19 using modern medicines. These types of interventions are considered simple, affordable, acceptable, and promising but need detailed investigations. They have also drawn the attention of all concerned in general and the WHO in particular to explore traditional medicine systems for the COVID-19 pandemic.

Several studies have reported the bioactivity of foods and herbs against the influenza virus and SAR-CoV-1. But clinical trials were conducted only to a limited extent. There is a variety of antimicrobial and antiviral essential oils. The essential oil vapors of *Cinnamomum zeylanicum* (cinnamon), *Citrus bergamia* (bergamot), *Cymbopogon flexuosus* (lemongrass), *Eucalyptus globulus* (eucalyptus), and *Pelargonium graveolens* (geranium) possess anti-influenza viral activity. Aerosolized tea tree (*Melaleuca alternifolia*) oil is known to inhibit airborne viral particles. Studies related to the potential of essential oil vapors are very limited. The bioactive compounds of *Cape jasmine, Chinese mahogany, Eucalyptus, Garlic, Ginger,* etc., were active against the influenza virus. Since there are only a limited number of allopathic medicines, the dietary therapy and herbal medicines are considered as potentially effective against COVID-19. But experimental validation is needed before their use is recommended (Panyod et al., 2020).

19.6 CONCLUSIONS

Traditional plant-based (herbal) medicines seem to be effective, affordable, and safe but to date there is not a single herbal preparation to fight with the Covid. These remedies were generally found to be immune modulator. Even the recently claimed cure 'Coronil' for COVID by the Patanjali Ayurved has been declared an immunity booster by the Ministry of AYUSH. There are thousands of plant species with general antiviral activity, but these need to be studied clinically to determine their potential against COVID-19.

REFERENCES

Ang L, Lee HW, Choi JY, Zhang J, Lee MS (2020a). Herbal medicine and pattern identification for treating COVID-19: A rapid review of guidelines. *Integrat Med Res* 9: 100407. (www.sciencedirect.com/science/article/pii/S2213422020300391)

Ang L., Song E, Lee HW, Lee MS (2020b). Herbal medicine for the treatment of Corona virus disease 2019 (COVID-19): A systematic review and meta-analysis of randomized controlled trials. *J Clinic Med* 9: 1583. (www.ncbi.nlm.nih.gov/ pmc/articles/PMC7290825/)

Deng Jia-gang, Hou X, Zhang T, Bai G, Hao E, Chu JJH, Wattanathorn J, Sirisa-ard P, Ee CS, Low J, Liu C (2020). Carry forward advantages of traditional medicines in prevention and control of outbreak of COVID-19 pandemic. *Chinese Herb Med* 12(3): 207–213. (www.sciencedirect.com/science/ article/pii/S1674638420300484)

Ekor M (2014). The growing use of herbal medicines: issues relating to adverse reactions and challenges in monitoring safety. *Frontiers in Pharmacology*, 4: 177 (https://doi.org/10.3389/fphar.2013.00177)

Gupta S, Sidhu MC, Ahluwalia AS (2017). Plant based remedies for the management of diabetes. *Current Bot* 87: 34–40. (https://updatepublishing.com/journal/index.php/cb/article/view/3169)

Heidary F, Varnaseri M, Gharebaghi R (2020). The potential use of Persian herbal medicines against COVID-19 through Angiotensin-Converting Enzyme-2. *Archive Clinic Infect Diseases* 5: e102838. (https://sites.kowsarpub.com/archcid/ articles/102838.html)

Kaur A (2012). *Medicinal Plants for Cancer Treatment*. M. Sc. thesis, Panjab University, Chandigarh.

Pandey MM, Rastogi S, Rawat AKS (2013). Indian traditional ayurvedic system of medicine and nutritional supplementation. *Evidence-Base Complement Alternat Med* 2013: 1–13. (https://pubmed.ncbi.nlm.nih.gov/23864888/)

Panyod S, Ho C, Sheen L (2020). Dietary therapy and herbal medicine for COVID-19 prevention: A review and perspective. *J Trad Complement Med* 10(4): 420–427. (www.ncbi.nlm. nih.gov/pmc/articles/PMC7260602/)

Sharma T, Sidhu MC (2017). Cytomorphological and preliminary phytochemical screening of *Eclipta alba* (L.) Hassk. *Int J Green Pharma* 11(1): S23–S32. (www.green pharmacy.info/index.php/ijgp/article/view/855)

Sidhu MC, Sharma T (2013a). A database of antidiabetic plant species of family asteraceae, euphorbiaceae, fabaceae, lamiaceae and moraceae. *Int J Herb Med* 1(2):187–199. (www. florajournal.com/archives/?year=2013&vol=1&issue=2&part=A&ArticleId=26)

Sidhu MC, Sharma T (2013b). Medicinal plants from twelve families having antidiabetic activity: A review. *America J PharmTech Res* 3(5): 36.

Sidhu MC, Sharma T (2014). Antihyperglycemic activity of petroleum ether leaf extract of *Ficus krishnae* L. On Alloxan-induced diabetic rats. *Ind J Pharma Sci* 76(4): 323–331. (https://pubmed.ncbi.nlm.nih.gov/25284930/)

Sidhu MC, Thakur S (2015). Documentation of antidiabetic medicinal plants in district Mandi of Himachal Pradesh (India). *International Journal of PharmaTech Research* 8(8): 164–169. (https://sphinxsai.com/2015/ph_vol8_no8/1/(164-169)V8N8PT.pdf)

Sidhu MC, Sharma T (2016). Meiotic and phytochemical studies of three morphotypes of *Solanum nigrum* L. from Punjab, India. *Ind Drugs* 53(4): 20–28. (www.indiandrugsonline.org/ issuesarticle-details?id=NTM1)

Tillu G, Chaturvedi S, Chopra A, Patwardhan B (2020). Public health approach of ayurveda and yoga for COVID-19 prophylaxis. *The J Alternat Complement Med* 26(5): 360–364. (www.liebertpub.com/doi/10.1089/acm.2020.0129)

Vellingiri B, Jayaramayya K, Iyer M, Narayanasamy A, Govindasamy V et al. (2020). COVID-19: A promising cure for the global panic. *Sci Total Environ* 725: 138277. (www.sciencedirect.com/science/article/abs/pii/S0048969720317903)

WHO (2020). WHO supports scientifically-proven traditional medicine. (www.afro.who.int/news/who-supports-scientifically-proven-traditional-medicine).

Ye Y, G-Champs T (2020). Guideline-based Chinese herbal medicine treatment plus standard care for severe corona virus disease 2019 (G-CHAMPS): Evidence from China. *Front Med* 7: 256. (www.ncbi.nlm.nih.gov/pmc/articles/ PMC7267028/).

Ziaei S, Heidari M, Amin G, Kochmeshki A, Heidari M (2009). Inhibitory effects of germinal angiotensin converting enzyme by medicinal plants used in iranian traditional medicine as antihypertensive. *J Kerman Univ Medical Sci* 16(2): 134–143. (https://jkmu.kmu.ac.ir/article_17291.html)

Zulcafli AS, Lim C, Ling AP, Chye S, Koh R (2020). Antidiabetic potential of *Syzygium* sp.: An overview. *The Yale J BiolMed* 93(2): 307–325. (https://read.qxmd.com/read/32607091/antidiabetic-potential-of-syzygium-sp-an-overview).

20 Role of Physiotherapy in Covid-19 Pandemic

Suraj Kumar,[1] Gowrishankar Potturi,[1] and Ranjeev Kumar Sahu[2]

[1]Department of Physiotherapy, Faculty of Paramedical sciences, Uttar Pradesh University of Medical Sciences, Saifai, Etawah, Uttar Pradesh, India
Ph: 7830337168, Email: surajdr2001@yahoo.com
[2]Babasaheb Bhimrao Ambedkar University, Lucknow, Uttar Pradesh, India
Email: rksahu2001@yahoo.com

CONTENTS

20.1 INTRODUCTION

Severe acute respiratory syndrome coronavirus 2 (SARS-CoV-2) is a highly contagious respiratory virus that spread rapidly among humans.[1] Early intubation is recommended when there is severe hypoxemia with SPO2 <90 mmHg.[2] Mortality during mechanical ventilation appears to be high.[3] Currently, 5% of COVID-19 cases require admission to ICU due to respiratory distress. The global current treatment options mostly followed for Acute respiratory Distress Syndrome (ARDS) due to COVID-19 pneumonia include high-flow nasal canula (HFNC), non-invasive ventilation (NIV), mechanical ventilation with lower tidal volumes, and continuous positive airway pressure (CPAP). There is high risk of aerosol dispersion during these techniques, provoking the risk of cross-infection, as most hospitals do not have negative pressure isolation rooms. There is also an acute shortage of mechanical ventilators globally.[4] Early intervention with chest physiotherapy can prevent the need for intubation in COVID-19 patients with respiratory distress. Chest physiotherapy focuses on airway clearance techniques, re-expansion of collapsed segments of the lungs, and maintaining adequate levels of oxygenation. These may delay or prevent the need for intubation and may facilitate early weaning and reduce the chances of re-intubation.

DOI: 10.1201/9781003358909-20

20.2 SHORT-WAVE DIATHERMY FOR COVIDPNEUMONIA

The bio-thermal therapy that uses deep tissue heating for various clinical conditions is a research-based approach that dates back 200 years ago.[4,5] Short-wave diathermy (SWD) for Covid is frequently used in infections and has a century-long track record of improving infection rates. It is frequently called the"physical therapy antibiotic." It has been shown to reduce death-rate 50% in pneumonia patients and to double the recovery rates. For COVID-19, the Portland Clinic of Holistic Health recommended the use of SWD for COVID-19 pneumonia to combat the infection and inflammation of the lungs, reduce mortality, and enhance recovery.[6] The benefits of SWD on pneumonia were first recognized by Frederick DeKraft and Byron Sprague Price of New York, about 1906.

In 1921, there was a severe pneumonia epidemic at the U.S. Marine Hospital in New York. A study was conducted on 41 patients with radiologically and laboratory diagnosed pneumonia cases, where 20 cases were given SWD and 21 were controls. The death rate was 17% in experimental (i.e., SWD group compared to 42% in control group)[7]. The effects of SWD can be summarized as follows:

1. Temperature: There was a drop in temperature resulting in conserving body energy and reducing pulse and respiratory rate.
2. Circulation: There was a lessening of cyanosis, which was more evident when lower lobes were affected. This effect is due to decrease in pain and improvement in the intrapulmonic circulation around the consolidated areas. It decreases the load on the right ventricle, resulting in improvement in the quality of pulse.
3. Respiration: The respiratory rate was reduced by an average of five per minute and respiratory grunt disappeared. All these effects contribute to improved pulmonary circulation in response to heat.
4. Pain: There was reduction in pain resulting in reduced anxiety levels and improved respiratory efforts.

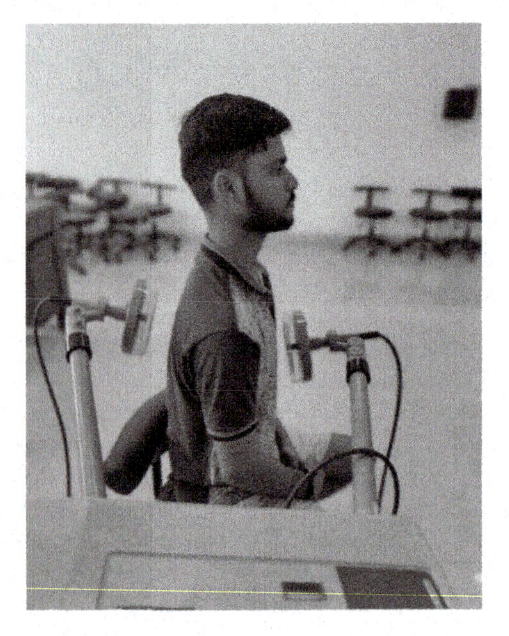

FIGURE 20.1 Technique of application of SWD.

A maximum of six to eight treatment sessions with SWD showed marked reduction in the size of inflammation and pulmonary opacity in radiographs. SWD applied through the chest in pneumonia patients provides a sense of wellbeing due to the *deus ex machine* principle. It has also shown to reduce bronchospasm, induce general relaxation, and reduce the viscosity of mucous aiding in proper drainage[7] (Figure 20.1).

SWD is a physical therapy antibiotic that has antimicrobial effects, possessing the capacity to reduce cyanosis, pain, dyspnoea, cardiac embarrassment, and induces restful sleep. The use of SWD in treating pneumonia has been successful since 1906. In the present pandemic, where 5% of the cases require mechanical ventilation due to development of pneumonia-like symptoms, SWD can be given at oligothermic dosages to combat the infection and respiratory distress. As there is a global shortage of ventilators, a risk of aerosol dispersion cross-infection, SWD with acute precautionary measures could be beneficial in reducing temperature, improving circulation, and reducing respiratory rate and pain. We recommend that RCTs be conducted for SWD to study its effect on cases of COVID pneumonia.

20.3 ROLE OF PHYSIOTHERAPY IN REDUCING COVID HOSPITAL-ACQUIRED ANXIETY AND DEPRESSION

COVID-19 is a serious pandemic the whole world is fighting. It is a serious burden on healthcare workers, administration, and government on critical care resources due to a large number of patients needing critical care.[8] It is estimated that around 5–15% of COVID-19 patients require critical care in COVID HOSPITAL.[9] An intensive care unit is also known as an intensive therapy unit of a hospital that caters to patients with severe or life-threatening illnesses and provides constant and close supervision from life-supporting equipment and medication.[10] The COVID HOSPITAL is a complex and stressful environment that is associated with unfavorable physical, psychological, cognitive, and functional consequences for patients.[4] It is often observed that patients who survive critical illness experience high rates of anxiety and depression that can persist months to years after hospital discharge.[11,12] COVID-19 patients exhibit a wide range ofpsychological stress that can impact the general health of the patient.[13] Anxiety, panic, stress, and depression have been linked to infectious epidemics.[14] Patients also have the risk of developing social stigma and xenophobia.[15] The role of physiotherapists in the management of anxiety and stress is becoming popular and recognized in current medical scenarios, yet the research base of this practice is limited.[16] Aerobic exercises and breathing exercises can be effective in reducing anxiety and depression.[17] RCTs on the role of physiotherapy on Covid hospital-acquired anxiety and depression in Covid-19 patients should be conducted to demonstrate that adequate medical care and physiotherapy in the form of aerobic activities, deep breathing exercises, and psychological support by verbal communication along with effective medicines can alleviate psychological symptoms.

The various techniques employed are as follows.

20.3.1 Active Cycle of Breathing Technique (ACBT)

ACBT is an active breathing technique performed by the patient, which helps in clearing the chest and improving lung function. The technique is performed in three steps:

1. Breathing control: This helps in relieving tightness in the chest, difficulty in breathing, and promotes relaxation. Patients are asked to breath in gently through the nose while keeping the shoulders relaxed and keeping eyes closed to concentrate on breathing.
2. Deep breathing exercises: Patients are asked to take a long, slow, and deep breath through the nose and to hold it for 2–3 seconds before breathing out gently and relaxed like a sigh.
3. Forced expiratory technique: Patients are asked to take a long breath in and to blow it out in a single phase through the mouth.

20.3.2 Proprioceptive Neuromuscular Facilitation of Respiration (PNF)

PNF techniques improve respiratory functions such as tidal volume, inspiratory reserve volume, expiratory reserve volume, inspiratory capacity, and vital capacity.

Patients are made to lie flat in supine. The therapist place his or her open hands on the lateral surface of the 8th, 9th, 10th, and 11th ribs bilaterally on the patient. The therapist then instructs the patient to take a deep breath, and at maximum inspiration the therapist asks the subject to hold the breath for five seconds while applying minimal manual resistance to the lower ribs bilaterally that is directed downwards and medially. Later patients are asked to breathe out maximally, at maximal expiration. The therapist then pushes the lower ribs bilaterally in downward and medial directions. The cycle isrepeated 10 times.

The PNF technique stimulates proprioceptive and tactile stimuli, which produce consistent reflexive responses on respiratory muscles leading to expansion of ribs, increased epigastric excursion, and respiratory depth and rate. It also properly aligns the respiratory muscles with respiratory rhythms.

20.3.3 Light Exercises

Patients are asked to repeat all active light exercises of limbs with 10 repetitions each three times a day (Figure 20.2).

Patients are given psychological support by verbal communication daily about healthy lifestyle and the importance of physiotherapy. Patient questions regarding COVID-19 are answered, and all efforts to remove any myths regarding COVID-19 are made.

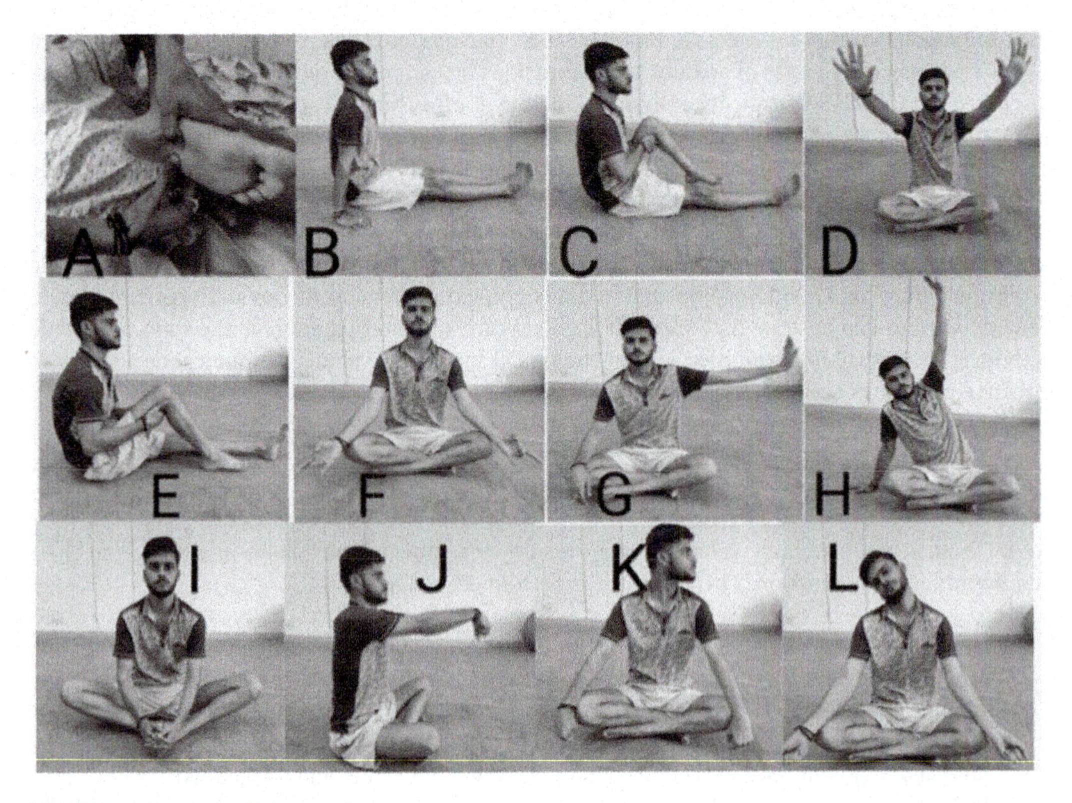

FIGURE 20.2 Active light exercises.

20.4 ROLE OF PHYSIOTHERAPY IN MANGEMENT OF DYSPNOEA AND IMPROVING SPO_2 LEVELS IN COVID-19 PATIENTS

Dyspnea is commonly defined as difficulty in breathing where a subject experiences breathing discomfort that consists of qualitatively distinct sensations that may vary in intensity.[18] Dyspnea is a highly debilitating symptom, which often leads to anxiety, depression, and exercise avoidance. It can worsen deconditioning leading to reduced health-related quality of life.[19] Onset of dyspnea along with noticeable drop in oxygen saturation is the key to distinguishing COVID-19 from other common illnesses. Researchers have observed that though many patients recover from mild symptoms of COVID-19 (nasal congestion, cough, fever, sore throat, diarrhea, abdominal pain, headache, myalgia, back pain, insomnia, fatigue), the onset of dyspnea occurring between day 4 to 10 of onset of symptoms can be discerned from other common illnesses.[20] Extensive inflammation of the bilateral respiratory bronchioles due to excessive activation of proinflammatory cytokines and chemotactic aggregation of t-lymphocytes at the site of inflammation may be the possible cause of dyspnea in COVID-19. In a RCT by the authors, there was a weak positive correlation between levels of dyspnea with age, strong negative correlation with SPO_2 levels in COVID-19 patients, and Neurophysiological Facilitation (NPF) improves in the levels of dyspnea and SPO_2 in COVID-19 patients.

The following techniques are used (Figure 20.3):

A. LU10 acupressure point stimulation
B. PNF in Respiration
C. NPF in Respiration

20.4.1 Neurophysiological Facilitation (NPF) of Respiration

NPF techniques produce reflex respiratory movement response by application of external proprioceptive and tactile stimuli that can alter the rate and depth of breathing. There are various NPF techniques. In the case of COVID we perform thoracic vertebral pressure, co-contraction of the abdomen, and intercostal muscle stretch techniques.

a) Thoracic vertebral pressure: The patient is positioned prone. A high manual pressure is applied to the thoracic vertebra in the region of T2-T5 and low manual pressure is applied to the

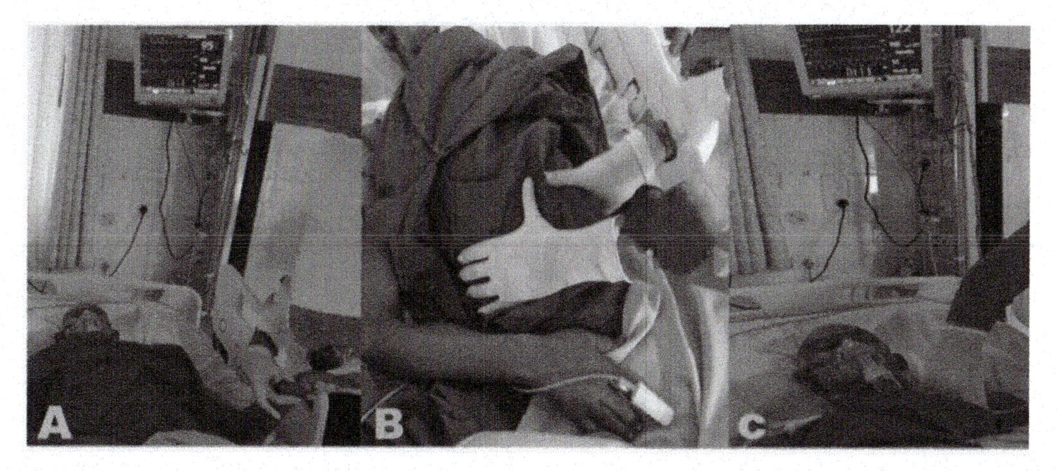

FIGURE 20.3 Techniques of chest physiotherapy.

region of T9-T12. The pressure at each vertebra level is given for 20 sec. This increases the rate and depth of breathing by dorsal root-mediated intersegmental reflex.

b) Intercostal stretch: The patient is made to lie flat in supine position. The limbs are positioned in neutral position. The therapist with his index and middle finger applies pressure over the intercostal muscles between the 2nd and 3rd rib bilaterally for 20 seconds. The pressure is directed caudally towards the next rib but not towards the vertebra and released. The cycle is repeated for 3 minutes. This stimulates the respiration by intercostal stretch reflex.
c) Co-contraction of abdomen: The patient is made to lie in supine position flat with limbs in neutral position. The therapist then applies moderate and firm pressure on the lower ribs and pelvis at right angles to the patient on the same side. The technique is repeated bilaterally one after the other side. The pressure is applied for 20 seconds. The cycle is repeated for 3 minutes alternatively. This technique activates the abdominal muscles by stretch reflex.

20.4.2 Acupressure

Acupressure is a method of massaging various acupuncture points with fingertips or knuckles. Acupressure has proved beneficial in patients with respiratory disease, which is associated with cough, dyspnea, and congestion.

We use the LU10 of the lung meridian, which is situated at the index finger. The point is stimulated with the therapist thumb for 20 seconds, 5 cycles in each session. This point is believed to increase oxygenation, decrease cough, and respiratory distress (Figure 20.4).

20.5 EXERCISES TO IMPROVE IMMUNITY DURING COVID-19 PANDEMIC

Physical inactivity is a major risk factor for many diseases and premature deaths.[21] Several studies have suggested that inactivity and poor nutrition are the actual causes of death.[22] Physical inactivity can lead to premature aging, obesity, cardiopulmonary diseases, musculoskeletal fragility, and depression. These problems are usually termed under the umbrella word "disuse syndrome."

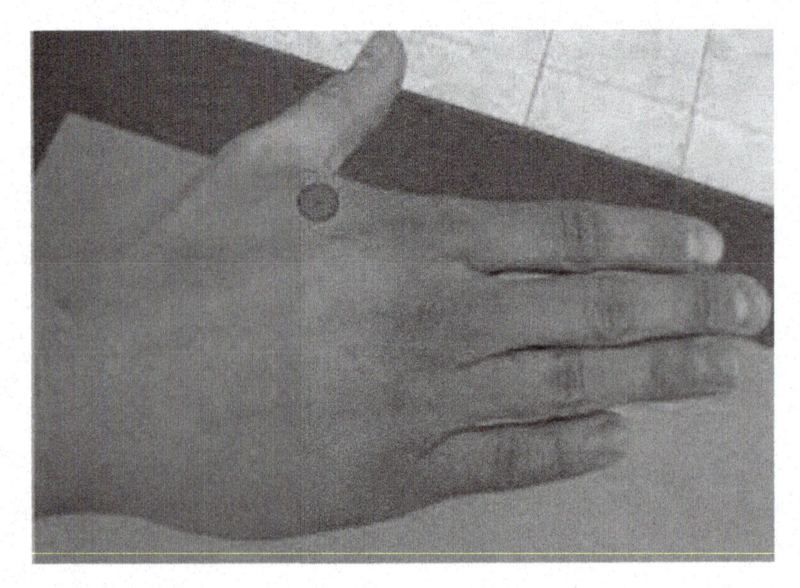

FIGURE 20.4 Acupressure point for improving oxygenation.

FIGURE 20.5 (1–13) Demonstrating various exercises to improve immunity.

The advantage of these exercises is that they can be performed without any sophisticated equipment, require minimal space, and can be performed in isolation. The following exercises are prescribed (Figure 20.5).

Exercise 1: VIPARITA KARANI WITH CHAIR

Procedure: Lie on the floor with hip and knee bending to 90° supported with chair against the gravity. Maintain this posture for 5–10 minutes. Concentrate on the breath during the whole posture.

Advantages: Improves the blood circulation and lymphatic drainage of upper part of the body, which enhances the immunity.

Precaution: Patients with glaucoma, hernia, and uncontrolled hypertension should take necessary precautions during this exercise.

Exercise 2: SHORT INDOOR BRISK WALK

Procedure: Walk indoors by swinging your arms during walking within your home or balcony for 5 minutes every hour in a day.

Advantages: Maintains healthy weight, prevents or manages various conditions including cardiopulmonary problems and type 2 diabetes, strengthens musculoskeletal system, elevates mood, improves balance and coordination, and improves immunity.

Precautions: Patients with elevated blood pressure, cardiomyopathy, valvular heart diseases, complex ventricular ectopy, and uncontrolled metabolic diseases should do walking with proper precautions and under medical practioner guidance.

Exercise 3 ASSISTED FISH POSE

Procedure: Sit on the floor or mat in long sitting position with legs out in front of you. Place a pillow or bolster just at the back of you. Slowly lower your back and shoulders onto the bolster. Straighten your arms out to the sides. Note that your hips should be on the mat or floor. Close your eyes and concentrate on thebreath for 3 minutes.

Advantages: Improves chest wall mobility and strengthens the respiratory system.

Precautions: Patients with recent head injuries, increased intracranial pressure, uncontrolled hypertension, low backache, and sciatica should take necessary precautions before doing this exercise.

Exercise 4: COBRA POSE

Procedure: Lie on your stomach on the floor or mat. Engage your abs by drawing your navel up and in towards your spine. Bend your elbows and place your hands down on the mat next to your ribs. Keep your thighs on the ground as you inhale to press into your palms and lift the chest off the ground. Keep a slight bend in your elbows while squeezing the shoulder blades together. Hold this pose for 5 breaths, and then slowly lower back down. Repeat the procedure 5 times a day.

Advantages: Stretches the muscles of shoulder, chest, and abdominal thus improves ventilator efforts, decreases stiffness of the lower back, strengthens the arms and shoulders, increases flexibility, invigorates the heart, and elevates the mood.

Precautions: Patients with recent shoulder or arm injuries, osteoporosis, advanced sciatica, or cervical spondylosis should perform this exercise with precautions and under strict medical practioner guidance.

Exercise 5: TREE POSE

Procedure: Stand in stride standing posture withyour feet hip-width distance apart. Make a namaste pose with your palms in front of your chest and engage your abs. Shift your weight on to the right foot and slowly lift the left foot off the ground and rest it on the back of the right calf muscle or inner thigh. Gaze to a point in front of you to maintain balance. Hold this position for 5–8 breaths and repeat by changing legs. Repeat this exercise 5 times a day.

Advantages: Tress pose stretches the muscles of the thigh, groin, and shoulders. It builds tone in the abdominals, calves, and shoulders.

Precautions: Patients experiencing headaches, insomnia, low blood pressure, dizziness, and vertigo should perform this exercise with caution.

Exercise 6: MOUNTAIN POSE

Procedure: Stand on your toes and take your hands above the head and stretch the whole body in the upper direction. Breathe in and out with control for 2 minutes.

Advantages: Stimulates nervous system, improves body balance and posture, strengthens the abdominal and gluteal muscles, and improves the function of respiratory and digestive systems.

Precautions: Patients experiencing headaches, insomnia, low blood pressure, dizziness, and vertigo should perform this exercise with caution.

Exercise 7: CHAIR POSE

Procedure: Try to sit on an imaginary chair with raised arms. Concentrate on your breath and maintain this posture for 2 minutes. Repeat 5 times a day.

Advantages: Stretches the shoulders and chest. Tones up the digestive system and cardiopulmonary system.

Precautions: Patients with sciatica, osteoporosis, hip arthritis, and knee arthritis should do this exercise with caution.

Exercise 8: MODIFIED SUKHASANA

Procedure: Sit on floor or mat with folded legs. Extend the sides of the body and lift the spine. Thenlift the arms above the head and slowly place them behind the back on the floor. Maintain this pose for 10 breaths and repeat for 5 times a day.

Advantages: Improves flexibility of upper spine, and allows optimal breathing and movement of "prana" (life force energy) throughout the body.

Precautions: Patients with shoulder injuries, low back pain, cervical spondylosis, and knee arthritis should do this exercise with caution.

Exercise 9: BRIDGING

Procedure: Lie straight on your back on floor or mat. Bend your hips and knees with feet flat on the floor. Raise your buttocks off the ground and straighten your spine while placing the arms at the side of the body. Maintain this posture for 10 breaths. Repeat 10 times a day. Do not use a pillow while performing this exercise.

Advantages: Strengthens the muscles of the lower limbs, paraspinal muscles, and chest, relaxes the wole body, reduces anxiety, improves digestion, and relieves insomnia.

Precautions: Patients with neck pain, low back pain, knee injuries, and shoulder injuries should perform this exercise with caution.

Exercise 10: BILATERAL SIDE STRETCHES

Procedure: Sit towards the front of a chair with your legs wide, knees and toes pointed out. Place your right hand on yourright thigh and lift the left arm towards the ceiling. On an inhale reach straight up then exhale to side bendover to the right. Hold for fivebreaths and repeat the same on the other side.

Advantages: Chest expansion, improves chest mobility, and improvedventilation.

Precautions: Patients with neck pain, low back pain, and shoulder injuries should perform these exercises with caution.

Apart from the above exercises, the following breathing exercises maybe beneficial in improving ventilatory capacity and efforts.

Exercise 11: ANULOM VILOM

Procedure: Sit in sukhasana or padmasana with spine straight. Using the right thumb, block the right nostril and inhale through the left nostril for 2 seconds. Now block both nostrils and hold the breath for 4 seconds. Repeat the same with opposite nostril. Repeat for 10 rounds.

Advantages: Helps in relieving anxiety and depression, warms up the body before doing other exercises, helpful in treating respiratory problems and high blood pressure.

Precautions: Cardiac patients, hypertensive patients, and pregnant women should not hold their breath while performing this exercise; they should just keep inhaling and exhaling.

Exercise 12: PURSED LIP BREATHING

Procedure: Sit in Sukhasana or Padmasana with spine straight. Relax shoulders as much as possible. Now inhale through your nose for 2 seconds, feeling the air move into your abdomen. Now purse your lips (make a small opening in the mouth) and exhale twice the time you have taken to breathe in. Repeat for 10 rounds.

Advantages: Relieves shortness of breath by slowing the breath rate, keeps the airways open for longer time, decreases work of breathing, and improves ventilation.

Precautions: Cardiac patients, hypertensive patients, and pregnant women should not hold their breath while performing this exercise; they should just keep inhaling and exhaling.

Exercise 13: DIAPHRAGMATIC BREATHING EXERCISE

Procedure: Lie straight on your back on floor or mat with a pillow under the knees and head. Place one hand on the upper chest and the other just below your rib cage. Breathe in slowly through the nose so that your stomach moves out against your hand. The hand on the chest should remain as still as possible. Repeat 10 times.

Advantages: It helps in strengthening the diaphragm, decreases work of breathing, and decreases oxygen demand.

Precautions: Patients with bronchial asthma, chronic bronchitis, reflux gastritis, or recent rib or chest injuries should perform this exercise with care.

Apart from physiotherapy, subjects in this research were also given a concoction made from various spices, Raj Nirwan Quath (RNQ), and mineral combination tablet, Raj Nirwan Bati (RNB). As a prophylactic measure, even healthcare workers are given RNQ. A study was conducted to analyze the prophylactic effects of RNQ. The study was conducted at a level-3 COVID-19 hospital in Uttar Pradesh. Among the 1767 healthcare workers of the Covid-19 Hospital, 1255 healthcare workers including physicians, surgeons, intensive care specialists, nurses, technicians (medical), physiotherapists, sanitation workers, and ward boys who worked in the COVID section of the university hospital were recruited for this study to analyze the prophylactic effects of RNQ and RNB. After the administration of RNQ and RNB, healthcare workers remained immune against COVID-19 and no healthcare worker reported any symptoms.

20.6 CONCLUSION

COVID-19 patients are usually affected from respiratory and psychological problems. They suffer from dyspnea, reduced respiratory rate as well as depth of respiration, weakness, anxiety, and depression leading to an increase in the risk of mortality. Physiotherapy interventions in the form of NPF, PNF, ACBT, acupressure, andaerobic activities improve the respiratory health of COVID-19 patients along with reducing levels of anxiety and depression. Early physiotherapy when given regularly can avoid the need for mechanical ventilation, and thus can prevent mortality in COVID-19 patients. Hence, early interventional physiotherapy can be incorporated intoscheduled treatment protocols of COVID-19 patients globally.

REFERENCES

1. del Rio C, Malani PN. Novel coronavirus-important information for clinicians JAMA, 323 (2020): 1039–1040.
2. Sohrabi C, Alsafi Z, O'Neill N, Khan M, Kerwan A, Al-Jabir A, et al. World Health Organization declares global emergency: a review of the 2019 novel coronavirus (COVID-19) Int J Surg, 76 (2020): 71–76.
3. Alhazzani W, Møller MH, Arabi YM, et al. Surviving sepsis campaign: guidelines on the management of critically ill adults with coronavirus disease 2019 (COVID-19) Intensive Care Med. 2020.
4. Surajkumar, Potturi G, Rajkumar, Dubey N, Agarwal A, Kumar A. A clinical review on low dose short wave diathermy therapy as a physical antibiotic on Covid-19 pneumonia. Arch Phys Med Rehabil 3 (2020): 062–067.
5. Huang-Wen H, Chihng-Tsung L. Heating in Biothermal Systems [Internet]. Convection and Conduction Heat Transfer. *InTech*. (2011). http://dx.doi.org/10.5772/19567

6. Jan Zbigniew Szopinski. Reflexive physical therapies. *The Biological Action of Physical Medicine* Academic Press (2014), pp. 73–222. https://doi.org/10.1016/B978-0-12-800038-0.00005-5.
7. Blackwood S. Shortwave diathermy (S.W.D.) in the treatment of unresolved pneumonia. S Afr J Physiother [Online] 36.3 (1980): 72–73.
8. Shang Y, Pan C, Yang X, et al. Management of critically ill patients with COVID-19 in COVID HOSPITAL: statement from front-line intensive care experts in Wuhan, China. Ann Intensive Care 10 (2020): 73.
9. Phua J, Weng L, Ling L, et al. Intensive care management of coronavirus disease 2019 (COVID-19): challenges and recommendations. Lancet Respir Med. 8(5) (2020):506–517.
10. Goran, SF. A Second Set of Eyes: An Introduction to Tele-COVID HOSPITAL. Crit Care Nurse. 30(4) (2010):46–55.
11. Hopkins RO, Weaver LK, Collingridge D, et al. Two-year cognitive, emotional, and quality-of-life outcomes in acute respiratory distress syndrome. Am J Respir Crit Care Med 171 (2005):340347.
12. Zoulay E, Pochard F, Kentish-Barnes N, et al. FAMIREA Study Group: risk of post-traumatic stress symptoms in family members of intensive care unit patients. Am J Respir Crit Care Med 171 (2005):987994.
13. Brooks SK, Webster RK, Smith LE, et al. The psychological impact of quarantine and how to reduce it: rapid review of the evidence. Lancet. 395(10227) (2020):912–920.
14. Wang C, Pan R, Wan X, et al. Immediate psychological responses and associated factors during the initial stage of the 2019 corona-virus disease (COVID-19) epidemic among the general population in China. Int J Environ Res Public Health 17(5) (2020):E1729.
15. Pompeo-Fargnoli A, Fargnoli A. The mental health impact of the COVID19 crisis: the battle ahead for inpatient survivors. Psychosomatics. Published online (2020).
16. Walker J, Shepherd W. Anxiety disorders: a nation-wide survey of treatment approaches used by physiotherapists. Professional Articles 7(10) (2001): 536–548.
17. Guszkowska M. Effect of exercise on anxiety, depression and mood. Psychiatr Pol 3 (2004):611–20.
18. Dyspnea. Mechanisms, assessment, and management: a consensus statement. American Thoracic Society. Am J Respir Crit Care Med 159 (1999):321–340.
19. O'Donnell DE, Ora J, Webb KA, et al. Mechanisms of activity-related dyspnea in pulmonary diseases. Respir Physiol Neurobiol 167 (2009): 116–132.
20. Cohen PA, Hall L, Johns JN, Rapoport AB. The early natural history of SARS-CoV-2 infection: clinical observations from an urban, ambulatory COVID-19 clinic [published online April 20, 2020]. Mayo Clin Proc. doi:10.1016/j.mayocp.2020.04.010.
21. Lees SJ, Booth FW, Sedentary death syndrome. Can J Appl Physiol 29 (2004): 447–460.
22. Mokdad AH, Marks JS, Stroup DF, Gerberding JL. Actual causes of death in the United States, 2000. JAMA 291 (2004): 1238–1245.

21 India's Healthcare Scenario during 2020

Prakash S. Lohar
MGSM's ASC College, Chopda Affiliated to KBC North Maharashtra University, Jalgaon, India

CONTENTS

21.1 INTRODUCTION

There are several challenges in the current state of healthcare in India. Some of these include inadequate reach of basic healthcare services, shortage of medical personnel, quality assurance, inadequate outlay for health, and most importantly insufficient impetus to research. During the Covid-19 pandemic, the healthcare sector in India has faced several challenges, which cannot be resolved by the government's healthcare schemes alone (Shing et al, 2020). The key is to get the private sector to participate, while the government continues to invest and enable. Technological innovations including artificial intelligence, big data, and digital tools of medical importance will improve the Indian healthcare sector.

21.2 HISTORY OF PANDEMICS

On 23 September 1896, the first "official" case of the bubonic plague in India was reported by an Indian physician in Bombay. The patient was suffering from high fever, large tumours, swollen lymph glands, and eventual gangrene in the extremities. Without treatment, the patient would succumb within days. Originating in Yunnan, China, the plague, caused by the bite of infected fleas in rodents, or contact with the carcass of an infected rodent in humans, travelled across the country to Hong Kong, from where it arrived in British India and the rest of the world, through trading ships. From thereon, the plague epidemic spread rapidly to major cities of India. Between the plague's arrival in 1896 and 1921, an estimated 12 million Indians lost their lives, compared with 3 million in the rest of the world combined (The Wire, 2020).

A century ago, an influenza pandemic killed between 50 and 90 million people across the world. It was reportedly quite common that a person falling ill in the morning could be dead by night. The severe acute respiratory syndrome (SARS) in 2003 eventually killed 774 people in 26 countries across five continents. In 2009, the H1N1 (swine flu) infected two billion people in six months (Van

DOI: 10.1201/9781003358909-21

et al, 2013). More than 11,000 fatalities and 30,000 infections occurred across 10 countries during the 2013–2016 due to West African Ebola outbreak.

As of 2019, the WHO estimated that one billion people lack access to basic care, and a further 100 million are pushed into poverty trying to access it. Considering the seriousness of global crisis in healthcare, there is a strong need for social entrepreneurship in health care. For success in health care, there is a need to turn the problem around from treatment to prevention. To bring a change, work has to be done at the community level. It is important for the government and industry to develop partnerships with a focus on improving coverage and providing access to quality healthcare services to people. A continuum of care systems also needs to be established by linking hospitals with health centres and with the community. The private-public partnership is the only way to solve India's healthcare problem. The starting point for this, as in any partnership, has to be mutual trust and a recognition that the solution has to be a win-win for both.

In India, the healthcare system consists of public and private healthcare service providers. However, most of the private healthcare providers are concentrated in urban India, providing secondary and tertiary care healthcare services. The public healthcare infrastructure in rural areas has been developed as a three-tier system of sub-centres (SCs), primary health centres (PHCs), and community health centres (CHCs) based on the population. India's healthcare scenario seems to be at a crossroads, where there are some positive achievements on the health indicator, but some serious shortcomings in dealing with health care of newborns to senior citizens. The country has been successful in eradicating polio and reducing epidemics caused by tropical diseases and controlled HIV to a large extent (Naveen et al, 2016). However, it still faces a huge economic burden due to the burden of communicable and non-communicable diseases (NCDs) and struggles to balance accessibility, affordability, and quality.

21.3 CHALLENGES IN THE HEALTHCARE SECTOR

The biggest challenge the healthcare sector is facing currently is a shortage of skilled medical workforce. There is one government doctor for every 10,189 people in India, whereas the WHO recommendation is 1:1000. While six states in India like Delhi, Kerala, Karnataka, Tamil Nadu, Punjab, and Goa have more doctors than the WHO norm, it is a highly imbalanced picture and most of them are unwilling to move to Bihar or Uttar Pradesh (UP), the states that suffer from an acute shortage of doctors. There needs to be a system where we can focus on training and upgrading medical workforce skills in the needed geographies. Here, the private sector can play a vital role in the skill development of doctors, nurses, and health workers. Across the healthcare spectrum, lack of infrastructure access, skilled professionals, quality, patient awareness, and use of health insurance are gaps.

For a healthy India, there is a need for preventive and/or promotive health: prevention, early detection, and treatment. A strong primary healthcare system will lead to a healthier India. In terms of affordability, health care is a calamity that can throw an afflicted family into the jaws of poverty. As per one estimate of WHO, 469 million people in India do not have regular access to essential medicines. Easy availability of essential drugs is critical for India's healthcare system. Indians are the sixth biggest out-of-pocket (OOP) health spenders in the low-middle income group of 50 nations, as per a May 2017 India Spend report. Around 70% of the overall household expenditure on health is on medicines, which is an important factor contributing to poverty. About 55 million Indians were dragged into poverty in a single year due to patient-care costs, according to a study by the Public Health Foundation of India (PHFI). Government spending on public health is just 1.3% of GDP in India (Statista Research Department, 2020). How to spend scarce tax rupees on health care must be considered carefully. World over, the trend is towards healthcare management, focusing on patient education and prevention.

To maximise benefits, it is important to establish a link among various health initiatives and also with related programmes like the National Health Mission. There is a need for uniformly pricing systems for various health interventions, including diagnostics and medicines and ensuring transparency. In 2019, there are caps declared by government on the prices of cardiac stents and orthopedics knee implants. It is always better to have a roadmap for a comprehensive policy formulation in consultation with the hospital sector and other stakeholders. The biggest issue that afflicts both patients and providers is the trust deficit between patients and healthcare service providers. Studies estimate that only two in five patients believe that hospitals act in their best interests.

21.4 GOVERNMENT HEALTHCARE SCHEMES

The Government of India has introduced several programmes to provide better care to its citizens. Ayushman Bharat – *Pradhan Mantri Jan Arogya Yojana* (PMJAY) in 2018, the non-communicable diseases programme, the communicable diseases programmes on tuberculosis (TB), malaria, HIV, and the National Tobacco Control Programme, the Pulse Polio campaign or the national immunisation programme, have done commendable work. However, more needs to be done. The Ayushman Bharat scheme is a radical shift as it reaches out to the poorest and the most vulnerable families, covering complex treatments, hospital stay costs, surgeries, and procedures (in both public and private hospitals). So far, 8,000 private hospitals have joined the programme but bigger, multi-specialty hospitals have yet to be established. PMJAY is the world's largest government healthcare scheme, and 50 crores marginalised beneficiaries have an opportunity to get access to hospital care (Oxfam, 2018). Health care and social security in general can no longer be the responsibility of a single department or ministry. Clean water and air, without which good health is not possible, depend on cropping practices, industrial regulation, pollution control, environmental protection, and law enforcement. It is not a single policy, but interconnected, multi-pronged thinking that is needed.

It is often said that quality, safety, and efficacy of a medicine is a must. There are two intertwined issues: one is how to ensure that best quality medicines are produced and that the integrity of the drug is maintained during the distribution process; second, determing how it can be ensured that quality medicines are affordable, so that they can be accessed when needed. The first issue requires strengthening of quality standards and their strict implementation so that there are not two different qualities of drugs – one for the developed regulated market, and the other, for the Indian market. To address the second issue, it is important to make quality generic drugs available to patients. Government and some start-ups have taken steps in this direction by opening a chain of retail pharmacies that dispense only quality generic drugs at a fraction of the cost.

Amidst various issues Pradhanmantri Bhartiya Janaushadhi Pariyojana (PMBJP) has come as a relief for many who are dependent on generics and cannot afford branded medicines. The presence of a large number of generic drug manufacturers in India means that essential drugs should be widely available across India. The Government of India has, through its Jan Aushadhi scheme, made a large-scale effort to make medicines available at affordable prices at pharmacies across the country. However, accessibility remains an overall challenge that needs to be tackled at various fronts. Primary healthcare centres need to be strengthened and incentives need to be given to generic medicine manufacturers, pharmacists, and supply chains. Public-private partnerships can provide some solutions to this issue. The role of the national regulator needs to be redefined to that of a facilitator. Technology and start-up incubators can play an important role here by creating new platforms for medicine supply and distribution.

21.5 IMPACT OF COVID-19 PANDEMIC ON HEALTH CARE SYSTEM IN INDIA

Coronavirus SARS-CoV2, responsible for Covid-19, emerged from a bio-lab in Wuhan (China) and put an extraordinary and unprecedented burden on health systems, organisations, and professionals (Li et al, 2020). Never before have we faced a global crisis of this magnitude, one that challenges every country's capacity to deliver healthcare services. Globally, 22 November 2022, there have been 635,229,101 confirmed cases of COVID-19, including 6,602,552 deaths and in India there have been 44,669,715 confirmed cases of COVID-19 with 530,591 deaths, reported to WHO.

The sudden explosion of the coronavirus pandemic all over the world has put serious pressure on healthcare facilities, which face multiple challenges and huge responsibility for treating Covid-19 patients. However, India's response to the pandemic must be appreciated since the nation increased production capability of the healthcare sector and built up systems to deal with the situation. As on 17 December 2021, there are a total of 23,680 COVID treatment facilities with 18,12,017 dedicated isolation beds (including 4,94,720 oxygen supported isolation beds) and 1,39,423 ICU beds including 65,397 ventilator beds (Press Information Bureau, 2022). The social discipline in the country, which stood out in comparison with many developed countries, must be appreciated.

All India Institute of Medical Sciences, New Delhi, is the apex medical institution of India and organised a response to the pandemic. There was no perfect strategy to fight this pandemic, and the strategies continued to evolve to learn from success and failures. India has been relying increasingly on quick antigen tests that can report false negatives as much as 50% of the time and its daily testing data does not specify what type of tests – antigen or the more sensitive real time-polymerase chain reaction tests – make up the total or whether they were conducted on symptomatic or asymptomatic people. Despite boosting antigen tests, India's 8% testing positivity rate is far higher than the WHO's 5% benchmark for controlling the outbreak.

The rapid spread and increase in the number of Covid cases has forced the world to look at alternate ways of living and operating both professionally and personally. The worldwide spread of the virus has raised numerous questions since December 2019. Covid-19 has made every country realise the need for a robust and agile healthcare system that can adapt to changing situations in a time-sensitive manner. The healthcare system needs to focus heavily on quality research and development, which will form the building block for a new innovative healthcare system.

Covid-19 has given India an opportunity to skip a full generation and transform healthcare and healthcare delivery using digital technology. Health care can learn so many things from new-age businesses and transform India. The pandemic has created a new urgency to close the gaps in providing healthcare access to remote areas of the country with the use of modern technology. This is where drones play a key role in providing pre-hospital emergency care and accelerating laboratory diagnostic testing and transport of vaccines, hematological products, and automated external defibrillators. In an overarching sense, virtual reality, big data, Internet of Things (IoT), telehealth, robotics, and genomics as areas that will define health care and exponentially impact its delivery to the masses (Zhewei et al, 2020).

Change is an inevitable fact of life. With the hope that the new decade will bring in the much needed positive change in Indian health care, the government, industry leaders, healthcare experts, doctors, and other stakeholders have set their goals for the future. Entrepreneurs in the medical industry have also begun prototyping and testing several of types of drones that can be used to enable access to rural areas for blood sample deliveries, tackling emergency situations, etc. Such innovation will help fill the gap caused by lack of standardised healthcare facilities across the country and bring timely medical services to previously inaccessible regions. Some of the areas of work include robotics where at the Surgical and Assistive Robotics Lab prototype devices such as hyperflexible surgical microscopes for neurosurgery, rehabilitative devices, and exoskeletons are being developed. Some innovations like machine learning and deep learning techniques for the analysis of medical images such as MRI, retina, histopathology slides for improved diagnostic support, ultrasound devices for real-time imaging in neurosurgery, mobile and internet-based solutions for self-help, and therapy for conditions such as depression will be the part of technological driven health care. Assistive solutions for people with visual disabilities such as tactile diagrams with audio readers that can be used for science education for children with visual impairment are also being developed.

The significance of telemedicine is of immense benefit to patients in remote locations (Damodharan, 2020). Offering convenience, it helps them to gain access to doctors without physical travel. This aids better management of chronic diseases and consistent post-operative monitoring. Wearable technology is aiding seamless and accurate health monitoring. For example, the advent of wearable devices supported by mobile technology can now allow a doctor to monitor a patient's vitals remotely. This technology has in-built patient monitoring devices that provide information on heart rhythm, blood pressure, breathing patterns, and blood glucose level.

21.6 TECHNOLOGICAL INNOVATIONS IN INDIAN HEALTHCARE SECTOR

In India, approximately about 5.8 million people die because of diabetes, heart attack, cancer, etc., each year. In other words, out of every 4 Indians, 1 has a risk of dying because of a non-communicable disease before the age of 70. According to the World Health Organization, 1.7 million Indian deaths are caused by heart disease (WHO, 2020). Technological innovations can improve our long-term health by tackling a number of issues in the medical field. Technology has already helped us develop improved medicines and drug dosage combinations, screen patients better, detect diseases early, perform complex surgical interventions, etc.

21.6.1 Artificial Intelligence

As a part of digital transformation, Artificial Intelligence (AI) is a major technological breakthrough for the medical space. The key categories of applications involve diagnosis and treatment recommendations, patient engagement and adherence, and administrative activities. AI allows for the creation of a personalised environment for both patients as well as healthcare providers (Thomas and Kalakota, 2019). Many healthcare experts anticipate that operationalising AI will result in a 10 to 15% increase in productivity over the next two years.

21.6.2 Big Data

Big data refers to complex and large data that cannot be handled by traditional data processing software; it is another area that will allow for preventive care. Next-generation sequencing (NGS) technologies, such as whole-genome sequencing (WGS), whole-exome sequencing (WES), and/or targeted sequencing, are progressively more applied to biomedical study and medical practice to identify disease- and/or drug-associated genetic variants to advance precision medicine (Collins and Varmus, 2015; Carter and He, 2016). Precision medicine allows scientists and clinicians to predict more accurately which therapeutic and preventive approaches to a specific illness can work effectively in subgroups of patients based on their genetic make-up, lifestyle, and environmental factors (Vassy et al, 2015). It will also allow for analytical solutions that will give insight into treatment viability, drug utilisation, and self-care programmes, specific to chronic conditions. The block chain will bring healthcare efficiencies by providing transparency in the process, eliminating intermediaries wherever possible, providing a guard against counterfeit drugs, and reducing unnecessary healthcare costs.

Rapid developments in mobile technologies, cloud computing, digital imaging, machine learning, and 3D printing have paved the way for breakthroughs in the development and adoption of healthcare technologies – from telemedicine to nanotechnology, lab-grown 3D organs to IoT and electronic health records to AI. Moreover, the use of data to build India-centric research (most of the research in the medical field is largely based on Caucasian samples) is possible only through digitalisation.

21.6.3 Digital Tools

The importance of digital tools in health care is significant. Apart from clinical decision support tools, digital tools can make clinical expertise available either remotely or through expertise embedded in medical equipment. The examples include digital pathology, tele-radiology, point-of-care diagnostic devices, tele-consultation, etc. The second issue is to keep millions of frontline health workers updated about the latest knowledge and skills, where e-learning academies and virtual classrooms can be of great help. Thirdly, these can help in enhancing the quality and effectiveness of care being delivered on the ground by paramedic staff like Accredited Social Health Activist (ASHA) workers through the use of appropriate apps on e-pads/mobile phones (Alexander, 2018).

21.7 CONCLUSION

Technological innovations related to medical imaging, data processing, digital tools, and big data derived from gene sequencing and personalised medicine will definitely improve efficiency in healthcare delivery in India. There is a need to collaborate with the larger fraternity and real-time challenges faced by the healthcare sector in India.

REFERENCES

Agarwal A., Nagi N., Chatterjee P., Sarkar S., Mourya D. 2020. Guidance for building a dedicated health facility to contain the spread of the 2019 novel coronavirus outbreak. *Indian J. Med. Res.* 151: 177–183.

Carter T.C., He M.M. 2016. Challenges of identifying clinically actionable genetic variants for precision medicine. *J. Healthc. Eng.* 231: 32–37.

Collins F.S., Varmus H. 2015. A new initiative on precision medicine. *N. Engl. J. Med.* 372: 793–795.

Damodharan D., Narayana M., Kumar C.N., and Math S.B. 2020. Telemedicine practice guidelines of India, 2020: Implications and challenges. *Indian J. Psychiatry*. 63 (1): 97–101.

Li Q., Guan X., Wu P., Wang X., Zhou L. 2020. Early transmission dynamics in Wuhan, China, of novel coronavirus-infected pneumonia. *N. Engl. J. Med.* 382: 1199–1207.

Naveen Thacker, Vashishtha V.M. and Thacker D. 2016. Polio Eradication in India: The Lessons Learned. *Pediatrics*. 138(4): 3–5.

Singh A., Deedwania P., Vinay K., Chowdhury A.R., Khanna P. 2020. Is India's health care infrastructure sufficient for handling COVID 19 pandemic? *Int. Arch. Public Health Community Med.* DOI: 10.23937/2643-4512/171004

Thomas D. and Kalakota R. 2019. The potential for artificial intelligence in healthcare. *Future Healthc J.* 6(2): 94–98.

Van Kerkhove M.D., Hirve S., Koukounari A. 2013. Estimating age-specific cumulative incidence for the 2009 influenza pandemic: A meta-analysis of A (H1N1) pdm09 serological studies from 19 countries. *Influenza Other Respi. Viruses*. 234: 34–37.

Vassy J.L., Korf B.R., Green R.C. 2015. How to know when physicians are ready for genomic medicine. *Sci. Transl. Med.* 57: 113–120.

Zhewei Ye, Zhang Y., Alessa T., and Goyal D. 2020. The internet of things: Impact and implications for health care delivery. *J Med Internet Res*. 22(11): e20135.

ONLINE DOCUMENTS

Alexander A. 2018. Real-time data sharing among rural health workers can save lives. www.livemint.com › Opinion

Oxfam, India. 2018. Fifteen Healthcare schemes in India that you must know about. www.oxfamindia.org

Press Information Bureau. 2022. Initiatives & Achievements-2021. Ministry of Health & Family Welfare, Government of India, 4 January 2022.

Statista Research Department. 2020. Public health expenditure as a share of India's GDP from financial year 2010 to 2018. www.statista.com

The Wire. 2020. How the Bombay Plague of 1896 Played Out. https://science.thewire.in/society/history

World Health Organization. 2020. The top 10 causes of death. www.who.int/news-room/fact-sheets/detail.

22 Healthcare in COVID-19 Era

Rupinder Kaur
BGJ Institute of Health, Panjab University, Chandigarh, India

CONTENTS

22.1 INTRODUCTION

Health care is about maintaining and improving health by prevention, diagnosis, treatment, or cure of illness or injury and rehabilitation. According to the WHO, "A well functioning healthcare system requires a financing mechanism, a well trained and adequately paid workforce, reliable information on which to base decisions and policies and well-maintained health facilities to deliver quality medicines and technologies." The disease can be physical or mental. Healthcare professionals are involved in providing health care at primary, secondary, and tertiary levels of care apart from home and community care. Depending upon the geographical barriers, socio-economical limitations, and personal limitations the usage of healthcare facilities differs. Essential primary care should be available to all as per the WHO irrespective of barriers. An efficient, robust healthcare system is an important part of the economic development of a country. India has approximately 24,855 primary health centers and 997 more are under construction; there are also 5,335 working community health centers and 354 are under construction as of 2019 data released by the Ministry of Health and Family Welfare Statistics Division, Rural Health Statistics 2018–2019 (1–3).

As per our health structure our preparedness for the pandemic is not up to the mark. In 1820, the cholera epidemic emerged in Asia, then in 1919 there was the flu pandemic, in 2003 the SARS epidemic, and in 2020 the COVID-19 pandemic are examples of major diseases that have caused us to think about public health, importance of sanitation, society, economy, and disease surveillance in new ways.

DOI: 10.1201/9781003358909-22

COVID-19, an infectious disease caused by a newly discovered coronavirus, spreads from person to person by droplets of saliva or discharge from nose, through cough, sneeze, fomite, surfaces, nosocomial transmission, and vertical transmission from mother to newborn, and has created not only healthcare crises but also economic crises.

In coming times we will be living with the novel coronavirus as we are living with other viruses such as the common cold, H1N1 (swine flu or Influenza A), polio, chicken pox, measles, and mumps. The present strategy is prevention is better than cure, so transmission control and awareness about personal and community hygiene is also important in such times to curtail morbidity and mortality. To tackle the virus, development of clearly defined integrated emergency response plan with information, faster, practical execution with identified initiatives in National Disaster Management Plan (NDMP) within the legal framework and upgrading of facilities are required.

22.2 AFTERMATH: BIOLOGICAL DISASTER

COVID required us to adapt ourselves by altering, making adjustments, revising, redesigning, and remodeling our healthcare response teams at primary care, secondary care, and tertiary care levels and to change these to more community-oriented health care rather than hospital-oriented care.

22.2.1 Changes in Primary Care

Primary care is health care provided day to day in the community as an initial approach for advice or treatment and acts as a coordinator for specialist care. It includes general care or prevention. Among preventive factors we can only emphasize on

- Behavioral factors
- Nutritional factors
- Social factors
- Mental health factors

22.2.1.1 Behavioral Factors: Means as an Individual What I Can Do to Prevent Infection

Maintaining physical hygiene – bathing, frequent handwashing (minimum five times a day). Proper hand washing reduces infection by 80%. We should wash hands before and after cooking and eating, using the toilet. It is advised that one should wash their hands after sneezing or coughing (even if a tissue was used). Kids should be taught to wash their hands after playing with toys. People handling animal waste should pay special attention towards the act of hand washing. It is a great idea to wash your hands as soon as you reach home with regular soap and water for 20 seconds reciting "Twinkle twinkle little star" following seven steps of handwashing:

1. Wet hands with running water
2. Apply soap on hands and spread it gently
3. Rub palms against each other in a circular motion
4. Interlace your fingers and move them back and forth
5. Rub the back of palms and fingernails
6. Clasp fingers of opposite hand individually
7. Rinse hands with clean running water.

Hand sanitizers and medicated soaps are not required in routine use for the general population. It is to be used when you are exposed to the outside environment. Drawbacks of sanitizers: not good for children as they tend to put their hands in their mouth, it is inflammable, and it does not clean dirt, dust, and grease.

- Removing shoes outside the entrance door of house: This is now an important part of daily routines that we need to incorporate into our life. Our ancestors used to follow this especially in central India and southern India. Whereever possible a shoe rack should be placed outside the entrance of house.
- Changing and washing clothes (soap and hot water) as soon as we reach home: Once you reach home you need to clean your hands thoroughly with sanitizer, keep your keys, etc., in sanitizer, clean mobile, go to washroom straight away without touching door handles, knobs etc., have a thorough wash from head to toe, soak your clothes in soap and warm water for 40 minutes then wash them and dry in sun, then iron your clothes for a thorough cleaning.
- Do not share personal items such as hairbrush, hats, scarves, toothbrush, towels, cups, glasses, and bottles as these can be sources of infection.
- Disinfecting surfaces with dilute bleach 1:100 solution (formed by mixing one part of bleach powder with 99 parts of water).
- When going out of your house wear clothing that covers maximum skin. Do not wear loose, flowing clothes, as they tend to gather infection. Cover head including ears with cap, dupatta (head cloth), or turban. Cover eyes with spectacles or goggles. Cover nose and mouth with homemade cloth mask because it can be washed with soap and hot water and reused while disposable masks need to be disposed of in biomedical waste bins as they are a biohazard (if thrown in house dustbin they become a source of infection for your family and for garbage collector/handlers). Never touch the outer surface of your mask after you have worn it and while removing as it may be infectious. Gloves, remove them inside out without touching the outer surface and dispose of them in biomedical waste bins (covered dustbins).
- Masks covering nose and mouth are now a major necessity. There are three main types of masks: Cloth mask – for all; surgical mask – triple-layered mask; and N95 mask – for healthcare professionals. These can be used as per the requirement of the individual.

Donning and doffing of mask is very important. Always tie the upper end of the mask first such that it fits snuggly over the nose, it should not be loose or tight. Then tie the lower end of the mask so that it covers the chin and snuggly rests on the face. Once tied securely it is not to be touched in any way.

- Benefits of cloth mask: This type of mask is recommended for the general public. For healthcare workers cloth masks should preferably be used with face shield or not be used. Decreases risk of infection. Can be washed (hot water and soap solution) and reused. Environmentally friendly (non-biohazard) and protects against asymptomatic patients.
- Risks of cloth mask: Not good for those who are exposed to many people everyday. Not effective for front-line workers such as healthcare workers, Covid patients and their medical team, police on duty, and volunteers who are in contact with numerous people throughout the day.

Surgical masks are routinely used in operation theaters. They are disposable and loose fitting with air leakage. They protect from splashes, sprays, droplets, and particles. They are good for people with Covid as the mask helps trap infectious respiratory infection. These masks can also be used for COVID caregivers and patients at home.

N95 masks are recommended for healthcare workers. They are tight fitting, in addition to protecting from splash, spray, droplets, and filters, including 95% of small particles, which includes bacteria and viruses. These masks come in many shapes and types. Some come with exhalation valve (decreases heat, humidity). They should fit properly to guarantee protection. Always perform a seal check after wearing an N95 mask. This mask is not recommended for children. In people with facial hair tight seal may not be possible so extra care is needed. Not to be used outside healthcare settings. Due to tight fit these masks are uncomfortable and stuffy.

Disposal: Closed lid waste bins lined with bin bags should be used exclusively for disposal of masks and gloves as they are a biohazard (if thrown in house dustbin they become a source of infection for your family and garbage collector.

22.2.1.2 Nutritional Factors

Diet should consist of fresh green leafy vegetables, fruits, fibers, adequate proteins, less sugars, oil, and spices. Drink lots of water and keep yourself well hydrated. Avoid preservatives and processed food. If you have any disease your diet should be as specified by your doctor and dietician.

22.2.1.3 Social Factors

Maintain social distance. Avoid handshakes, hugs, and kisses and limit group sizes. Restrict entry of people in a room/office depending on size of room. Research has shown that following social distancing reduces chances of infection drastically. Wait for your turn patiently – at vendor, bank, market, hospital, etc. When visiting your doctor try to contact him/her by phone and visit only in person if necessary. Contact other people by phone instead of visiting.

22.2.1.4 Mental Health Factors

We need to take good care of mental health by keeping in touch with family, friends, coworkers, etc. Follow the advice of experts in health care and government and seek mental health services if required. Physical distancing, pre-existing health conditions, unemployment, decreasing earnings, and keeping a business alive can all affect mental health.

22.2.1.5 Support Children and Elders in Family and Surroundings

Special care should be given to children and elders so that they feel safe. Create a schedule for the whole family such that it consists of a well-regulated routine that involves regular wake-up time, morning prayers, yoga or exercise, meals, play time, reading, listening to music, watching a movie, etc. Teach them precautions, educate them about the pandemic to create awareness, and try to be a role model. Try to involve children and the elderly in group video calls with friends and family.

22.2.1.6 Lockdown

Preventive measures to be implemented and followed by public/private institutions and establishments in case of lockdown to avoid COVID-19 infection

- Decrease burden of disease
- Create interventional policies and improve infrastructure. Designate hospitals, assign hospital beds, and procure equipment, diagnostic kits, essential medicines, etc.
- Train and assign human resources to manage crises
- Pay attention towards early signs of COVID-19 symptoms, public information, education, prophylactic measures
- Devise new policies as needed.

22.2.2 Changes in Secondary Care

Secondary care refers to specialized care required by a patient.

22.2.2.1 Specialty Care

These are the patients who can't be treated at dispensaries and need to go to hospitals. We should first call the concerned doctor and on his advice visit a fever clinic or you can call a helpline number who will visit and assess your requirement for hospital visit or you can consult a physician by telemedicine in case following symptoms are seen.

- Fever, tiredness, breathlessness with or without fever
- Nasal congestion, runny nose, sore throat
- Repeated shaking with chills
- Aches pains
- Marked fatigue
- Anosmia (lack of smell or taste)
- Lack of appetite, nausea, diarrhoea
- Gangrene of extremities (blue toes or fingers)
- Pain or pressure on chest
- Confusion or inability to wake up fully
- Blue lips or face.

Continue your routine medicines for BP, heart, sugar, thyroid, liver, kidney, etc. Routine wellness checkups are not required. In case of emergency visit the hospital.

22.2.2.2 Isolation

Isolation is a means of avoiding contact with a person infected with COVID-9 or a person showing COVID-19 symptoms. Isolation room should be a self sufficient single room with attached toilet. Its floor should be mopped with hypochlorite solution at least twice a day and fixtures should be cleaned every 4 hours. The patient should have separate utensils that should be cleaned separately from other dishes. Preferably one person with training (or can be trained) should be assigned as caregiver.

Types of isolation include:

a) Home isolation is recommended for all with or without history of symptoms and with history of travel including travel from affected country, state, district, or locality. As per government guidelines 14 days of self-quarantine is required.
b) Hospital isolation is recommended for positive Covid cases and for those patients that have co-morbidities such as diabetes mellitus, hypertension, cardiovascular disease, chronic kidney disease, liver disease, cancer, etc.

22.2.3 Changes in Tertiary Care

Tertiary care basically comprises specialized hospital care to patients referred from primary and secondary care providers. For example, patients requiring cardiac care, neurology care, nephrology care, hepatic care, cancer patients, etc.

In the COVID era, there is a huge responsibility to ensure safety of doctors, nurses, paramedics, and support staff working in hospitals. Hospitals should be segregated into COVID, suspected COVID, and non-COVID areas so that adequate precautions can be taken. Special preparedness and modifications are required for emergency patients, cancer patients, organ transplant patients, dialysis patients, neonatology, and trauma patients. Strategies to handle non-COVID emergency and nonemergency cases in such a way that deaths of these cases do not rise due to lack of healthcare facilities should be developed. The number of patients with TB, HIV, HCV, diarrhea, typhoid, common cold, and malaria has increased (4).

22.2.4 Physical and Mental Wellbeing of Healthcare Workers

Keeping healthcare staff motivated, involving them in suggestions on improvement, addressing the flexibility of roles, addressing their fears and problems regarding their profession even at mental and family level. There is an acute shortage of healthcare workers and employee turnover that needs to be addressed urgently (5).

As per a study conducted by a UK-based organization on physicians' health, it was found that 76% complained of lack of sleep, 66% complained work stress, 41% had considered quitting medical profession, 39% had problems with long working hours, 35% said they could not give adequate time to their patients, 87% could not keep up with medical advances being released, and 55% could not give enough time to their health (6,7).

As per the WHO, recommended Doctor:Population ratio should be 1:1000 while in India it is currently around 1:1615. Nurse:patient ratio should be 1:2 as per the WHO for ill, but stable patients, but in India it is at 1:483. India has a shortage of 600,000 doctors and 2 million nurses as per data from 2019. This deficit can partly be compensated for by active use of telemedicine services throughout the country. Due to technology advances in the past decade timely care to remote patients timely and at cost-effective rates is possible now. Artificial intelligence and use of robotics are emerging as major game changers for prompt, convenient, and accessible health care. This will also decrease the hospital bed requirement and readmissions, thus helping to decrease overall cost to patient and country as a whole. Frequent and timely updating of healthcare needs, infrastructure, SOPs, treatment protocols, scientific evidences, research needs, preventive strategies for both healthcare professionals, and community (8,9).

22.3 TECHNOLOGY UPGRADES IN HEALTH CARE

"Healthcare facilities at your doorstep" is becoming more and more common, especially for elderly populations. Innovative technology at door steps in the form of apps or websites so that data and information collected at the field is stored at a central server to train primary healthcare workers to augment healthcare workforce with minimal damage to human lives and improved healthcare scenario with shift from hospital to community-focused care.

Aligning patient needs: Managing emergency visits by telemedicine conferencing by emergency doctor and then accordingly educating a family member towards the needs of the patient and visiting the emergency will help in decreasing patient flow, referring nonurgent cases to OPDs and specialized care patients to respective specialists, thus decreasing contacts and increasing preparedness and patient-centered care (10).

Telehealth (Figure 22.1) also helps to reduce white coat fears, encourages people to get routine medical checkups without being exposed to infection, and allows posthospitalization or postoperative follow-ups to be done. Routine medicine reviews, services related to pregnancy and after delivery, etc., can also be addressed through certified doctors and health authorities via telemedicine (2) (Figure 22.2).

The pandemic led to lots of changes in the healthcare sector. One of the most important amendments was the inclusion of telemedicine in the Indian Medical Council (professional, conduct, etiquette and ethics) Regulations, 2020 appendix 5 at 3.8.1– consultation by Telemedicine.

22.3.1 Present State of Telemedicine in India

Telemedicine comes under the combined jurisdiction of the Ministry of Health and Family Welfare and Department of Information Technology. Government of India under the Ministry of Health and Family Welfare Telemedicine division has set up a National Telemedicine Portal on e-health for linking medical colleges with the National Rural Telemedicine Network through the National

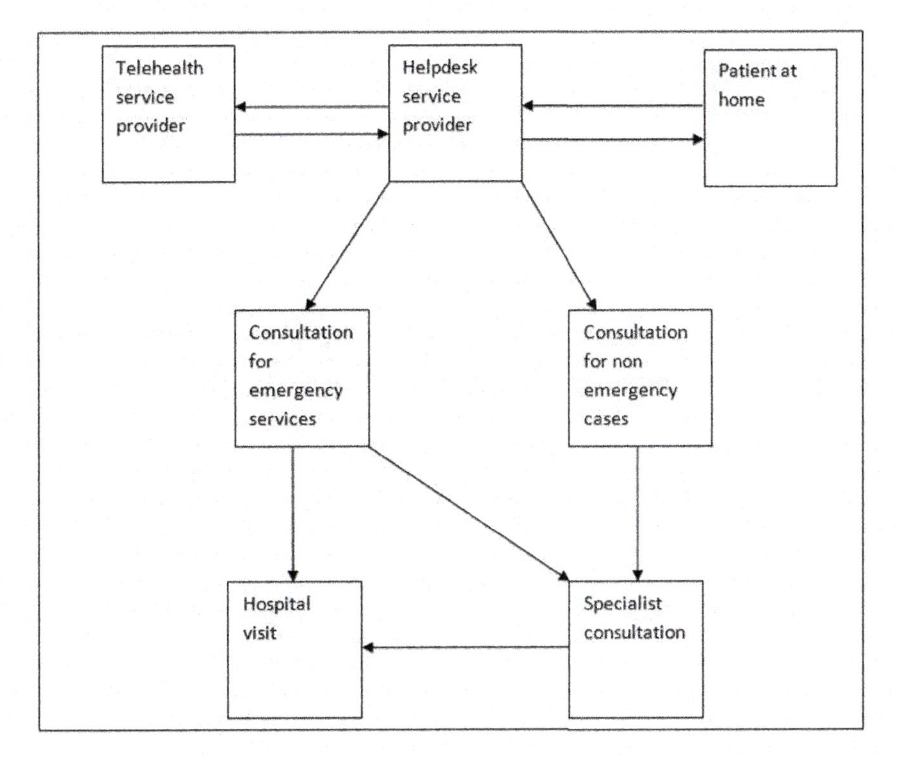

FIGURE 22.1 Pictoral diagram showing how telehealth works.

Medical College Network (NMCN). The National Digital Health Authority of India (NDHAI) and National e-Health Authority (NeHA) aim at developing cost-effective high-quality health services for the country. Various other telemedicine projects in India include:

- Electronic health records under MoHFW started in 2013 and were upgraded in 2016.
- National Rural AYUSH Telemedicine Network.
- Village Resource Centre developed by ISRO are learning centers and provide connectivity to specialty hospitals.
- AROGYASHREE is a mobile telemedicine conglomerate for hospitals and specialists and rural mobile units for situations such as the COVID-19 outbreak.
- ECG jackets are being worked developed to continuously monitor ECG without hospitalization.

India needs to engage stakeholders in implementing and integrating ehealth technologies to improve services and access by providing more funds and planning strategy guidelines so that ehealth reaches the maximum number of people.

It has been reported that 74% of Americans are using telemedicine and are satisfied with their consultations. Most of them found that it provided them 24/7 remote health care in the convenience of their homes. Emergency care can also be handled safely and effectively at home to a certain extent.

Health insurance companies are including treatment costs for 'pandemic diseases' in their portfolio and some of have started recognizing telemedicine as a modality for treatment even in developing countries like India (11,12).

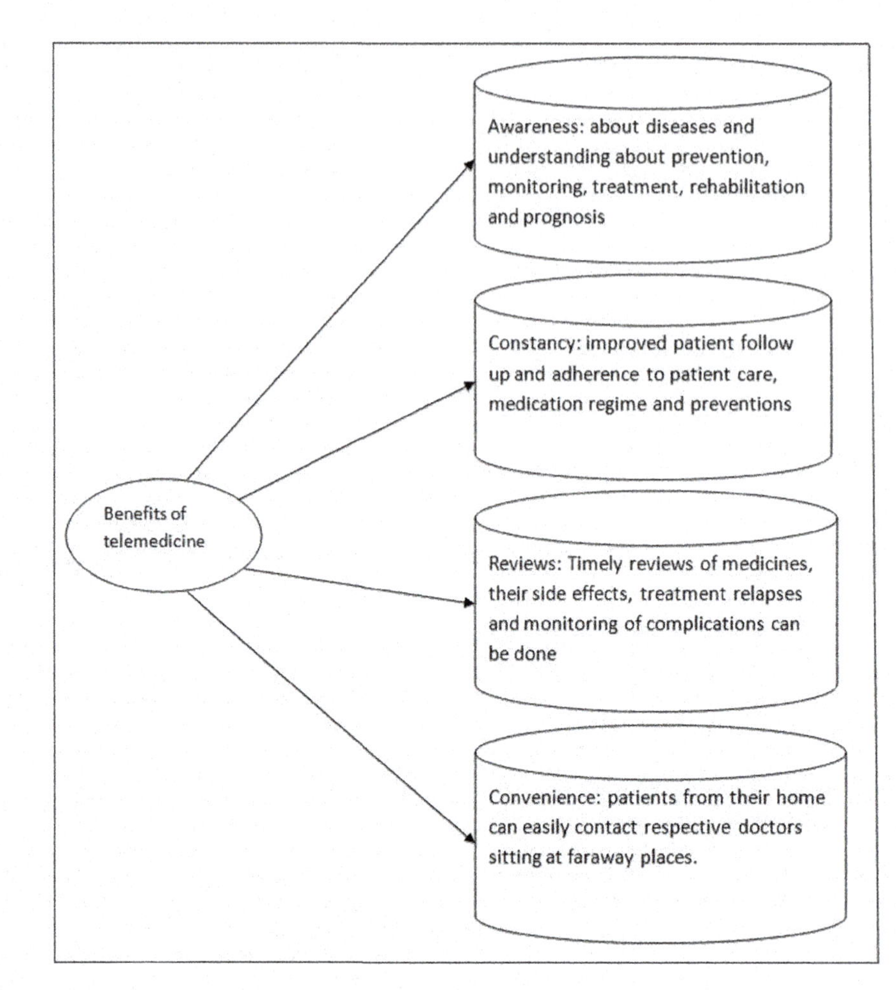

FIGURE 22.2 Diagram showing benefits of telemedicine.

22.4 CHALLENGES FOR HEALTH CARE

a) Medico-legal effects due to use of telemedicine: Telemedicine still is a new entity in India, so lots of medico-legal issues regarding jurisprudence may arise due to difference in the interstate laws like if a patient sitting at Himachal Pradesh contacts a doctor from Chandigarh by telemedicine and if some medico-legal problem arises then whether the case to be filed in Himachal Pradesh or Chandigarh. Similarly state laws may also differ in certain cases.

b) Impact on health services of non-communicable diseases: It is a challenge to provide health care to patients who regularly need it such as patients on dialysis, patients with pre-existing diabetes mellitus, hypertension, coronary artery disease, hepatic failure, and those at high risk due to pre-existing ailments. Limited emergency treatment and elective surgeries has decreased the quality of life of the general population.

c) Re-skilling and protection of healthcare workers: Better coordination of global governance systems for health care, lack of expertise to deal with such pandemics, strained health care worldwide due to rapidly increased demand on health care and healthcare workers has overstretched healthcare delivery systems. Thus, making amendments in healthcare systems to meet the needs of patients and staff is required.

d) Health assessment of all healthcare workers: Frequent assessment of healthcare workers for physical and mental wellbeing and providing needed resources is also important. Protocols should be adhered to to minimize loss of workforce.
e) Realigning and remodeling of hospital, office, and residential architecture so that air flow can be maintained to avoid infections.
f) There is dire need for realigning and strengthening the existing government, public, private healthcare infrastructure by maximizing its utility to the masses with least expenditure. There should be increased stress on in-house manufacturing of medical equipments, medicines, and other medical supplies as these will be in short supply due to disruption in supply chains (13).
g) Underfunding, overarching: Health care is allocated very meager funds by our government, and this needs to be enhanced.
h) Non-clinical work: Healthcare professionals have to spend around 22% of their time dealing with administrative and paperwork as per a survey (14). This loss of professional time leads to lower wages and patient care.

The COVID–19 pandemic has blurred the divide between developed and developing nations. Many of the highly developed nations did poorly as compared to some developing nations and adopted some practices prevalent in developing countries. There are many lessons to be learned from the pandemic.

REFERENCES

1. "Health care" www.en.m.wikipedia.org/wiki/Health_care [Google Scholar]. Assessed on 15th June 2020
2. MoHFW guidelines www.mohfw.gov.in [Google Scholar]. Assessed on 29th June 2020
3. "Ministry of health and family welfare statistics Division, Rural health Statistics 2018-19" www.main.mohfw.gov.in/sites/default/files/Final%2520RHS%25202018-19_0.pdf. Assessed on 22nd June 2020
4. N Subramanian, Director – Medical Services, Indraprastha Apollo Hospitals, New Delhi "Managing tertiary care hospitals during COVID-19 pandemic." www.expresshealthcare.in/amp/blogs/managing-a-tertiary-care-hospital-during-covid-19-pandemic/420792/ Assessed on 18th June 2020
5. "Skilled health workforce in India does not meet WHO recommended threshold – Medical Xpress" https://medicalexpress.com/news. Assessed on 24th June 2020
6. Cimpean and Drake, "Health and work spotlight on mental health" 2011. Labour Force Survey 2014. www.assets.publishing.services.gov.uk/government/uploads/system/uploads/attachment_data/file/677542/Infograghic.png. Assessed on 22nd June 2020
7. David Cutler, PhD, "The good and bad news of health care employment" 2018. www.scolar.harvard.edu/cutler/publications/good-and-bad-news-health-care-employment. Assessed on 22nd June 2020
8. KD Rao, "Composition and distribution of the health workforce in India: estimates based on data." www.rguhs.ac.in. Assessed on 24th June 2020
9. CDC Corona virus (COVID-19). https://pubmed.ncbi.nlm.nih.gov [Google Scholar]. Assessed on 24th June 2020
10. Gary William Shannon, Rashid Bashshur. "History of Telemedicine: Evolution, Context and Transformation" 2009, Google Books, www.books.google.co.in/books/about/History_of_Telemedicine.html? Assessed on 15th June 2020
11. www.mdportal.com/education/history-of-telemedicine 2015. Assessed on 15th June 2020
12. F.Kennedy St, Cambridge. www.jamanetwork.com/channels/health-forum/fullarticle/2764547. Assessed on 18th June 2020
13. Economic Times "India facing shortage of 600,000 doctors, 2 million nurses: Study" 2019. www.m.economictimes.com/industry/healthcare/biotech/healthcare/india-facing-shortage-of-600000-doctors--2-million-nurses-study/amp_articleshow/68875822.cms. Assessed on 24th June 2020
14. S Woolhandler and DU Himmelstein. 2014. Administrative work consumes one-sixth of U.S. physicians' working hours and lowers their career satisfaction. *International journal of health services : planning, administration, evaluation, 44*(4), 635–642. https://doi.org/10.2190/HS.44.4.a

23 Role of the World Health Organization in Combating the COVID-19 Pandemic

Brijesh
Advocate, Punjab and Haryana High Court, Chandigarh, India
Email Id: brijeeeesh@gmail.com
Phone Number: +91-8195949182

CONTENTS

23.1 INTRODUCTION

The Coronavirus Disease-2019 pandemic brought the complex health systems of the world to catastrophe. The pandemic has spawned a healthcare crisis that reveals a deep underlying problem in global health systems. The pandemic ramifications are not merely limited to human infection and deaths but also have associated social consequences (Lidia, 2020). The Secretary-General of the United Nations, Mr Antonio Guterres, aptly stated that "COVID-19 is the greatest test since World War II [...] it is more than a health crisis, it is a human crisis [...] The pandemic continues to unleash a tsunami of hate and xenophobia, scapegoating, and scaremongering" (Yeo, 2020). Statements like these point towards the cynicism of the member nations in this situation (Kamla & Brijesh, 2020). In the absence of bold and spirited actions by nations, the number of cases has escalated to millions, health systems have collapsed, economies have plumetted, and people are in despair, with the poorest being hit the hardest (Maira, 2020).

The World Health Organization (WHO), which was already under fire of critics for its poor handling of the Ebola Outbreak in 2014, is now being criticized for its organizational lethargy, absence of decisive leadership, bureaucratic indolence, underfunded programs, and inability to evolve to meet the needs of the 21st century (Sharma, 2020). There have been various concerns of the global community regarding the WHO's functioning as many African and Asian countries still depend upon the WHO for their basic health needs. These concerns were further amplified after U.S. President Donald Trump declared that the United States of America should stop funding of the WHO.

DOI: 10.1201/9781003358909-23

23.2 ABOUT THE WORLD HEALTH ORGANIZATION

The WHO, headquartered in Geneva, Switzerland, is a specialized agency of the United Nations responsible for maintaining international public health. It has six semi-autonomous regional offices and 150 field offices worldwide (World Health Organization, 2020).

The objective of the WHO has been enshrined in Article 1 of the WHO Charter, reproduced hereunder:

Article 1

From now on, the World Health Organization (called the Organization) shall be the attainment by all peoples of the highest possible level of health.

The fundamental functions of the WHO, as defined by its constitution, can be broadly divided into three categories: (1) normative functions, which includes negotiating international conventions and agreements, guidelines, and non-binding recommendations standards and recommendations; (2) guiding and coordinating functions, which includes various health packages for poor and the underprivileged, and its disease-specific programs; and (3) research and technical cooperation functions, which includes disease eradication and prevention (Walt, 1993).

23.3 WORLD HEALTH ORGANIZATION AND EPIDEMIOLOGY

The WHO's constitution identifies epidemiological and statistical services as its core function under Article 2(f). Furthermore, under Article 2(g), the constitution mandates that the Organization "stimulate and advance work to eradicate epidemic, endemic and other diseases." In the financial year 2012–2013, the WHO identified five broad areas for the devolution of funds. Two of those five areas were related to communicable diseases: to decrease the "health, social and economic burden" of communicable diseases, and the second to combat HIV/AIDS, malaria, and tuberculosis, in particular (World Health Organization, 2012).

In the recent past, the WHO has played a commendable role in spreading communicable diseases. In the 1970s, the WHO played an essential role in controlling malaria. It helped countries procure and distribute mosquito nets and insecticide sprays to prevent the spread of malaria (World Health Organization, 1978). The WHO, in 1988, launched a Global Polio Eradication Initiative to eradicate polio in association with the United Nations International Children Emergency Fund (UNICEF), Rotary International, the U.S. Centers for Disease Control and Prevention (US CDC), and various other organizations. The WHO was successful in reducing polio cases by 99% (World Health Organization, 1982). In the last 30 years, which is successful in bringing down the number of deaths from tuberculosis. The practices advocated by the WHO have saved over seven million lives. Notwithstanding its achievements, the WHO was criticized for its failure in containing the Ebola virus in 2014.

Recently, there has been a swing in the WHO's primacies due to the surfacing of organizations like the US CDC. Organizations like UNCDC possess better resources and expertise for global information systems. Notwithstanding the same, the WHO still plays a crucial role in regulating procedures, integrating information systems, and ensuring the health statistics' trustworthiness (Taylor, 2002). Furthermore, many countries still rely on the WHO's guidelines to form their Standard Operating Procedures during emergencies.

23.4 WORLD HEALTH ORGANIZATION AND THE COVID-19 PANDEMIC

In light of recent events and circumstances, it can be asserted without any doubt that the WHO has been unsuccessful in assessing the health hazards posed by the novel coronavirus, which first appeared in the Chinese city of Wuhan. The statement corroborates that the organization initially claimed no "human to human transmission" of the virus, notwithstanding the growing evidence to the contrary. Furthermore, the WHO has been lambasted for its initial advisories condemning restrictions on trade and travel, and its initial disinclination to declare COVID-19 as a global pandemic. Its mistakes indeed led to wastage of crucial time to contain the virus's spread and prepare for the pandemic (Kapoor, 2020).

To the shock of the global community, the WHO, on January 14, 2020, tweeted that "preliminary investigations conducted by the Chinese authorities have found no clear evidence of human-to-human transmission of the novel coronavirus (2019-nCoV) identified in Wuhan, China." From the empirical reality around us, we know that the claim is false, and it is believed that the same was known to the authorities in Wuhan. Furthermore, the same was informed to the WHO by Taiwan as well, which the WHO ignored. One day before this tweet was uploaded, a woman who travelled to Thailand from Wuhan was positive. It was informed that the woman never visited the seafood market, which is linked with the outbreak of the virus. The same suggests that, by then, the virus was already spreading within Wuhan (Tufekci, 2020).

In the recent past, the global community's concerns regarding global health were further amplified when Mr Donald Trump, the President of the United States of America, declared that the U.S. would stop funding the operations of the WHO (BBC, 2020). It is believed that defunding the WHO was a tactic played by the Trump administration to create a camouflage to cover its failures in combating the COVID-19 pandemic. The same may yield catastrophic results. Notwithstanding that the WHO has failed in the run-up to this pandemic, we must understand that the WHO functions include more than containing the COVID-19 pandemic. The majority of third-world nations depend upon the WHO to sustain their basic health needs and infrastructure.

The WHO's failure can be attributed to its structural defect as it has not been conceived to function independently. Instead, it is designed to act upon the whims and fancies of the nations who fund it and appoint its head. The people's Republic of China has played a pivotal role in electing the current leadership of the WHO. Presently, we need an accurate diagnosis of the conundrum. Global leaders must be inclined towards reform, not abolition of the WHO as it plays a crucial role in well-functioning global health systems.

23.5 GLOBALIZATION AND HEALTH

We are in the third decade of the 21st century, and by now, globalization has become a part of our everyday life. There are mixed opinions of researchers regarding globalization. For some, it holds a prospect of a new and a brighter future. While for others, it represents a threat that needs to be confronted and counteracted (Bettcher & Lee, 2002). Globalization poses various challenges to global health, and there are concerns regarding the egalitarian distribution of the health infrastructure. On the other hand, globalization has provided an opportunity for health systems to improve as it has facilitated the transfer of medical and public health knowledge from one part of the globe to another (Yach & Bettcher, 1998).

Traditionally, WHO, governments, and non-government organizations were in charge of the activities related to public health. In the contemporary era, the World Bank and the World Trade Organization have a significant influence on public health. New alliances are being formed to regulate the direct and indirect consequences on health (Walt, 1998). Due to globalization, the spread of infectious diseases has been accelerated. The rapid spread of COVID-19 is an empirical reality of the same. The COVID-19 pandemic has intensified glaring inequalities in global health systems (Lee, Sridhar & Patel, 2009). The WHO must play an essential role in ensuring that those very countries are not excluded from most of the global economy's benefits.

In the recent past, a boundary-less world was propagated by hyperglobalizers. They believe that the boundaries obstruct the flows of trade, finance, and people. With the spread of COVID-19 pandemic, we have realized that boundaries between the countries are good. At least for global health

(Maira, 2020). Different countries are at different stages of economic and social development and have different compositions of resources. Therefore, they must follow different paths to progress. The sub-systems with complex systems must have boundaries around them, as per the systems' theory. The boundaries must be permeable so that the sub-systems can maintain their integrity and evolve (Alter, 2018).

23.6 PLURALISM IN GLOBAL HEALTH

After the second world war and the formation of the United Nations and the WHO, only a few international organizations influenced the global agenda. It is only since the last two decades that institutions like The World Bank have strongly influenced global health priorities (London, 2017).

Pluralism has led to the global health agencies' rupture and an increasingly disintegrated, inept, *ad hoc*, and incongruent international health plan. The same has created a leadership gap for an overarching convening and coordinating role. Given this complex system, the WHO is mandated by its constitution to coordinate *inter se* between the member nations (Marshal, 2005). It is the only organ of the United Nations that has been empowered to develop and implement international health norms and standards. The benefits of cooperative supra-national action on global health issues, while numerous, could, however, be hampered by a shift in the WHO's budgetary allocations and policy priorities away from global normative development toward operational work at the country level (Taylor, 2002).

23.7 CONCLUSION

Thomas S. Kuhn aptly remarked in his work "Structure of the Scientific Revolutions" that societies' prevalent power structures always oppose ideas (Kuhn, 1962). The prevalent power structures determine which ideas are worthy of admission, and which ideas are not worth consideration. The prevalent power blocks do not want outsiders to dilute their influence in the organization. The establishment always resists change. Therefore, fundamental reform of ideas and institutions in human societies are always difficult until a crisis. Therefore, reforms must be brought in the WHO structure, and it must be made more independent.

On the other hand, President Trump's declaration to defund the WHO is extremely dangerous, given this pandemic situation. The present pandemic needs to contained globally, including in poor counties that are entirely dependent upon the WHO for their primary health needs, since it is the only international organization whose mission, reach, and infrastructure are suitable for this. The United States contributed about 15% of the total WHO's current funding, and the already stretched-thin organization may not be able to make up for the deficit quickly.

On the other hand, we need to recognize that corruption and shortcomings have engulfed the leadership of an organization deeply flawed. Notwithstanding the same, the WHO is still the jewel of the global health community. The WHO employs thousands of dedicated and selfless medical staff in 194 countries, and even now it is leading the global fight against COVID-19 pandemic.

The organization needs to be restructured, and the first order of business is to make sure that it is led by health professionals who are given the latitude to be independent and the means to resist bullying and pressure, and who demonstrate spine and an unfailing commitment to the Hippocratic Oath when they count most.

On examining diseases in a historical context, we can see how communities handle a disease such as a plague, and how its problems depend mainly on how the community understands that particular disease. As such, despite diseases not being the result of social construction, the response mechanisms depend on how society reacts. The United Nations and its bodies have their limitations because they are structurally designed to be not independent. Instead, they work on the whims and fancies of the nations that fund it. All the nations must unite their efforts in fighting the COVID-19 pandemic to save humanity on this planet before it is too late.

REFERENCES

Alter, S. 2018. *In Pursuit of Systems Theory for Describing and Analysing Systems in Organization.*

BBC. 2020. *Coronavirus: US to halt funding to WHO, says Trump.* April 15. Accessed May 14, 2020. www.bbc.com/news/world-us-canada-52289056.

Bettcher, D. & Lee, K. 2002. "Globalisation and Public Health." *Journal of Epidemiology and Community Health* 56 (1): 8–17. doi:https://doi.org/10.1136/jech.56.1.8

Kamla & Brijesh. 2020. "Extra-Health Ramifications of the COVID-19 Pandemic and Role of the U.N. World Focus." *World Focus* 65 (2): 60–65.

Kapoor, Kriti. 2020. *From passivity to pandemic: A timeline of Dr WHO's failure.* 04 06. Accessed May 14, 2020. https://theprint.in/opinion/passivity-pandemic-timeline-who-failure/396049/

Kuhn, T.S. 1962. *The Structure of Scientific Revolutions.* Chicago: University of Chicago Press.

Lee, K., Sridhar, D. & Patel, M. 2009. "Bridhing the Divide: Global Governance of Trade and Health." *Lancet* 373 (9661). doi:https://doi.org/10.1016/S0140-6736(08)61776-6

Lidia, K. 2020. "The world community expects the World Health Organisation to play a stronger leadership and coordination role in pandemic control." *Frontiers in Public Health* 470–476.

London, A.J. 2017. "The pluralism of coherent approaches to global health." *The Hatings Center Report* 47 (5): 26–27. doi:https://doi.org/10.1002/hast.766

Maira, Arun. 2020. "Pathways to a more resilient economy." *The Hindu*, May 06: 8.

Marshal, P.A. 2005. "Human righrs, cultural pluralism, and international health research." *Theoratical Medicine and Bioethics* 26 (6): 529–557. doi:https://doi.org/10.1007/s11017-005-2199-5

Sharma, M. 2020. "The WHO's relevance is fading." *The Hindu*, August 10: 8.

Taylor, AL. 2002. "Global governance, international health law and WHO: looking towards the future." *Bull World Health Organ* 80 (12): 975–980. doi:12571727

Tufekci, Zeynep. 2020. *The WHO Shouldn't Be a Plaything for Great Powers.* 04 16. Accessed May 14, 2020. www.theatlantic.com/health/archive/2020/04/why-world-health-organization-failed/610063/

Walt, G. 1998. "Globalisation of International Health." *Lancet* 351 (9100): 434–437. doi:https://doi.org/10.1016/S0140-6736(97)05546-3

Walt, G. 1993. "WHO under stress: implications for health policy." *Health Policy* 24 (2): 125–144. doi:https://doi.org/10.1016/0168-8510(93)90030-s

World Health Organization. 1978. *Malaria.* Accessed November 22, 2020. www.who.int/en/news-room/fact-sheets/detail/malaria

World Health Organization. 1982. *Global Polio Eradication Initiative Welcomes the Organisation for the Islamic Conference Decision to Step Up Effort to Eradicate Polio.* Accessed November 22, 2020. www.who.int/mediacentre/news/releases/2003/pr78/en/

World Health Organization. 2012. *World Health Organisation Programme Budget.* New York: World Health Organization.

World Health Organization. 2020. *Who Are We?* Accessed May 14, 2020. www.who.int/about/who-we-are/history

Yach D. & Bettcher D. 1998. "The globalisation of public health: threats and opportunities." *American Journal of Public Health* 88 (5): 735–744. doi:https://doi.org/10.2105/ajph.88.5.735

Yeo, Peter. 2020. *U.N. Response to COVID-19 Pandemic.* Accessed May 13, 2020. https://betterworldcampaign.org/un-response-to-global-pandemic/

24 Environmental Safeguards in Running a Level-3 COVID-19 Facility

The SGPGIMS Way

Rajesh Harsvardhan

Department of Hospital Administration, SGPGIMS & Medical Superintendent & Executive Registrar, SSCI&H, Lucknow, India

CONTENTS

24.1 INTRODUCTION

Prior to the COVID-19 pandemic, there were safeguards in place as prescribed by the Environment Protection Act, 1986 and as amended in 1991, and its rules and regulations.[1] Further, chemical waste, hazardous waste, solid waste, and bio-medical waste were addressed with their proper management protocols in place too.[2]

The coronavirus took 325 days (Table 24.1) in India to reach the 10 million (one crore) mark, since the first case was reported on January 30, 2020 but the strategies to curb and curtail the spread of the COVID-19 pandemic were already in various stages (from planning to execution)[3] and being incorporated at three tiered governance system (centre, state/union territory, and at district level) in India with premier hospitals invoking their policies and protocols in line with the above. The same was the case at SGPGIMS.[4]

24.2 COVID-19 PREPAREDNESS AT SGPGIMS

The emergency situation was sensed in time by the top leadership at SGPGIMS with the pressing need to provide immediate care for incoming COVID-19 patients.[4] An initial facility with 10

DOI: 10.1201/9781003358909-24

TABLE 24.1
India's Journey in COVID-19 Cases

India's journey in COVID-19 cases

Date	Case/s	Number of Days
30th Jan.2020	01	-
14th Mar.2020	100+	44
28th Mar.2020	1000+	14
13th Apr.2020	10000+	16
6th May.2020	50000+	23
18th May.2020	100000+	12
2nd Jun.2020	200000+	15
12th Jun.2020	300000+	10
20th Jun.2020	400000+	08
26th Jun.2020	500000+	06
1st Jul.2020	600000+	05
6th Jul.2020	700000+	05
10th Jul.2020	800000+	04
13th Jul.2020	900000+	03
16th Jul.2020	1000000+	03
6th Aug.2020	2000000+	21
22th Aug.2020	3000000+	16
4th Sep.2020	4000000+	13
15th Sep.2020	5000000+	11
27th Sep.2020	6000000+	12
10th Oct.2020	7000000+	13
28th Oct.2020	8000000+	18
20th Nov.2020	9000000+	22
19th Dec.2020	10000000+	29

Source: [3,13,14]

ventilator beds and 4 isolation beds was set up on March 20, 2020 and later the 210-bed (later augmented to 246) Apex Trauma Center was converted into Rajdhani Corona Hospital (RCH) on the directive of the Chief Minister of Uttar Pradesh on April 16th, 2029. Due to the pressing needs of the time another 36-bed ICU unit was activated as RCH 2.[5]

To deal with the pandemic clinical management protocols were published and revised based on approved guidelines by medical centres[6] and even at SGPGIMS.[4] Cautious management of bio-medical waste (BMW) was a key concern and an approved protocol by the Hospital Infection Control Committee of SGPGIMS[7] for providing an environment where patient and healthcare staff would feel secure.

24.3 KEY THRUST AREAS

This chapter discusses the ways SGPGIMS addressed BMW, sanitization and housekeeping, dead body management, healthcare worker training and safety through containment/preparedness plan implementation, Covid-19 screening, operational management related with execution or delivery of healthcare services by commissioning of central control room, and continuity of healthcare delivery by telemedicine services.

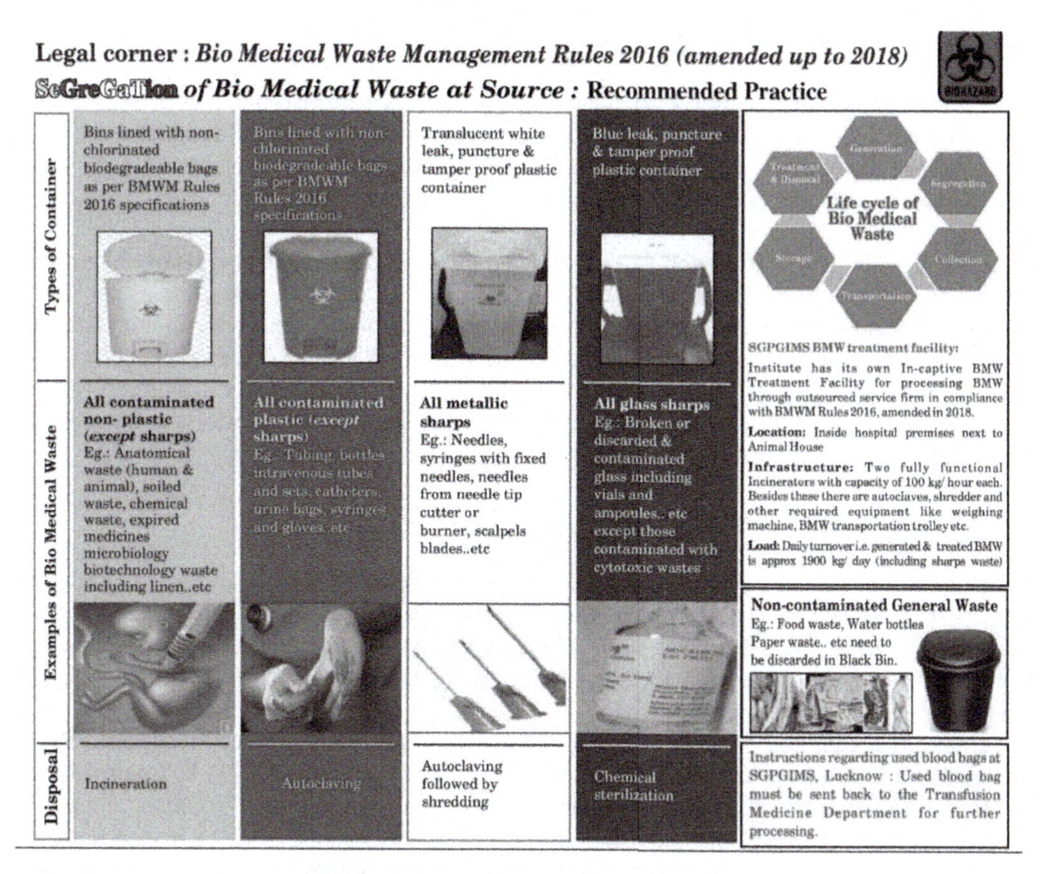

FIGURE 24.1 Synopsis of BMW Management Rules (as amended until 2018).

Source: [8]

24.3.1 Bio-Medical Waste Management Protocol

The guidelines as prescribed in BMW Rules (Figure 24.1) were already being followed at SGPGIMS. The major change that COVID-19 brought was based on Central Pollution Control Board (CPCB) Guidelines for COVID-19 related waste management as and when revised and other IEC materials to aid in reducing spread of the virus.[3,6]

To equip the already in-service manpower, the following steps taken were:

1. Personnel protection as advised by the government.[6]
2. Repeated, structured, and customized (based on audience characteristics) training schedule for resident, nursing staff, housekeeping, sanitation and BMW worker staff based on module of IPC (infection prevention and control).[2,3,6] This was later translated to hands-on training of staff during class with practical session after each class. Their actions were then monitored by designated staff and prompt corrections made as needed.
3. Supervisor rounds were made a norm and hierarchical authorities were designated for troubleshooting as and when required. The Duty Officer, COVID Control Room, RCH (1&2), SGPGIMS was the first point of contact for all such issues.
5. Audio-visual, information education, and communication (IEC) materials[2,3,6] were strategically placed and exhibited donning/doffing area and nursing stations. WhatsApp was also used as a tool for communication.

For collection in ward/ICU/laboratory/radiology units:

1. Inside: Double-lined, earmarked colour-coded BMW bins were placed in earmarked locations.
2. Outside: COVID-19 labelled dual lined and tied BMW was kept in designated locations (storage area) which was collected in COVID-19 labelled trolley[2] minimum thrice a day through a earmarked pathway (ramp) only.

The designated BMW collection workers entered the premises only with regularly disinfected trolleys after donning in full PPE. The end disposal of collected waste was done as per CPCB guidelines.[9,10] Additionally as SGPGIMS had its own functional incinerators the waste was incinerated, after autoclaving to render them completely harmless – *a step beyond prescribed norms*. All the waste thus generated inside the RCH was considered as infected, hence treated accordingly. The shift-wise data of collected waste was maintained meticulously, which was later uploaded on the district portal on a daily basis.

24.3.2 Sanitization and Housekeeping Protocol

In line with the approved protocol by the Hospital Infection Control Committee of SGPGIMS workers were trained on the aspects mentioned above. The housekeeping staff was also trained on dead body packaging[11] involving a three-layered system after sealing all body openings and removing of attached tubing with 1% hypochlorite spray and later transfer to dead body bag for transfer to morgue.

Further, as a contingency measure a Rapid Response Team (RRT) was formed, tasked with responding to any contingency, acting on the command of the Duty Officer, COVID Control Room, RCH. The RRT was also tasked with regular environmental sanitization of premises around the RCH but within its periphery only on a daily basis.

A log of all such sanitization activities was maintained and confirmation was obtained through on-site inspection/telephonic communication on a regular basis. Similar interspaced and periodic sanitization drive was also conducted across the entire 550-acre SGPGIMS facility with 1% hypochlorite solution for safety of all healthcare workers living on hospital premises.[6]

Later, when outdoor patient activities were started the same protocol was extended to functional Outdoor Patient Departments (OPDs).[6]

24.3.3 Dead Body Management Protocol

In line with the approved protocol by the Hospital Infection Control Committee of SGPGIMS7[6,11] workers were trained on the aspects mentioned above.

The RRT staff was tasked with transfer of dead bodies from outside the entry door of inpatient areas via dedicated pathways (common to BMW transfer) to morgue after donning in full PPE.

Instructions regarding dead body management were displayed in the working areas of the respective stakeholders involved in dead body management.

24.3.4 Healthcare Worker (HCW) Safety

HCWs are at the forefront of the fight against COVID-19 pandemic and are hence exposed to the risk of infection. In addition to the HCWs deployed in the designated COVID-19 treatment areas, we recognize that the HCWs in areas that are not designated COVID-19 treatment areas are also at significant risk of exposure to COVID-19 due to inadvertent admission of COVID-19 patients in those clinical care areas. SGPGIMS recognizes that it is of utmost importance to protect the HCWs during this pandemic, to prevent collapse of healthcare services being provided by the institute. Measures adopted to protect the HCW include administrative, engineering, and safe work practices.[12]

In view of COVID-19 infections among HCW at SGPGIMS, a Containment Committee was also constituted under the Department of Hospital Administration, SGPGIMS, a team of Healthcare Professionals & Workers are assigned the task of conducting contact tracing activity for COVID-19 positive cases reported from the non-COVID areas of SGIPGIMS.

24.3.5 COVID-19 Screening

As a precautionary measure single entry points to non-COVID areas were designated at which COVID screening is being done. Mandatory compliance is required for wearing face mask, hand hygiene, and maintaining social distancing throughout the institute. Daily records of COVID screening by respective departments are maintained.

24.3.6 Central Control Room

In the wake of the COVID-19 pandemic, to ensure appropriate operational management of non-clinical/administrative concerns related with execution/delivery of healthcare services of the institute, a Central Control Room was established by the Department of Hospital Administration. The Central Control Room is manned by Senior/Junior Residents of the Department of Hospital Administration designated as Duty Officers. This department is primarily responsible for inter-sectoral coordination between different service verticals and for any concern related to non-clinical/administrative/operational management of RCH-I/II, holding areas, quarantine areas, and non-COVID areas.

24.3.7 Continuity of Healthcare Delivery by Telemedicine Services

The telehealth system framed the foundation of the Electronic Outpatient Dept. Clinic (eOPD) and Electronic COVID Care System.

The eOPD network facility started on May 11, 2020 at SGPGIMS in which 26 departments could provide tele-followup of old/registered patients 6 days a week. A total number of 41,705 patients benefited from using this system until September 2020.

The eCCS network was initiated at SGPGIMS on May 1, 2020 with the aim to provide decentralised care of patients suffering with COVID-19/non-COVID-19 disease. Teleconsultation and remote clinical decision support were provided from the panel of specialist doctors sitting at Telemedicine, SGPGIMS to a remote location in the state of U.P. The total number of patients that benefitted from this system was 4062.

24.4 CONCLUSION

COVID-19 tested the managerial acumen at the institute, but has been addressed through collaborative teamwork of all stakeholders at SGPGIMS right from bottom to top.

REFERENCES

1. MoEFCC. (September 08, 2020). *Rules and Regulations, Environment Protection.* Retrieved from MoEFCC, GoI: http://moef.gov.in/rules-and-regulations/environment-protection/
2. MoEFCC. (September 08, 2020). *Environment, Waste Management.* Retrieved from MoEFCC, GoI: http://moef.gov.in/environment/waste-management/
3. Organization, W. H. (January 06, 2021). *WHO Coronavirus Disease (COVID-19) Dashboard.* Retrieved from WHO: https://covid19.who.int/

4. SGPGI. (2020). *SGPGI Recommended Protocols*. Retrieved from SGPGI: www.covid19sgpgi.com/recommended-protocols
5. Kapoor, P., Behari, A., Pande, S., Bhatnagar, A., Katharia, R., Rehman, K. P., & Yadav, K. (June 2020). Creation and Activation of RCH 2. *SGPGIMS Newsletter July–Sep 2020*, p. 5. Retrieved January 06, 2021, from www.sgpgi.ac.in/PR/news_letter_Jul_Sep20.pdf
6. MOHFW. (2020). *Home*. Retrieved from MoHFW: www.mohfw.gov.in/
7. HICC, S. (April 09, 2020). *SGPGI Recommended Protocols*. Retrieved from SGPGI: https://fc3f4a71-2ea1-491c-bdc0-cf254a782475.filesusr.com/ugd/dcdd21_39ea710654454567a1162c8e6b4d239a.pdf
8. Aggarwal, A., Harsvardhan, R., Mishra, R., Goel, A., Mehrotra, A., Kushwaha, R., & Mehrotra, A. (March 2019). Legal Corner. *Hospital Infection Control Newsletter*. Lucknow, U.P., India: Ganpati Printers & Packagers.
9. Central Pollution Control Board, N. D. (2020, July 17). *COVID-19 Waste Management* . Retrieved from CPCB, MoEFCC: https://cpcb.nic.in/uploads/Projects/Bio-Medical-Waste/BMW-GUIDELINES-COVID_1.pdf
10. Central Pollution Control Board, N. D. (2020). *COVID-19 Waste Management* . Retrieved from CPCB, MoEFCC: https://cpcb.nic.in/covid-waste-management/
11. MoHFW. (March 15, 2020). *Guidelines on Dead Body Management*. Retrieved from Home, MoHFW: www.mohfw.gov.in/pdf/1584423700568_COVID19GuidelinesonDeadbodymanagement.pdf
12. Containment/Preparedness Plan of COVID-19 for Health Care Workers of SGPGIMS (2020)
13. Narain, Nikhil. (2020). *Starting in January, How India's Covid-19 Trajectory Escalated from 1 to 10 Lakh Cases*. Noida: News 18 India.
14. Asrar, Shuja. (2020). *India's 10 Million Coronavirus Cases Explained in 10 Charts*. New Delhi: Times of India.

Index

D

E

F

G

H

I

J

K

L

M

N

T